Mineral Engineering Conference (MEC 2018)

IOP Conference Series: Materials Science and Engineering
Volume 427

Zawiercie, Poland
26 - 29 September 2018

Editors:

Marcin Lutynski
Tomasz Suponik

ISBN: 978-1-5108-7336-0
ISSN: 1757-8981

Printed from e-media with permission by:

Curran Associates, Inc.
57 Morehouse Lane
Red Hook, NY 12571

Some format issues inherent in the e-media version may also appear in this print version.

Printed by Curran Associates, Inc. (2018)

Institute of Physics
Dirac House, Temple Back
Bristol BS1 6BE UK

Phone: 44 1 17 929 7481
Fax: 44 1 17 920 0979

techtracking@iop.org

Additional copies of this publication are available from:

Curran Associates, Inc.
57 Morehouse Lane
Red Hook, NY 12571 USA
Phone: 845-758-0400
Fax: 845-758-2633
Email: curran@proceedings.com
Web: www.proceedings.com

Table of contents

Volume 427

Mineral Engineering Conference
26–29 September 2018, Zawiercie, Poland

Accepted papers received: 7 September 2018
Published online: 17 October 2018

Preface

Mineral Engineering Conference

Peer review statement

Papers

The use of selected structural tests in the determination of phase composition of industrial wastes based on the example of Zn-Pb slags

I Jonczy.....1

Trends in global copper mining – a review

S Pietrzyk and B Tora.....12

Development of dry coal gravity separation techniques

W Mijał and B Tora.....22

The role of β-diketones in the extraction of cobalt (II) and nickel (II) ions

K Witt and E Radzyminska-Lenarcik.....30

Copper recovery from model chloride solution using polymer inclusion membranes with 1-decyl-2,4-dimethylimidazle

E Radzyminska-Lenarcik and K Witt.....41

Influence of dextrin on beneficiation of components from copper flotation concentrate

A Bakalarz, M Duchnowska and R Kubik.....52

Influence of foam height on salt flotation of copper-bearing shale

T Ratajczak.....60

Analysis of distributions of various coal types properties by means of statistical methods

T Niedoba, P Pięta and A Surowiak.....65

Studies on Polish copper ore beneficiation in Jameson cell

P Pięta, T Niedoba, A Surowiak, O Şahbaz, C Karagüzel and Ö Canieren.....70

Determination of flakiness index (FI) and shape index (SI) using open source digital image analysis software

K Oleksik, K Ostrowski and D Mrocheń.....82

Application of kinetic, incentive, limits and arrheniusan curves for flotation characterization

J Drzymala.....89

Modelling of oil agglomeration of dolomite by the use of an artificial neural network optimized using Optimal Brain Damage algorithm

M Kamiński and A Bastrzyk.....95

Application of a vision systems for assessment of particle size and shape for mineral crushing products

D Krawczykowski.....104

Investigation of post-industrial pollutions' immobilization in a hydraulic self-solidifying clay-cement binder

A Stempkowska, Ł Wójcik, P Izak, M Staszewska and J Mastalska-Popławska.....109

Impact of HPGR pressing force on the level of product recovery

D Saramak.....119

Influence of HPGR operation on the reduction of Bond's working index of crushing product

D Saramak and Z A Iwanów.....126

Investigations of zinc recovery from metallurgical waste

D Saramak, D Krawczykowski and T Gawenda.....133

Investigations over the selection of reagent dose and speed of rotor in laboratory process of ore flotation for HPGR crushing products

K Oleksik, D Saramak and A Krawczykowska.....138

Management of waste from energy production – waste combustion in Poland

A Uliasz-Bocheńczyk and P Bąk.....146

Monitoring of coal jig operation using a radiometric meter with a variable time of measurement

S Cierpisz and J Joostberens.....153

Water content in froth products depending on the salinity level of pulp flotation

K Tupek-Murowany and A Młynarczykowska.....164

Biomineralization of calcium carbonate controlled by biomolecules produced by *Bacillus* sp.

A Bastrzyk, I Polowczyk, KJ Legawiec, J Chojniak, K Paraszkiewicz and G A Płaza.....169

Image histogram as a tool for a coal stream homogeneity evaluation

A Heyduk, J Joostberens, A Nowrot, J Pielot and K Fabiańczyk.....175

Testing the reduction properties of coke produced from a mixture of hard coal and rubber wastes

P Matusiak and J Kozioł.....183

Technology of fine coal grains recovery from the mining waste deposits

D Kowol and P Matusiak.....191

Development of research work in the air-water spraying area for reduction of methane and coal dust explosion hazard as well as for dust control in the Polish mining industry

D Prostański.....203

Preliminary assessment of new approach to scale-up the laboratory results of flotation reagents tests based on mineralogical analysis

M Krzemińska and P Kurzydło.....216

Hydrometallurgical processing of serpentine ore

E A Abdel-Aal, D El-Sayed, A K Ismail and F I El-Hosieny.....223

A new methodology of floatability tests on Polish copper ores based on grain size-density fractions

A Krawczykowska.....236

Effect of curing time on the dissolution of a secondary copper sulphide ore using alternative water resources

V Quezada, L Velásquez, A Roca, O Benavente, E Melo and B Keith.....244

Analysis of process of grinding efficiency in ball and rod mills with various feed parameters

D Foszcz, D Krawczykowski, T Gawenda, E Kasińska-Pilut and W Pawlos.....254

Increasing efficiency of technological process by limiting impact of corrosive environment on operation of spiral classifiers

M Czekajło and K Żakowski.....265

Analysis of the working chamber size influence on the effectiveness of grinding in electromagnetic mill

D Krawczykowski, D Foszcz, S Ogonowski, T Gawenda and M Wołosiewicz-Głąb.....273

New technology used to prepare ore for flotation based on impact comminution technique

A Wieniewski, B Skorupska and M Osoba.....287

Assessing the benefits of automatic grinding control using PST technology for true on-line Particle Size Measurement

R Maron, C O'Keefe and J Sepulveda.....296

Substitution of magnetite in dense medium separation by Zinc-Lead waste

D Gajda, M Lutyński and M Kujawska.....305

The behavior of oxidized copper during the sulphide Cu ore flotation in KGHM Polska Miedź S.A.

A Grotowski.....312

Analysis of improving the quality of coal slurry based on the use of regression models

A Manowska and M Mazurek.....322

Sorption rate of CH_4 and CO_2 in coal at different pressure ranges

M Lutynski and L Zavanella.....332

The testing of waste from the installation of salt debris leaching

T Suponik, P Pierzyna, D Franke and Z Adamczyk.....340

IOP Conf. Series: Materials Science and Engineering **427** (2018) 011001 doi:10.1088/1757-899X/427/1/011001

MINERAL ENGINEERING CONFERENCE

MEC 2018

September 26th- 29th, 2018

Zawiercie, Poland

Preface

The Mineral Engineering Conference MEC 2018 was held in Zawiercie, Poland and gathered over 100 participants. It followed the tradition of earlier conferences where mineral engineering science and industry can meet and exchange ideas. MEC has become one of the most highly acclaimed meetings in the field of mineral engineering in Europe. The MEC2018 for the third time is consisting of 14th International Conference on Non-Ferrous Ore Processing (ICNOP), 55th Symposium of Physicochemical Problems of Mineral Processing and 23rd International Conference of Mineral Processing.

The Conference welcome ceremony hosted vice-rector of The Silesian University of Technology prof. Bogusław Łazarz, Vice President of the Management Board of KGHM Polska Miedz S.A. Radosław Stach and senator of the Republic of Poland - Adam Gawęda. Plenary session was devoted to 20th anniversary of copper ore enrichment plants of KGHM Polska Miedz SA and consisted of 4 speeches given by notable guests such as Prof. Tsuyoshi Hirajima, Prof. Andrzej Wieniewski, Dr Andrzej Grotowski and Miroslaw Kidoń from the Sierra Gorda SCM.

Seven sessions of the conference were devoted to Physicochemical Problems of Mineral Processing, Mineral Processing and Hydrometallurgy. One technical session was devoted to companies presenting their solutions and products for the mineral industry.

The conference series contains only selected articles but it gives an overview of current situation of mineral engineering and related fields.

Editors

Dr. Marcin Lutyński, DSc, Eng.

Dr. Tomasz Suponik, DSc, Eng.

IOP Conf. Series: Materials Science and Engineering **427** (2018) 011002 doi:10.1088/1757-899X/427/1/011002

Peer review statement

All papers published in this volume of *IOP Conference Series: Materials Science and Engineering* have been peer reviewed through processes administered by the proceedings Editors. Reviews were conducted by expert referees to the professional and scientific standards expected of a proceedings journal published by IOP Publishing.

IOP Conf. Series: Materials Science and Engineering **427** (2018) 012001 doi:10.1088/1757-899X/427/1/012001

The use of selected structural tests in the determination of phase composition of industrial wastes based on the example of Zn-Pb slags

I Jonczy

Silesian University of Technology, Faculty of Mining and Geology, Akademicka 2, 44-100 Gliwice, Poland

iwona.jonczy@polsl.pl

Abstract. The article presents the results of phase composition analyses of Zn-Pb slags based on selected structural tests. Based on these tests it has been found that the application of several complementary research techniques may provide reliable information regarding the presence of individual phases. In view of obtaining qualitative characteristics of the components of slags, the phase identification tests by means of X-ray diffraction were found to be precise. Based on them, also the percentage of the individual crystalline phases and the percentage of the amorphous substance have been determined. Studies using electron scanning microscopy and X-ray spectral microanalysis allowed for the determination of the chemical composition of the individual components. In that case, the X-ray spectral microanalysis was found to be more precise and it allowed for the identification of substitutions of elements occurring in small amounts in the crystalline phases, i.e. the rare earth elements.

1. Introduction

Steel production is one of the branches of industry that produces significant amounts of wastes, including slags. For years the slags were deposited to dumps. Today, however, seeking the possibilities of reusing the slags i.e. in production of various aggregates that serve as an additive to cement is increasingly common [1-8].

There are also known testes to recover metals from metallurgical wastes; very interesting are researches conducted in China – process of coal-based direct reduction followed by magnetic separation was conducted to extract metallic iron and separate Pb and Zn from lead slag [9]. A study on the recovery of copper, cobalt and zinc from copper smelter and converter slags by roasting with sulfuric acid has been conducted by Turkish scientists. In their experiment they show that a total of: 88% of copper, 87% of cobalt, 93% of zinc, and 83% of iron were extracted in 2 h of roasting at 150°C and 3:1 acid/slag ratio [10].

These type works are determined by the necessity to conduct diverse research of slags, especially mineralogical and chemical analyses – as due to the phase composition, smelter slags are one of the most differentiated groups among waste materials. To exemplify this, among the slags produced in the processing of Zn-Pb ores in Bukowno, Kucha and Jędrzejczyk [11] have distinguished i.e. augite, omphacite, kirschsteinite, diallage, chromite, magnetite, wustite, periclase, Fe and Al carbides, phosphides and silicides as well as metallic phases of Fe and Pb. Among the slags from copper production, Muszer [12,13] has distinguished metallic phases, sulphides, arsenides, antimonides and oxides rich in Ag, Pb, Cu, Ni, Co, Sb, As, Zn, Fe, Cr. The following, on the other hand, are dominant in the phase composition of steel production slags from the Gliwice-Łabędy dump: iron oxides represented by wustite, magnetite, hematite and pericase, perovskite, melilites, dicalcium silicates and pyroxenes [14,15].

In the world, research of metallurgical slags are conducted, among others in Portugal [16], in the United States of America [17] and the Czech Republic [18], as also by the Russian scientists, who were conducted tests on the vanadium slag samples. The evolution of the compositions of spinelide, silicate, and metallic phases was investigated [19]. Babenko et al. [20] describe a phase composition of magnesia converter slag in which they distinguished larnite, brownmillerite and ferruginous periclase, with a series of solid solutions of magnesiowustite, magnesioferrite and calcium ferrites.

The diversity of the phase composition of slags results in the fact that each of them should be considered separately. Excessive generalizations are not allowable as the phase composition of slags changes depending on the steel production process, the type of the furnace charge material, the use of additives and fluxing agents.

Phases that crystallize in the conditions of a metallurgical furnace are identified as materials which are formed as a result of natural geological processes. Their chemical composition, however, is often much richer than their natural counterparts. They may also contain impurities, namely various elements which form substitutions in their internal structure.

One should also note that phase components of slags form highly diversified forms which are the result of the variable crystallization conditions. The intensity of the slag cooling process has an impact on the size and form of the crystals. In slags, well-formed crystals of mineral phases, the crystallization of which proceeded in the conditions of slow setting may be found as well as fine crystallites in glaze mass – if the crystallization process was interrupted due to sudden cooling of the slag.

This paper presents the results of the mineralogical analysis of Zn-Pb slags conducted using selected research methods. In view of the diversified and variable phase composition of slags, we have aimed at exhibiting the importance of the selection of research methods in such a study.

2. The scope and methodology of research

The range of the study incorporated highly weathered Zn-Pb smelter slags that have been deposited to the dump in Ruda Śląska-Wirek over a hundred years ago. The liquidation works conducted for a certain period and the extraction of material from the central part of the dump has made the acquisition of samples from the base of the dump possible. 10 samples of wastes, 2 kg each, have been collected. The samples were subject to microscopic observation and described. 3 of the samples were selected for further laboratory research. Before commencing laboratory analyses, the samples were dried and reduced by quartering. A part of each of the samples was ground in a bead mill, while the other part served as preparations for the microscopic study – 2–3 preparations were made from each of the samples.

Within the study, identification of phases using X-ray diffraction method was applied. Moreover, electron scanning microscopy and X-ray spectral microanalysis were used.

X-ray structural tests were conducted in the Institute of Ceramics and Building Materials, Refractory Materials Division in Gliwice, Structural, Thermal and Thermomechanical Testing Laboratory. The analysis of the phase composition using X-ray diffraction (XRD) was conducted using a PANalytical XPERT PRO MPD powder diffractometer with an X-ray lamp with a Cu anode, a graphite monochromator at the diffracted beam side, X'Celerator semiconductor strip counter and a spinner. X-ray diffractograms were made for the range of angles of 5–80° 2Θ at lamp operational parameters of 35 mA and 45 kV. To identify the phase composition, HighScore software by PANanalytical as well as the ICDD PDF-4+ (2011 issue) database were used. The quantitative analysis of the phase composition was conducted using the Rietveld Method with the SIROQUANT software. The following corrections were applied in the analyses: the effect of anomalous X-ray scattering and the microabsorption. The elementary cell parameters were determined using the Rietveld method only for the phase components in case of which the diffraction spectrum was well-formed. The results of the determinations of the elementary cell parameters were compared with catalogue data of the ICCD PDF 4+2012 database to identify solid solutions.

The tests using electron scanning microscopy were conducted in the Institute of Ceramics and Building Materials, Refractory Materials Division in Gliwice with the use of a microscope equipped with an electron gun in the form of Schottky emitter (accelerating voltage of 200 V–30 kV) and a Wide Field Optics optical system comprised of three electron lenses. Microscopic images were obtained using detectors: backscattered electrons detector BSE (2.0 nm resolution at 30 kV), secondary electrons (SE) detector (1.2 nm) working in high-vacuum mode, LVSTD low vacuum secondary electron detector (1.5 nm) and a CL cathodoluminescence detector (radiation range of 350–650 nm).

The X-ray microspectral analysis was conducted in the Institute of Non Ferrous Metals (Department of Material Engineering and Powder Metallurgy) in Gliwice, using an X-ray JXA 8230 microanalyser manufactured by JEOL. The analyses were conducted using metallographic specimens dusted with a thin layer of carbon to transfer the electric charge. X-ray mapping of elements was conducted using Energy Dispersive Spectroscopy. Local quantitative analyses of selected grains were conducted.

3. Results

3.1. X-ray diffraction

Tests conducted using the X-ray diffraction method allowed for a precise determination of the phase composition of slags, both in terms of quality and quantity.

It has been exhibited that the amorphous substance is dominating in the composition of the analyzed Zn-Pb slags, apart from which the crystalline phases occur and are represented by: oxides (hematite, magnetite, corundum, solid solution of spinel, hercynite) and silicates (mullite, quartz, cristobalite, anorthite, melilite, sanidine, sekaninaite) as well as phases which were subject to secondary crystallization in the dump, among which the following were distinguished: carbonates (calcite, agagonite), sulphates (gypsum, epsomite, hexahydrite) and hydroxides (ferrihydrite). The results of the analysis have been presented in table 1.

The amorphous substance constitutes the main component of slags and its content varies in the range from 45.00 to 74.60%. The formation of such significant amounts of amorphous substance is related to the rapid setting of slag during its release to the so-called slag pits. Additionally, such slag may be sprayed with water which increases the rate of the setting process. In such conditions, a material resembling glaze is formed.

Amorphous substance rarely occurs in the entire volume of the slag melt, but it usually co-exists with the crystalline phase. In such substances, domains of crystalline phase appear (known as crystallites in mineralogy), mixed with the domains of amorphous phase. By changing the conditions of liquid cooling, it is possible to change the proportions of one phase to another in a relatively wide range. In case of smelter slags, however, the crystallites constitute a small percentage among other components surrounding them. The crystallization commences in the conditions of the furnace, but the process is suddenly interrupted during the cooling of the slag melt, which results in the formation of fine crystals surrounded by amorphous substance. Their clear identification is impossible in e.g. microscopic observations in thin section, however tests using X-rays provide a reliable image of the presence of crystalline phases. In case of the tested Zn-Pb slags, among the crystalline components identified using such a method, the finding of the solid solution of spinel characterized by variable chemical composition was interesting. In sample No. 2 the spinel was found to have an iron impurity, while the spinel found in sample No. 3 contained substitutions of zinc. The X-ray diffraction method also provides good results in terms of the identification of hydroxides, the determination of which in

IOP Conf. Series: Materials Science and Engineering **427** (2018) 012001 doi:10.1088/1757-899X/427/1/012001

microscopic studies may be difficult due to the microcrystalline structure. The presence of one of the iron hydroxides – ferrihydrite – was found in the tested slags.

Table 1. Phase composition of slags based on X-ray diffraction.

No. of sample	Phase composition	Quantitative participation (%)
	Amorphous substance	74.6 ± 0.5
1	$CaSO_4 \cdot 2H_2O$ – Gypsum	1.3 ± 0.1
	$Al_6Si_2O_{13}$ – Mullite	15.0 ± 0.2
	Fe_2O_3 – Hematite	1.8 ± 0.1
	Fe_3O_4 – Magnetite	1.6 ± 0.1
	Al_2O_3 – Corundum	1.4 ± 0.1
	SiO_2 – Quartz	1.3 ± 0.1
	$CaCO_3$ – Calcite	1.8 ± 0.1
	$CaCO_3$ – Aragonite	1.1 ± 0.1
	Amorphous substance	56.5 ± 0.5
2	Ca $[Al_2Si_2O_8]$ – Anorthite	17.1 ± 0.2
	Fe_2O_3 – Hematite	1.6 ± 0.1
	Fe_3O_4 – Magnetite	5.7 ± 0.1
	$CaSO_4 \cdot 2H_2O$ – Gypsum	2.5 ± 0.1
	$Al_6Si_2O_{13}$ – Mullite	5.5 ± 0.2
	$(Fe,Mg)Al_2O_4$ – Spinel solid solution	2.8 ± 0.1
	$(Ca,Na)_2(Al,Mg,Fe^{2+})[(Al,Si)SiO_7]$ – Melilite	0.7 ± 0.1
	$FeAl_2O_4$ – Hercynite	6.9 ± 0.1
	SiO_2 – Quartz	0.8 ± 0.1
	Amorphous substance	45.0 ± 1.0
3	SiO_2 – Quartz	10.2 ± 0.5
	$KAlSi_3O_8$ – Sanidine	3.7 ± 0.3
	$MgSO_4 \cdot 7H_2O$ – Epsomite	10.0 ± 1.0
	Fe_2O_3 – Hematite	1.2 ± 0.1
	Fe_3O_4 – Magnetite	1.0 ± 0.2
	$CaSO_4 \cdot 2H_2O$ – Gypsum	2.1 ± 0.2
	$(Fe^{+2}, Mg)_2Al_4Si_5O_{18}$ – Sekaninaite	Below the limit of determination
		1.0 ± 0.1
	$(Mg,Zn)Al_2O_4$ – Spinel solid solution	Below the limit of determination
	$Fe_{1.44}O_{32}(OH)_{3.68}$ – Ferrihydrite	6.3 ± 0.4
	SiO_2 – Cristobalite	19.0 ± 1.0
	$MgSO_4 \cdot 7H_2O$ – Hexahydrite	

3.2. Scanning electron microscopy

The tests using electron scanning microscopy made it possible to identify the components of the slags as well as to exhibit their chemical composition (figure 1, table 2).

IOP Conf. Series: Materials Science and Engineering **427** (2018) 012001 doi:10.1088/1757-899X/427/1/012001

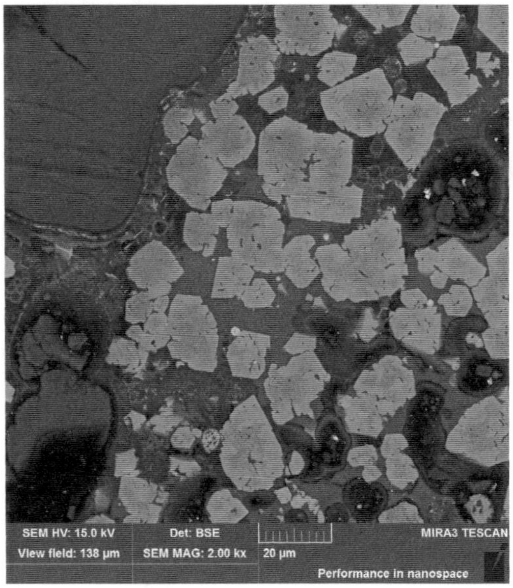

Figure 1. Microphotography of Zn-Pb slag; scanning electron microscopy.

Table 2. Chemical composition of Zn-Pb slag phases; points of analyzes according to figure 1.

Element	No. of analysis (% mass.)				
	1	2	3	4	5
O	58.33	57.53	63.92	75.82	74.32
Na	-	-	0.19	-	-
Mg	1.56	1.93	14.93	-	-
Al	14.53	23.29	0.89	-	-
Si	2.15	-	19.04	-	0.18
P	-	-	-	-	-
S	-	-	-	12.42	13.16
K	-	-	-	-	-
Ca	0.20	-	0.14	11.53	11.89
Ti	0.69	0.29	0.12	-	-
Fe	22.54	16.96	0.78	0.23	0.45
Total	100.00	100.00	100.00	100.00	100.00

Explanations: 1−2 hercynite, 3 forsterite, 5−6 gypsum

The microphotography and the analyses presented in table 2 have been supplemented by the EDS spectra, which are the dependence of counts vs. the radiation energy. The spectra allow for the identification of elements constituting components of a given phase and – in a quite evident manner – for exhibiting the chemical composition in the analyzed micro areas (figure 2).

IOP Conf. Series: Materials Science and Engineering **427** (2018) 012001 doi:10.1088/1757-899X/427/1/012001

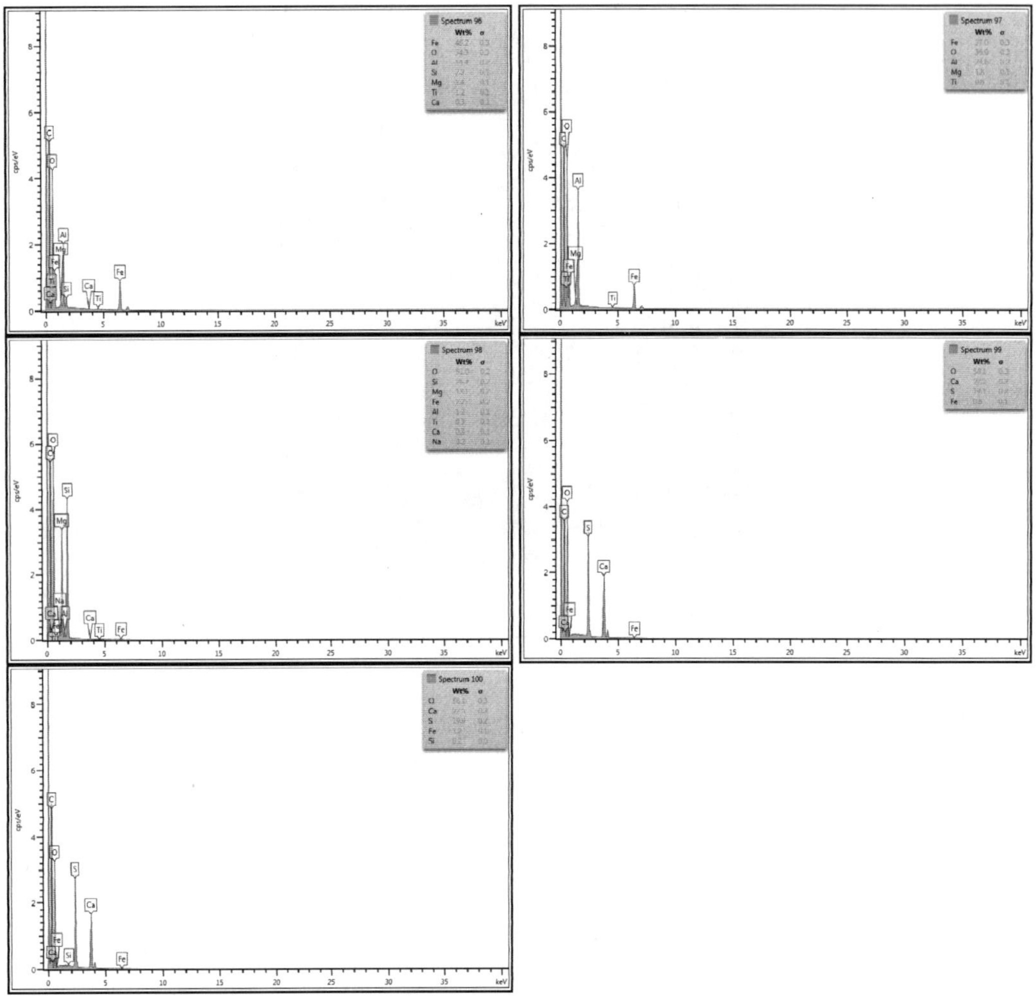

Figure 2. EDS spectrums; scanning electron microscopy (from the left: EDS spectrums from 1 to 5 point – points according to figure 1).

Based on the analysis conducted using electron scanning microscopy, the presence of most of the components identified using the X-ray diffraction was found, i.e. the hercynite (figure 1, table 2, analysis 1-2), as well as forsterite which was not identified in X-ray structural tests (figure 1, table 2, analysis 3). This may be explained by the form of the forsterite, which constitutes very fine crystallites which result in low results in the phase identification using X-ray diffraction. In such case, the reflections of the better crystallized phase components occurring in larger amounts may overlap the forsterite peaks.

Observations using electron scanning microscopy have also allowed for the assessment of the structure of the individual phases, including the morphology, dimensions and their mutual co-occurrence. Figure 1 presents numerous fine euhedral hercynite crystals accompanied by fine crystallites of fosterite and agglomerations of gypsum filling the pores formed in the slag.

Microphotographs of fractures or the surface of slag are helpful in preparing morphological descriptions of crystalline phases. Especially the structure of secondary phases filling the pores and fissures formed in the slag during the setting process were well observable. Figures 3 and 4 depict gypsum crystal tiles.

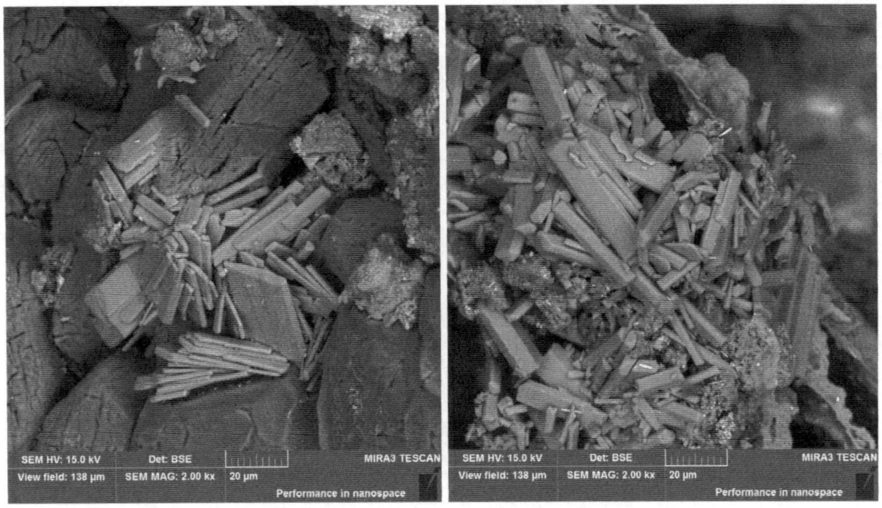

Figures 3, 4. Tabular habit of gypsum; scanning electron microscopy.

Tests conducted using the electron scanning microscopy as well as the studies conducted using X-ray spectral microanalysis, described in a subsection below, allow for the creation of element maps exhibiting the distribution of elements at the analyzed area of the sample.

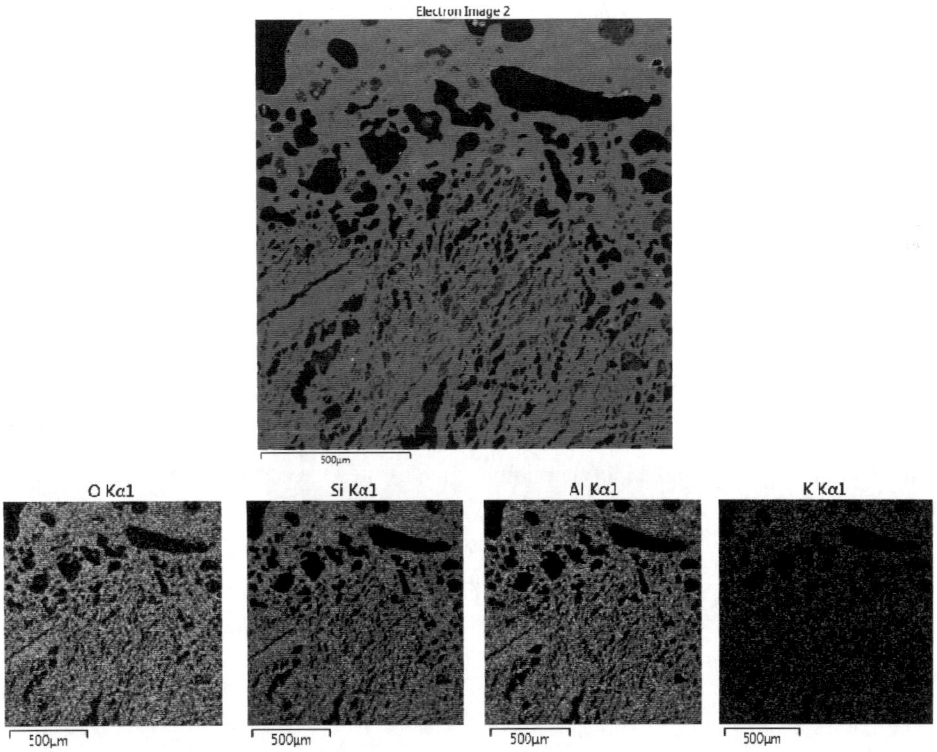

Figure 5. Maps of main elements in the Zn-Pb slag amorphous substance; scanning electron microscopy.

Figure 5 presents the surface distribution of the main elements occurring in the amorphous substance separated in the tested Zn-Pb slags, where silicon, aluminum and potassium were dominant besides oxygen.

3.3. X-ray spectral microanalysis

The tests conducted using the X-ray spectral analysis have confirmed the presence of phases in the analyzed slags that have been identified using the X-ray diffraction method. Additionally, based on these tests iron sulphides have been identified, which were not found using other methods (figure 6, table 3).

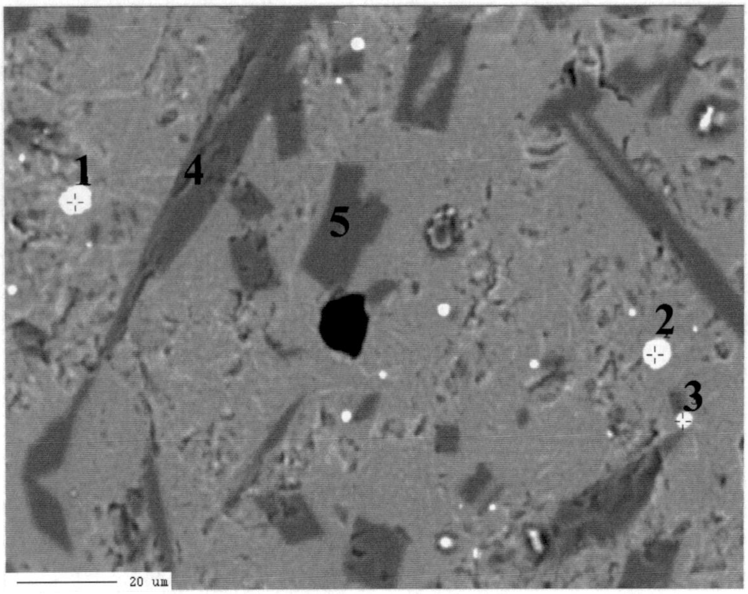

Figure 6. Microphotography of Zn-Pb slag; X-ray spectral microanalysis.

Iron sulphides occur in the form of fine spherical inclusions surrounded by amorphous substance. In their chemical composition, the content of iron is from approx. 52 to over 58%, while the content of sulphur varies between 37 and 39%. Such a composition of elements suggests the presence of FeS pyrrhotite, where, theoretically, the content of sulphur may reach 37.67% and the content of iron reaches 62.33% [21].

Based on the X-ray spectral microanalysis, also the presence of impurities was identified, namely the elements that occur in slags in low amounts, such as the rare earth elements. It has been found that the elements representing the lanthanides, such as La lanthanum, Ce cerium, Dy dysprosium as well as Ir iridium, considered to be a lanthanide, constitute substitutions in the internal structure of sulphides, mostly mullite.

IOP Conf. Series: Materials Science and Engineering **427** (2018) 012001 doi:10.1088/1757-899X/427/1/012001

Table 3. Chemical composition of Zn-Pb slag phases; points of analyzes according to figure 6.

Element	No. of analysis (% mass.)				
	1	2	3	4	5
O	-	-	-	46.948	46.453
Si	-	-	-	12.238	11.892
Al	-	-	-	40.451	41.250
Ca	-	-	-	0.015	0.011
Mg	-	-	-	0.019	-
Fe	52.985	58.112	58.850	0.042	0.039
Ti	-	-	-	0.139	0.142
Ni	6.777	4.088	3.372	-	0.003
As	-	-	-	0.015	-
Cd	0.036	-	-	0.019	0.014
Co	0.436	0.378	0.310	-	0.004
Cr	0.002	-	0.002	0.009	0.022
Zn	0.038	-	-	-	0.027
W	-	0.014	-	-	0.006
Tl	0.159	0.127	0.145	-	0.014
S	39.180	37.052	37.117	-	0.002
La	0.210	0.093	0.083	-	0.073
Ce	-	0.079	0.069	0.023	-
Dy	0.134	0.050	0.023	0.044	-
Ir	0.041	0.005	0.029	0.037	0.046

Explanations: 1–3 iron sulphide, 4–5 mullite

4. Discussion

The identification of phase components using X-ray diffraction has proven to be one of the most precise techniques used in qualitative analysis of the phase composition of slags. Based on it, it was also possible to determine the percentage of each of the components in a given sample. The application of this technique also allowed for the identification of solid solutions of oxide phases and the identification of two kinds of polymorphous calcium carbonate – calcite and aragonite.

The identification of hydroxides, which may manifest in micro- or cryptocrystalline forms, usually poses a difficulty in the analysis of the phase composition of slags. Microscopic observations, i.e. in thin plates, do not provide the possibility to identify the individual hydroxides unequivocally. Their identification due to the presence of the OH groups by means of the X-ray spectral microanalysis, on the other hand, is not possible. Favourable conditions may thus be achieved using X-ray structural analysis – the presence of ferrihydrite was found in the analyzed slags. A good result was also obtained in earlier studies using Mössbauer spectroscopy to identify iron hydroxides. Depending on the type of tested Zn-Pb waste material different iron phases have been shown: in the weathered slag four different products of the iron oxidation were detected: divalent $Fe(OH)_2$, mixed valence magnetite, Fe_3O_4 and three valence $Fe(OH)_3$, α Fe_2O_3 and γ-Fe_2O_3 compounds, in the vitrified material metallic forms of iron are dominated [22,23].

The tests using electron scanning microscopy and X-ray spectral microanalysis allowed for the determination of the chemical composition of the individual phases and the amorphous substance. The results were more precise and detailed, however, in case of tests conducted using the X-ray spectral microanalysis. These tests served the identification of lanthanide and platinium group elements substitutions in the sulphide phases. These elements are considered critical materials, which were defined in 2008 as „minerals that are subject to the risk of fluctuation or restriction of supply, the deficit of which may have severe consequences to the entire economy" by the committee for Minerals Critical for the Economy of the United States [24,25].

The conducted analyses of the mineral composition of slags indicate the necessity to apply several complementary research methods. This may be exemplified by the inclusions of iron sulphides that were identified using the X-ray spectral microanalysis, but were not found using other methods, e.g. using the X-ray diffraction method. The microphotographs show that the iron sulphides are in the form of fine inclusions surrounded from all sides by the amorphous substance, which is the main component of slags. This is why the identification using the X-ray diffraction method could have been hindered. It may be assumed that the low peaks resulting from sulphides were overlapped by the reflections of other components. A similar situation was noted in observations using electron scanning microscopy – this technique, however, has identified the presence of forsterite.

All the examples provided above clearly show that conducting a mineralogical study of smelter slags requires the application of several complementary research methods. This is caused by the variable and differentiated phase composition of slags as well as the specific structure of the phase components, most often occurring as fine crystallites covered by the amorphous substance.

5. Conclusions

Based on the conducted research, the following conclusions have been made:

- It has been exhibited that the application of several complementary research techniques is necessary in the determination of the phase composition of slags, as only the application of the full spectrum of tests may provide reliable information regarding the presence of the individual phases in the slags. Some phases, e.g. iron sulphides, forsterite and ferrihydrite were found in the Zn-Pb slags using only one of the applied research methods.
- In case of the tested Zn-Pb slags and in view of obtaining qualitative characteristics of the components of slags, the phase identification tests by means of X-ray diffraction were found to be precise. Based on them, also the percentage of the individual crystalline phases and the percentage of the amorphous substance have been determined.
- Analyses using electron scanning microscopy and X-ray spectral microanalysis allowed for the determination of the chemical composition of the individual phases. In that case, the X-ray spectral microanalysis was found to be more precise and it allowed for the identification of substitutions of elements occurring in small amounts in the phase components, i.e. the rare earth elements.

References

[1] Baricová D, Pribulová A, Demeter P, Bulko B and Rosova A 2012 *METABK* **51(4)** p 465
[2] Borell M 2005 *Proc. Int. Conference of Mining, Environment, Metals, Energy Recovery* (Sweden) p 1
[3] Chycki A, Sobierajski S, Gramała J and Kubacz N 2001 *Rudy i Metale Nieżelazne* **46(12)** p 615
[4] Fidancevska E, Vassilev V, Hristova-Vasileva T and Milosevski M 2009 *Journal of the University of Chemical Technology and Metallurgy* **44(2)** p 189
[5] Iacobescu R I, Koumpouri D, Pontikes Y, Şaban R and Angelopoulos G 2011 G *U.P.B. Sci. Bull., Series B* **73(1)** p 1454
[6] Rai A, Prabakar J, Raju C B and Morchalle R K 2002 *Construction and Building Materials* **16(8)** p 489
[7] Sobierajski S 2002 *Rudy i Metale Nieżelazne* **7** p 334
[8] Tajchman Z, Tora B and Nowakowski K 2001 *Gospodarka Surowcami Mineralnymi – Mineral Resources Management* **17** p 295
[9] Wang Ch, Li K, Yang H and Li C 2017 *ISIJ International* http://dx.doi.org/10.2355 /isijinternational.ISIJINT-2016-683] dostęp 21.03.2018
[10] Arslana C and Arslan F 2002 *Hydrometallurgy* **67 p** 1

[11] Kucha H and Jędrzejczyk B 1995 *Mineralogia Polonica* **26(2)** p 75
[12] Muszer A 1996 *Fizykochemiczne Problemy Mineralurgii* **30** p 193
[13] Muszer A 2006 *Physicochemical Problems of Mineral Processing* **40** p 89
[14] Jonczy I 2014 *Archives of Metallurgy and Materials* **59(2)** p 481
[15] Jonczy I 2016 *Archives of Metallurgy and Materials* **61(1)** p 61
[16] Mateus A, Pinto A, Alves L C, Matos J X, Figueiras J and Neng N R 2011 *Environment and Waste Management* **8(1-2) p** 133
[17] Yildirim I Z and Prezzi M 2011 Hindawi Publishing Corporation Advances in Civil Engineering [https://doi.org/10.1155/2011/463638]
[18] Ettler V, Legendre O, Bodénan F and Touray J-C *The Canadian Mineralogist* 2001 **39(3)** p 873
[19] Smirnov L A, Rovnushkin V A and Smirnov A L 2015 *Russian Metallurgy (Metally)* **3** p 191
[20] Babenko A A, Krivykh L Yu, N V Mukhranov, V V Levchuk, S A Remigo, and M V Savel'ev 2012 *Steel in Translation* **42(2)** p 114
[21] Mineralogy Database [https://www.mindat.org/] dostęp 21.03.2018
[22] Jonczy I and Stanek J 2013 *International Journal of Nuclear Research* **58 (1)** p 127
[23] Jonczy I, Fornal P and Stanek J 2015 *Archives of Foundry Engineering* **15(3)** p 21
[24] Blaschke W, Witkowska-Kita B, and Biel K 2015 Annual Set The Environment Protection *Rocznik Ochrona Środowiska* **17** p 792
[25] Communication from The Commission to the European Parliament, The Council, The European Economic and Social Committee and The Committee of The Regions on the 2017 list of Critical Raw Materials for the EU (Bruksela, 13.09.2017 r., COM (2017) 490 final)

IOP Conf. Series: Materials Science and Engineering **427** (2018) 012002 doi:10.1088/1757-899X/427/1/012002

Trends in global copper mining – a review

S Pietrzyk[1] and B Tora[2]

[1] Department of Non Ferrous Metals, AGH University of Science and Technology, Krakow, Poland

[2] Department of Mining and Geoengineering, AGH University of Science and Technology, Krakow, Poland

tora@agh.edu.pl

Abstract Copper and its alloys are used for diverse applications that are required for us to attain a specific standard of life. Continuous copper production and use are needed for the development of the society. Future copper demand will be covered mainly by the discovery of new deposits and mining. However, technological improvements and efficient designs intended to limit copper consumption will be equally important, similarly to recycling, as part of the activities associated with circular economy. Current data indicate that the global extraction of copper reached 20.2 million tons in 2016, while the output from metallurgical plants attained the level of 19.0 million tons in the same year. However, the refined copper production output increased to 23.3 million tons in 2016, including 3.9 million tons obtained from recycling. One can ask the question: Can we run short of copper, considering the present 6% increase and projected further increase of copper extraction and use? This paper is trying to find an answer to this question, based on such factors as copper resources and their long-term availability, global distribution of identified and projected copper resources, global extraction rates, and the trends in copper extraction capabilities.

1. Global copper resources

Usually, the projection of the availability of a given metal in the future is based on the evaluation of the existing resources in comparison to mining capabilities.

International classifications (United Nations Framework Classification (UNFC)) distinguish [1,2] the extractable and non-extractable resources, comprising non-industrial and subeconomic ones, as well as economic ones that have not been included in industrial or non-industrial resources (figure 1).

The class of "Reserves" includes the resources that can be extracted in an economically feasible manner, omitting losses or waste, and taking into account depletion. the term "Resources" includes geological assets, excluding the "Reserves". According to the United States Geological Survey (USGS), the global copper reserves are presently estimated at ca. 720 Mt, while the remaining resources are estimated at ca. 5.600 Mt (figure 4). The latter figure contains the identified and undiscovered quantities that are estimated at ca. 2.100 Mt and 3.500 Mt, respectively [3].

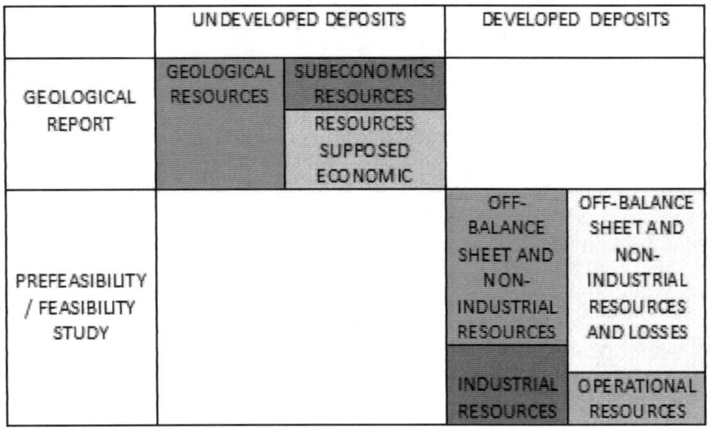

Figure 1. Classification of resources in international systems [1, 2].

The estimation of the undiscovered copper resources, based on the status of geological knowledge, was provided by USGS in 2013. Undiscovered copper resources were estimated, assuming two basic forms of occurrence, being so typical that they constitute 80% of identified resources. Those are porphyry deposits (60%) and sedimentary ore deposits (20%) [4]. The distribution of such resources is presented in figure 2 [4].

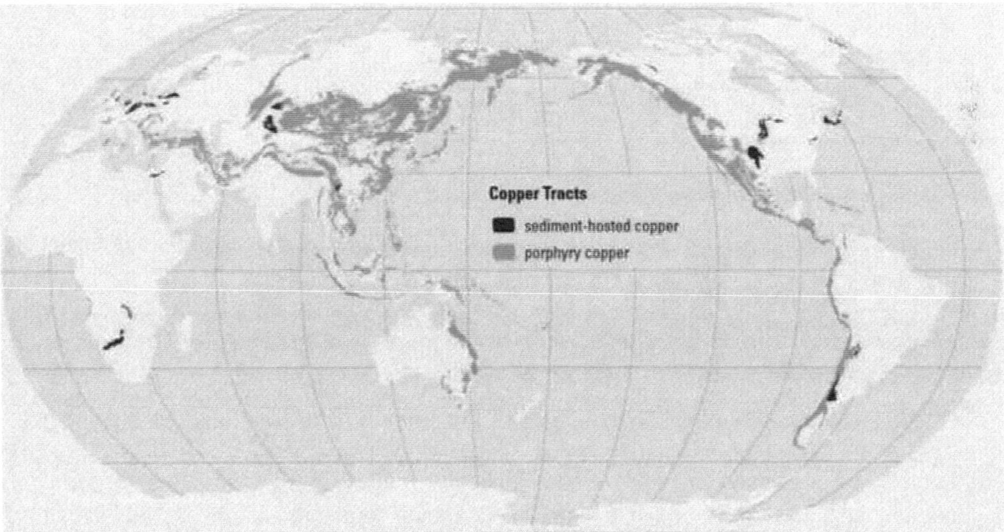

Figure 2. Distribution of main global copper resources (without those from sea bottom) [4].

The estimated average amounts of undiscovered porphyry and sedimentary resources amount to 3.100 and 400 Mt, respectively, giving 3.500 Mt of copper on the global scale (not qualified as

IOP Conf. Series: Materials Science and Engineering **427** (2018) 012002 doi:10.1088/1757-899X/427/1/012002

extractable deposits or resources). In total, the copper resources (both undiscovered and identified) are estimated at 5.600 Mt (figures 3, 4, and 5) [3, 5].

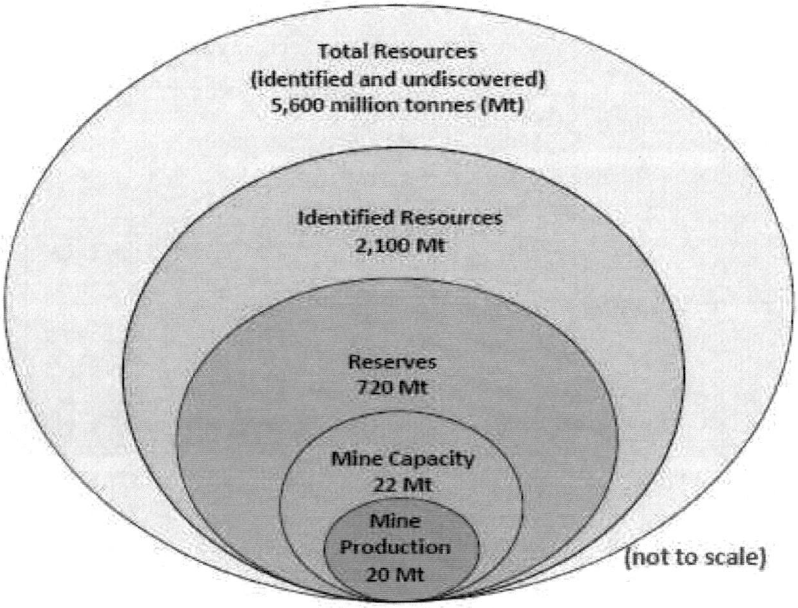

Figure 3. Global resources and mining output [3].

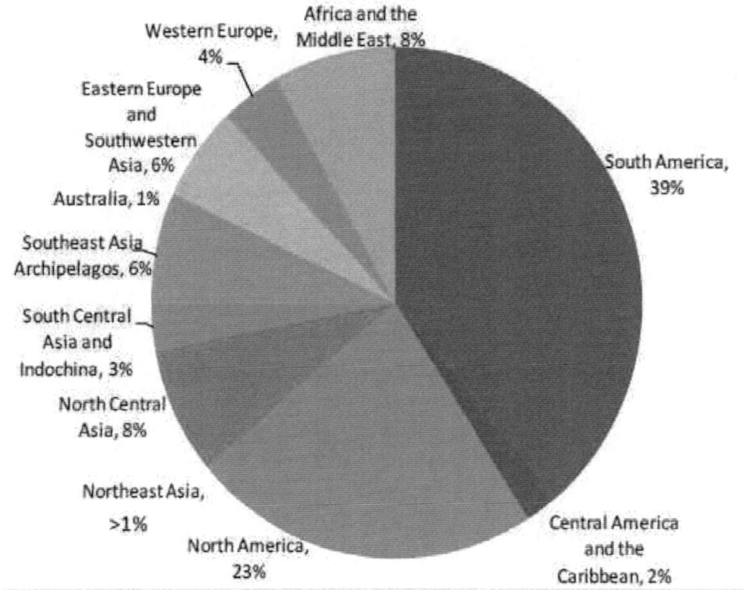

Figure 4. Distribution of identified copper resources [3].

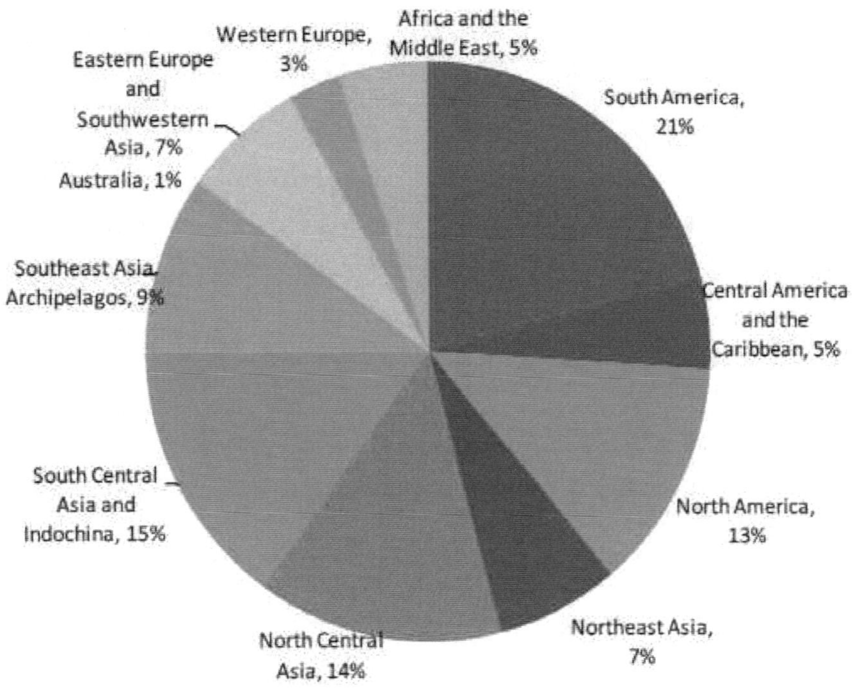

Figure 5. Distribution of undiscovered copper resources [3].

The latter value excludes the copper resources occurring at sea bottom. The geological recognition of such resources will increase the figures in both groups: reserves and resources. We should also take into account the average Cu content in the earth's crust down to the depth of 3.3 km (estimated as a limit of future mining operations), and that amounts to ca. 85 ppm. This parameter additionally increases the resource content to the astronomical figure of 300.000 Mt [6].

2. Global copper mining

Pure copper, in the form of natural copper, is rarely found and it constitutes about 1% of all copper compounds. The remaining copper resources occur in the form of sulphide ores in about 90% and oxide ores in about 9%. Copper is recovered from the metallic minerals occurring in more than 160 compounds. The main copper minerals are the following: chalcosine, bornite, chalcopyrite, digenite, covellite, cuprite, malachite, djurleite, anilite, and idaite. The basic production of original copper starts with the extraction of natural copper ores in copper mines. There are three basic methods of copper ore extraction: open-pit, underground, and leaching mining. Open-pit mining is a dominating form on a global scale.

Figure 6 presents the global trends in copper mining in 1900-2016. As we can notice, the output reached 20.2 Mt in 2016 and showed more than a 3% increase. We also witnessed the appearance of a new copper production technology in the 1970's, consisting in copper extraction and leaching in heaps, with the delivery of the mineral in the form of electrolyte (the so-called solvent extraction and electrowinning, or the SX-EW method).

IOP Conf. Series: Materials Science and Engineering **427** (2018) 012002 doi:10.1088/1757-899X/427/1/012002

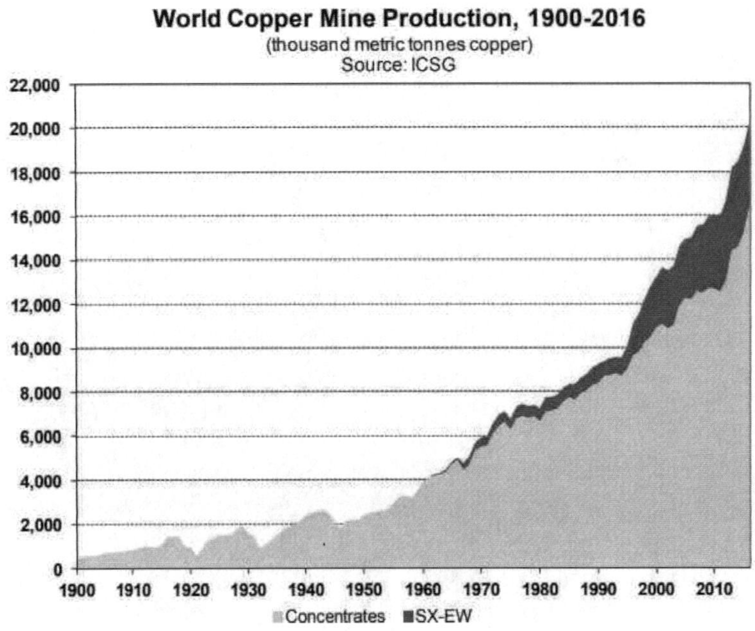

Figure 6. Global copper mining [3].

Figure 7 presents the changes of continental shares in copper ore mining.

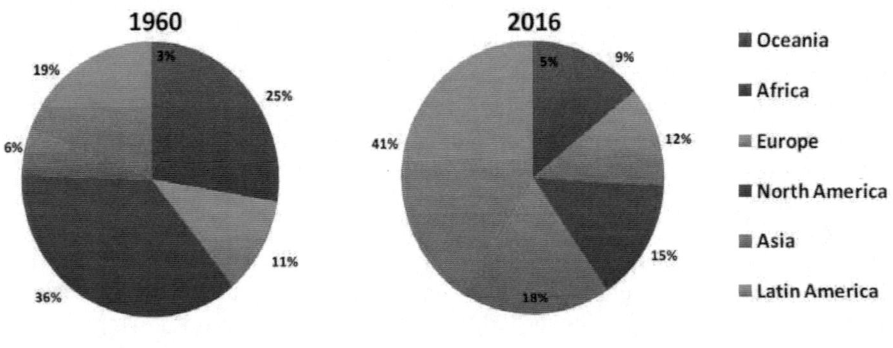

Figure 7. Continental shares in copper mining [3].

We can observe a fast increase of copper production in South America, mainly in Chile and Peru, with a 41% of share in global copper production. A strong increase from 6 to 18% also occurred in Asia, mainly in China. A drop was noted in North America, from 36 to 15%. Chile is the largest copper ore producer, representing the level of 5.5 Mt in 2016 (figure 8). The Escondida Chile is the largest copper mine, with the output of 1.270 Mt (applying the concentration and SX-EW methods).

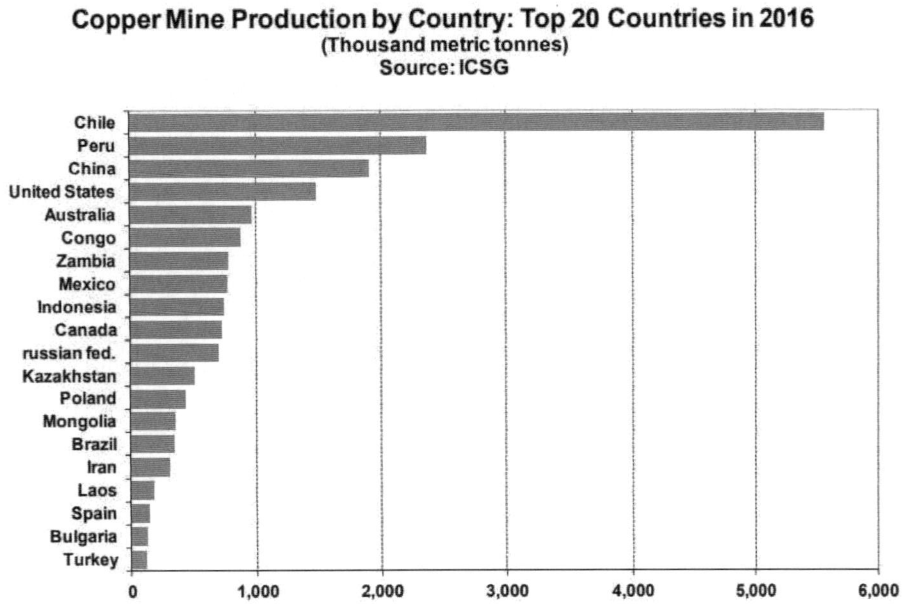

Figure 8. The largest global copper mining countries [3].

3. Global metallurgical copper production and copper applications

The original copper production consists in obtaining the metal from natural ores. We can distinguish two production methods: pyro- and hydrometallurgical processes. The former consists in high-temperature melting of copper concentrates, containing 20-40% of Cu, obtained by ore enrichment. Copper is melted into the form of cake, or an alloy of sulphides, mainly Cu_2S and FeS, containing 50-75% of Cu. The next stage involves converting processes, changing melted cake into the so-called blister copper (with 98.5-99.5% of Cu). Subsequent stages include fire refining and casting in the form of ingots or anodes designed for further electrorefining copper processes. Copper cathodes are the final product, with high 99.99% purity.

There is an alternative method of copper production called the SX-EW method. It is applied to poor oxide ores, with low copper content, as well as some sulphide ores. The method includes leaching (Solvent Extraction, SX) and electrowinning (EW).

Refined copper is the output product of both methods, with a comparable purity class. It is estimated that the SX-EW method was used to produce 16% of the total global refined copper output in 2016, and the remaining proportion originated from the pyro-metallurgical processes.

However, there is still another source of raw material available for the production of copper: copper scrap. That resource is subject to recycling. Scrap is divided into new (post-production scrap), occurring as a result of ore processing, and old scrap (post-consumer scrap) collected from scrapped products. Scrap constitutes valuable and increasingly significant raw-material in the recycled copper production, in the times of circulation economy implementation. It was estimated that recycled copper constituted about 17% of the global refined copper output in 2016 (figure 9).

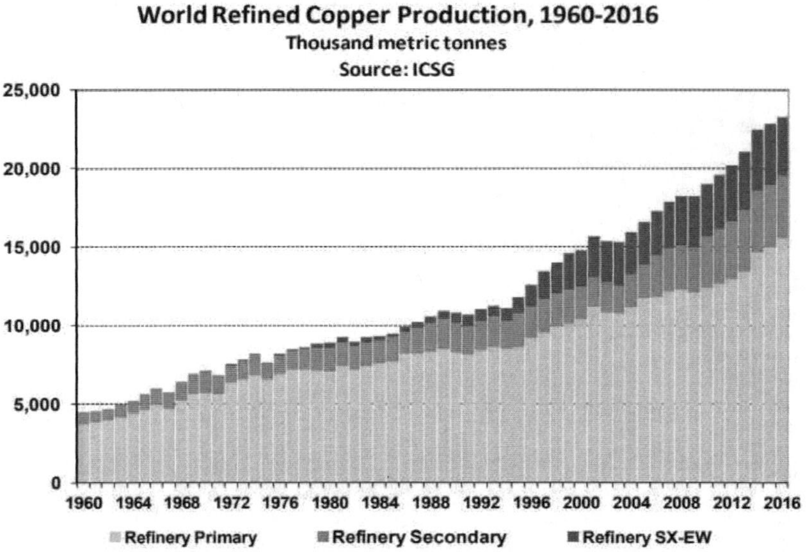

Figure 9. Global output of refined copper, 1960-2016 [3].

The production of blister copper reached the level of about 19 Mt in 2016. China was the largest producing country of blister copper and anodes in 2016 (7.2 Mt). The total refined copper output amounted to 23.3 Mt in that year, including 3.9 Mt obtained from recycling (recycled refined copper).

Copper consumption is continually increasing, together with the development of global economy, and it is expected that the trend will continue because more and more people and societies will aspire to a higher standard of life, with the access to electricity, sewage systems, and modern appliances.

Wide-ranging copper applications rely mainly on the metal's properties, e.g. high electrical and thermal conductivity, high workability and ductility, resistance to corrosion, and antibacterial effect. The latter feature is becoming more and more significant.

Nearly 90% of the global copper output is used in electrical devices and communication systems. Copper represents the highest electrical conductivity after silver, and that causes that it is widely used (40%) in electricity generation and electricity distribution systems by both industrial and individual consumers, in an efficient and safe manner. 12.5% of copper is used for the construction of electric and electronic equipment (electric circuits, cables, contacts etc.) [10,11]. Another 12.5% is used in the transportation industry (e.g. in the form of copper cable bundles with the highest purity, applied in trains, airplanes, trucks etc., conducting current from batteries to various control devices, lighting, on-board computers, or satellite navigation systems).

Another 20% of the whole copper output is used in the building industry (sewage systems inside the buildings, roof decking, or facade panels). Copper allows to produce light and durable structures that do not require maintenance, since the material is natural, nicely looking, and recyclable. The remaining 10% of copper output is designed for coins, sculptures, jewellery, musical instruments, kitchen utensils, and other consumer products [11].

Besides, a different application is coming back to copper consumption, with anti-microbiological touch surfaces designed for hospitals or public facilities.

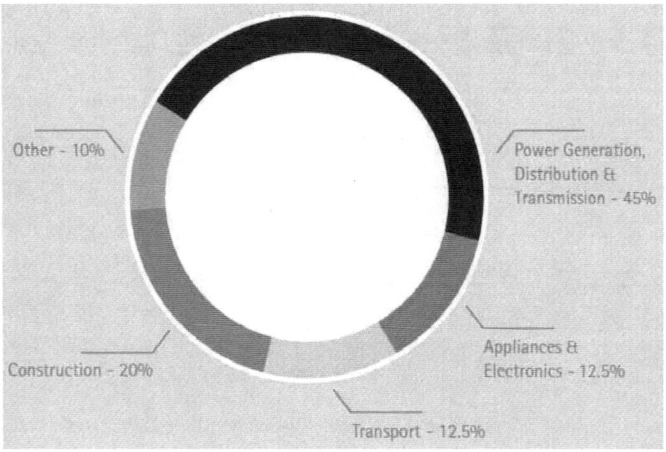

Figure 10. Copper applications [7].

4. Projections of the development of copper mine capabilities

It is estimated that copper extraction will reach 25.9 Mt in 2020 of which 19% will be obtained by the SX-EW method. That will be equivalent to a 10% increase in comparison to 2016 when the global extraction capability reached 23.5 Mt.

It is envisaged that the mining capabilities will be increasing by 2.9% a year, with the implementation of new capabilities in the existing mines (figure 11).

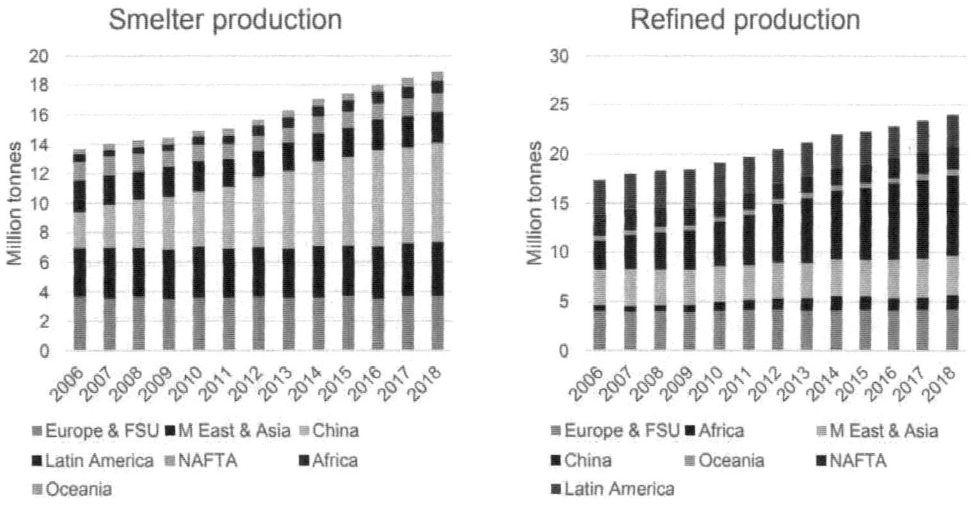

Figure 11. Trends in copper extraction [3].

What is a good indication of copper mining trends is the relationship between mining output and mining capabilities, called the proportion of using production capabilities. The global indicators of using copper mine production capabilities amounted to about 86% in 2016, and that shows some reserves, allowing for extraction increase.

Another good indicator of the status of copper resources, in comparison to copper extraction capabilities, is the comparison of the figures relating to resources and mining output (figure 12) [8].

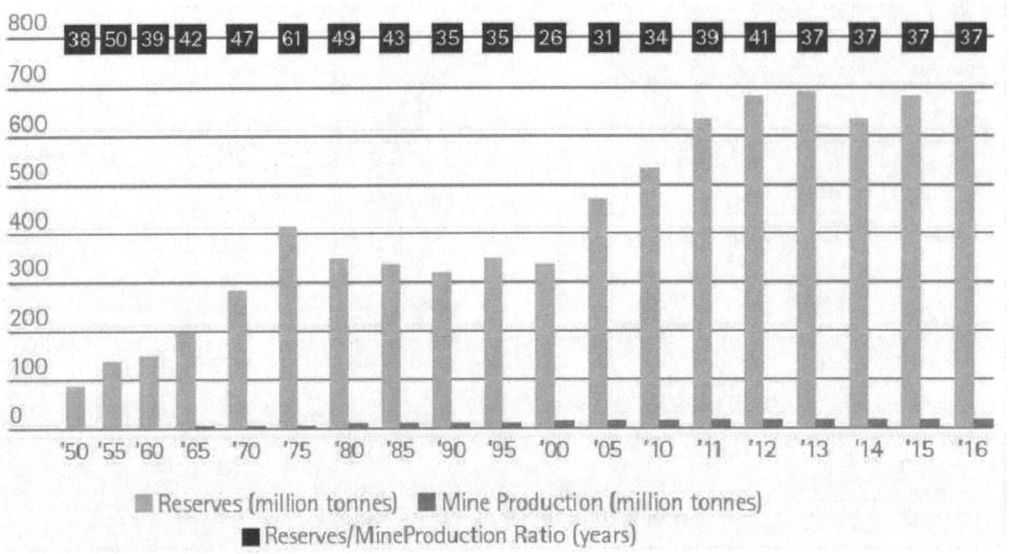

Figure 12. Historical copper reserves and output in Mt [8].

5. Conclusions

Based on the data presented above, we can conclude that there are very promising trends in the development of copper mining. The current copper resources and ore mines' capabilities, as well as the technologies of the production of primary and refined copper provide good prospects for further regular increase of copper production in the future.

However, certain limitations exist as well. Those can affect the growing trends. One of the reasons of such an option is the fact that the high-quality copper ores become rare and difficult to discover. The success of the resource exploration requires long-term and continued capital investments at the initial stages of resource evaluation which is charged with a high risk of losses. The presently operated copper mines struggle with a number of problems, starting with the exhaustion of quality ores and the necessity to extract deeper and deeper deposits and ending with the growing expectations of the governments, regulatory authorities, and the local communities.

The Wood Mackenzie consulting agency expects that a possible drop in ore mining can reach even 17% [9]. The analysts estimate that the mining output may be dropping by 230 kt a year after 2019. The global increase of demand for copper is estimated at 2.1% a year, or 521 kt of new supply. Consequently, the deficit can increase in case of ore mining decrease.

New high-class copper ore deposits are required to meet the growing global demand for copper. The increase of copper ore extraction can be enforced in the coming decades by such new industries as electric car production, power generation from renewable resources, or infrastructure development in growing and emerging economies.

The electric car production is increasing fast. Hybrid cars consume 40 kg of copper each: twice as much in comparison to conventionally or natural gas-powered cars. Electric cars use 90 or more kilograms of copper, which is three or four times more than in the case of traditional cars. It is expected that the global fleet of electric cars will increase from the present one million vehicles to about 140 million by 2035 [11].

The increase of primary copper production can be greatly supported by renewable energy. Renewable technologies, based on wind or solar energy displayed a nearly 50 time increase in the recent decade. The share of solar energy in global energy generation increased more than twice in only three years, although it still constitutes a small part of the total energy balance. From now up to 2040, 8 billion dollars will be spent on the global renewable resources, or about two thirds of all the expenditures for energy. The renewable energy systems require 4 to 12 times more copper than the traditional energy generation plants using fossil fuels. The growing demand for solar energy will require from 7 to 10 million tons of copper until 2030 [11].

Another hope of copper mining is associated with the growing economies that create new demand for copper. It is expected that the technological development and the positive demographic growth only in Asia will create an additional demand for more than 30 million tons of copper by 2030.

References

[1] Nieć M 2009 Polska i międzynarodowa ramowa klasyfikacja zasobów (UNFC) złóż kopalin stałych i węglowodorów– podobieństwa i różnice *Górnictwo Odkrywkowe* **50** (2-3) pp 50-7

[2] Definitions for mineral resources and reserves from U.S. Department of the Interior 2001 U.S. Geological Survey *Mineral Commodity Summaries* (Washington, D.C.: GPO)

[3] World Copper Fact book 2017, *International Copper Study Group*, Lisbon, Portugal

[4] https://www.usgs.gov/media/images/global-copper-map-0

[5] Edelstein D L 2013 Copper, metals and minerals *U.S. Geological Survey Minerals Yearbook 2011* **1**, pp 20.1–20.25

[6] Kesler S E and Wilkinson B 2008. Earth's copper resources estimated from tectonic diffusion of porphyry copper deposits *Geology* **36(3)** pp 255-8

[7] The Long-Term Availability of Copper 2017 *International Copper Association* http://copperalliance.org/wordpress/wp-content/uploads/2018/02/ICA-long-term-availability-201802-A4-HR.pdf

[8] Copper Mineral Commodity Summary 2017 *United states Geological Survey* (USGS) https://minerals.usgs.gov/minerals/pubs/commodity/copper/mcs-2017-coppe.pdf

[9] The market fundamentals of copper *Northern Fox Resources* https://www.thenorthernfox.com/our-markets/

[10] Pietrzyk S and Tora B 2017 Processing of Non – Ferrous Metals Secondary Raw materials in Poland – Trends, Opportunities and Threats *Inżynieria Mineralna – Journal of the Polish Mineral Engineering Society* **2(40)** pp 81-92 DOI: 10.29227/IM-2017-02-9

[11] Wieniewski A and Myczkowski A 2017 Achievements and Participations in the Development of Mineral Processing and Waste Utilization if the Institute of Non Ferrous Metals (IMN) in Gliwice *Inzynieria Mineralna – Journal of the Polish Mineral Engineering Society* **2(40)** pp 15-24 DOI: 10.29227/IM-2017-02-2

IOP Conf. Series: Materials Science and Engineering **427** (2018) 012003 doi:10.1088/1757-899X/427/1/012003

Development of dry coal gravity separation techniques

W Mijał and B Tora

Department of Environmental Engineering and Mineral Processing, Faculty of Mining and Geoengineering, AGH University of Science and Technology, Mickiewicza 30, 30-059 Cracow, Poland

walmij@agh.edu.pl

Abstract. Dry separation method was used at the beginning of 20th century to separate coal separation process. First industrial applications of dry separation took place in the middle of 1930s. At that time air separator Frazer-Yankey, based on aero suspension was used in in Pittsburgh and Lundale, West Wirginia coal preparation plants. Later, some more constructions like RAW air jig, air table "V" type Birtley, air jig type Kirkup, air table "V" type Peale-Davis or air jig type Bruay-Soulary were constructed. Fast development of wet technologies of processes, especially jigging and heavy dense medium separation caused the discontinuation of the further development of dry separation methods and replacement of dry preparation plants with plants based on wet methods of enrichment. At the end of 20th century and beginning of 21st century dry separation method started once again to be widely used in the world. New types of equipment like FGX vibrating air table, TFX air jig or KAT process technology were introduced to wide applications in Asian countries. The paper describes few constructions of old air separators and new ones, used nowadays. It contains comparison of the separation results obtained by means of various equipment and shows possibility to separate coal in laboratory scale air jig.

1. Introduction

Many different kinds of enrichment methods are used in coal preparation, like for example gravity separation processes conducted in heavy dense media, in jigs or on shaking tables or physico-chemicals processes like flotation process. One of the methods used in coal beneficiation process is dry method. This method was popular at the beginning of the 20th century. Then it was successfully used in the United States (1919). Separator Frazer-Yancey was used in USA and it was working as a heavy dense medium (air and sand suspension). The first big coal preparation plant was built in 1930 in Pittsburgh. First type of this device was equipped with rigidly placed working bed of small angle. The working bed consisted of sieve with very small mesh and big working chamber was situated above of it. 5 coal preparation plants with this technological solution were built in the early 1930s in USA. The biggest separator in this plants was used in Lundale, West Wirginia with total capacity of 200 Mg/h. In the middle of 1930s Frazer-Yankey separator was modified and the option with moving working bed was added. At the same time dry separation method came also to Europe where it was used in England (1925), followed by Belgium, Germany or Poland (1928). The rule of the dry method is based on separation of the material by ascending or pulsatory air stream. Dry enrichment is usually applied in locations where water shortages occur causing that there is not enough water to conduct wet processes as well in a harsh climate due to the possibility of freezing of the separation products after being separated in a water medium. Nowadays, this method is used in China, Mongolia, India, USA or Russia. The raw materials that can be enriched by means of this method are mainly hard coals with a large proportion of coal-fired or tailings fractions and for lignite (hard types).

Recently, a developing time started for new equipment for dry coal enrichment. It occurred especially in China (CFX, TGX or FGX) and Korea (KAT Process, a dry coal preparation process developed by Korean Institute of Geoscience and Mineral Resources). The air vibrating table type FGX became very popular which is used in United States, Turkey, India, South Africa and a dozen other countries. In 2012 the Institute of Mechanized Construction and Rock Mining purchased a vibrating air separator type FGX-1 through the company WARKOP Sp. z o.o. (this company represent Tangshan Shenzou Machinery Co. Ltd. on the Polish market) [1, 2, 3, 4, 5, 6].

2. Dry separation equipment

Dry beneficiation techniques can be divide into two main categories: beneficiation on air table and separation by air jigs. Air tables work by using ever-rising stream of air. Work plate can also vibrate to increase the accuracy of separation. Obtaining of the products depends on construction of the table. Good examples of old separators of this type are air table “V”, type Birtley or air jig type RAW.

Air table “V”, type Birtley consists of two symmetrical parts (with divided work plate) getting V shape. Both parts work independently of each other. Each of the halves has a slope in a direction perpendicular to the axis of the table. The working plate consists of steel net with riffles of decreasing height related to the feed loading inlet. The air chamber is located under the net and the air is supplied to it by the fan through a pipe. The air pressure can be regulated in each part of the air table. Under the influence of the shaking action of the table and blasting air raw material is separated into three products: concentrate, middlings and tailings. Heavy particles (tailings) are transported on the surface of the net by riffles and are unloaded at the end of working plate. Concentrate is unloaded at the beginning part of the table.

The RAW air type jig is one of the precursors of air jigs currently used. Its construction is based on box equipped with steel sieve arranged in a staircase-like way. Each subsequent sieve is thinner than sieve before (3 parts of sieves, together). The box of air jig is located on steel construction by spring levers (as a result the box shakes). The air chamber is located under the sieve and is supplied by pulsatory air stream. Raw feed is separated on upper part of sieve by air pulsation during working plate motion. Firstly, fine coal secondly coarse coal and lastly middlings and tailings are separated. Clean coal and middlings are separated at the bottom of the box. Tailings are unloaded at the end of the last sieve to a belt conveyor.

In the past some works on constructing air separator with aero suspension were also performed As a load sand or magnetite were used. One of the devices used to separate coal by aero suspension was Frazer-Yankey separator used in USA in 1930s. The sand was used here as a load. The comparison of the results of coal separation in dense medium separator with aero suspension and typical dense medium liquid are shown in table 1. The research was conducted by W.I. Korszunov in USSR [5, 6, 7, 8, 9].

Table 1. Comparison of the results of coal separation by heavy liquid and aero suspension [7].

Type of coal	Separation in heavy liquid with density 2.0 g/cm^3				Separation in aero suspension (concentration of the solid phase 33.4%)			
	Light product		Heavy product		Light product		Heavy product	
	Yield [%]	Ash [%]	Yield [%]	Ash [%]	Yield [%]	Ash [%]	Yield [%]	Ash [%]
Anthracite	91.2	10.7	8.8	83.4	92.2	10.8	7.8	90.7
Coking coal	69.6	15.0	30.4	80.5	72.6	17.4	27.4	82.2

Results showed that raw feed of coal can be separated by aero suspension with similar efficiency comparing to separation in heavy dense medium. Very good results were obtained after separation of anthracite coal. Aero suspension helps to get final product with ash content >90% in tailings and around 10.8% of ash in concentrate. This result was better than in case of separation in heavy liquid. It showed that separation in air separators can be efficient.

After many years construction of air tables was modified and now two good constructions exist: vibrating air table, type FGX from China and TFX air jig for small particles and KAT (Korean Advanced Technology) technology developed by Korea.

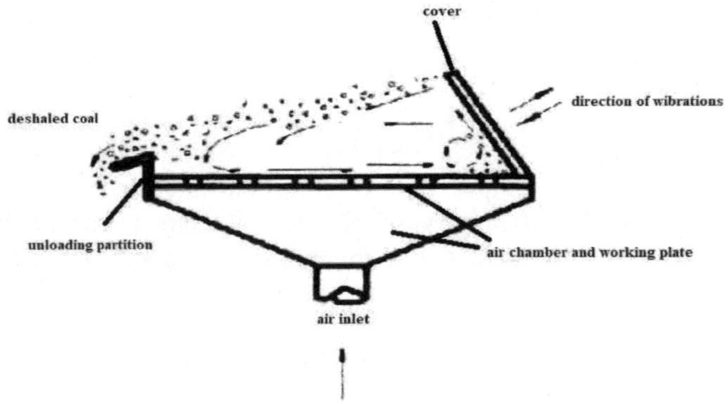

Figure 1. Material distribution on FGX working plate.

Vibrating air separator FGX consists of a funnel feed, dosing feeder, perforated work plate, vibrator, air chambers, dust removal module and a mechanism that allows to change the angle of inclination of the working plate and the frequency of vibrations. The feed is fed on the installation by means of a vibratory feeder and then it goes to a working plate inclined at various lateral and longitudinal angles set in a vibrating motion by means of a vibratory drive.

In order to ensure air supply under the working plate the air chambers are located which are fed by a centrifugal fan. The fine coal material forms a fluidized bed (air-solid slurry) as a result of contact with air. Thanks to it, individual particles can fall relatively to each other dependably on their size or density. Under the influence of combined forces: air current and vibrations the coal bed is raised and becomes stratified according to the density. The lighter material is suspended on the surface of the fluidized bed and the particles of higher density sink deeper. An additional phenomenon is the liquefaction effect resulting from the interaction between small particles. This phenomenon improves the efficiency of the separation of coarse fractions. Fine material located on the surface of the layer tends to slide over its surface and falls continuously under the influence of gravity through the partition at the edge of the plate (dumping of enriched coal). The heavy material falls to the bottom of the layer and is moved towards the waste collection point (gangue). Figure 1 shows the distribution on FGX working plate.

KAT (Korean Advanced Technology) table is a type of pneumatic table separator. In this separation process the differences in density between grains in ROM coal are used. Figure 2 shows the separation scheme. The denser material falls down on deck to form a lower layer and moves to the oscillating direction of the table deck by the influence of eccentric motion. The upper layer of light particles roll down with help of airflow and shaking table to the lower part of the deck and accumulates along the blocking wall. When the level of accumulated light particles exceeds the height of blocking wall it starts to overflow. This specific design of KAT table generates an autogenetic medium of light particles along the lower end of the deck thus increases a product quality. KAT table separates coal into 3 different fractions. The light fraction-coal, the heavy fraction-tailing and the middling.

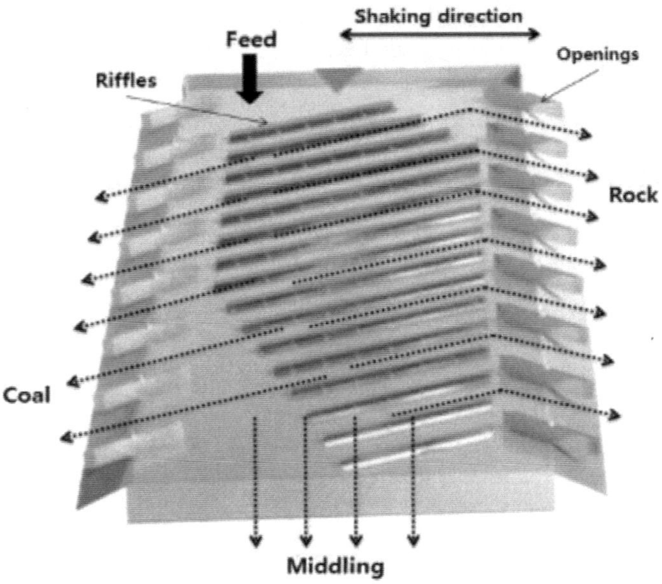

Figure 2. KAT air table deck – trajectory of separated products [13].

As an example of modern air jig a TFX air jig will be described. The dry cleaning jig operates on a principle which is basically similar to that of a wet cleaning jigging machine. The fine size material on the deck tends to become gradually loosened and stratified under the combined effect of vibration force and the pulsation produced by the upward air-stream coming from deck bottom. The heavier material gradually sinks down to bed bottom while the lighter material gradually floats up to the top of the bed. During the proper jigging process, a table material bed is finally formed. The heavier material at bed bottom is discharged through the discharge unit while the other materials move forward to go through further separating process under the effect of vibration force. The TFX air jig unit is shown in figure 3 [10, 11, 12, 13, 14].

Figure 3. Air jig TFX unit [14].

3. Results

To show the quality of dry separation all results from TFX, FGX and KAT will be compared with result from laboratory scale air jig created in Silesian University of Science and Technology. This

IOP Conf. Series: Materials Science and Engineering **427** (2018) 012003 doi:10.1088/1757-899X/427/1/012003

small jig have a periodic activity, working chamber consists of rings, dust extraction system is attached to the bottom of working chamber. To collect the separation products whole working chamber is dismantled into smaller parts. Every ring is a one of final products (usually 3 products: clean coal, middlings and tailings).

Table 2. Results of separation by using KAT Process technology for coal in particle size fraction 1-5 mm [13].

Test	Clean coal		Middlings		Tailings	
	Yield [%]	Ash [%]	Yield [%]	Ash [%]	Yield [%]	Ash [%]
1	57.20	16.00	20.40	44.50	22.40	86.10
2	53.40	9.40	22.90	37.90	23.70	88.70
3	58.30	10.70	14.00	37.30	27.60	84.00

Table 3. Results of separation by using KAT Process technology for coal in particle size fraction 5-10 mm [13].

Test	Clean coal		Middlings		Tailings	
	Yield [%]	Ash [%]	Yield [%]	Ash [%]	Yield [%]	Ash [%]
1	41.10	13.30	18.20	52.90	40.70	88.60
2	34.50	6.90	32.80	53.40	32.70	89.80
3	36.70	7.70	26.10	46.10	37.20	89.60
4	44.40	8.40	18.90	56.80	36.60	87.00
5	47.80	10.60	13.50	66.90	38.60	84.70

KAT Process technology for separation of coal particle size fractions 1-5 mm and 5-10 mm showed that specific construction and motion of working plate can help to get final tailings with high quality. Most of separated tailings have ash content higher than 85%. The stone separated by the dry stone method can find application in the building materials industry or engineering works. Concentrate can be used as final product in coal power plants or by regular customers at home.

Table 4. Results of separation by using TFX air jig for coal in particle size fraction 13-0 mm [14].

Coal	Raw coal ash [%]	Clean coal		Middlings		Tailings	
		Yield [%]	Ash [%]	Yield [%]	Ash [%]	Yield [%]	Ash [%]
Coking coal	19.75	77.49	9.10	7.50	28.43	15.01	70.41
Power coal	42.53	58.73	27.26	21.93	54.88	19.39	73.47
Power coal	35.85	60.97	20.71	20.90	46.27	18.13	74.73

Table 4 shows that air jig can be used for coking coal or steam coal as well. Final products fulfill the requirements of users. Results show that the air jig can help to get good results during separation of coking and steam coals. There is still some amount of coal in tailings - the ash content is too low <75%. Concentrate of ash content higher than 25% is unacceptable.

Table 5. Results of separation by using laboratory air jig for coal in particle size fraction 2-20 mm [15].

Test	Clean coal		Middlings		Tailings	
	Yield [%]	Ash [%]	Yield [%]	Ash [%]	Yield [%]	Ash [%]
1	77.40	10.60	11.40	17.50	11.20	78.20
2	67.70	11.70	16.20	20.10	16.10	46.10

Table 6. Results of separation by using laboratory air jig for coal in particle size fraction 12-6 mm [15].

Test	Clean coal		Middlings		Tailings	
	Yield [%]	Ash [%]	Yield [%]	Ash [%]	Yield [%]	Ash [%]
1	72.11	6.01	9.10	12.39	18.80	67.47
2	74.90	8.91	10.60	14.59	14.50	68.62

Tests on laboratory scale jigs show that dry separation is a good process to get final products of low ash content. In particle fraction 20-2 mm ash content level in concentrate is between 10-12%. Ash content in tailings shows that coal particles can occur in heavy product.

Table 7. Results of separation by using vibrating air table type FGX for coal in particle size fraction 20-0 mm [16].

	Parameter	Feed	Concentrate	Middlings	Tailings
1.	Ash content [%]	31,7	21,2	29,5	80,5
2.	Moisture content [%]	9,6	9,0	8,4	5,3
3.	Sulphur content [%]	0,56	0,62	0,68	0,44
4.	Calorific value [kJ/kg]	17 151	21 558	18 885	1 860
5.	Yield [%]	100	77	3	18

Table 7 showed that coal in particle size fraction 20-0 mm can be enriched with good results. Concentrate can be used by commercial sector (ash content around or a little bit more than 20% and good calorific value >21000kJ/kg). Coal-fired boilers in Polish energetic sector are adapted for using coal of higher content of ash. Middlings can be sold to energetic sector or can be used to produce energetic mixes.

4. Conclusions

Dry separation is one the old methods used before introduction of wet processes such as beneficiation in jigs, heavy dense medium or flotation process. Among the advantages of this method can be mentioned that they:

- Eliminate water and sludge circulation from processing plant. As an example the solution showed by scientists from China University of Mining and Geology and Tangshan Shenzhou Manufacturing Company can be presented (figure 4):

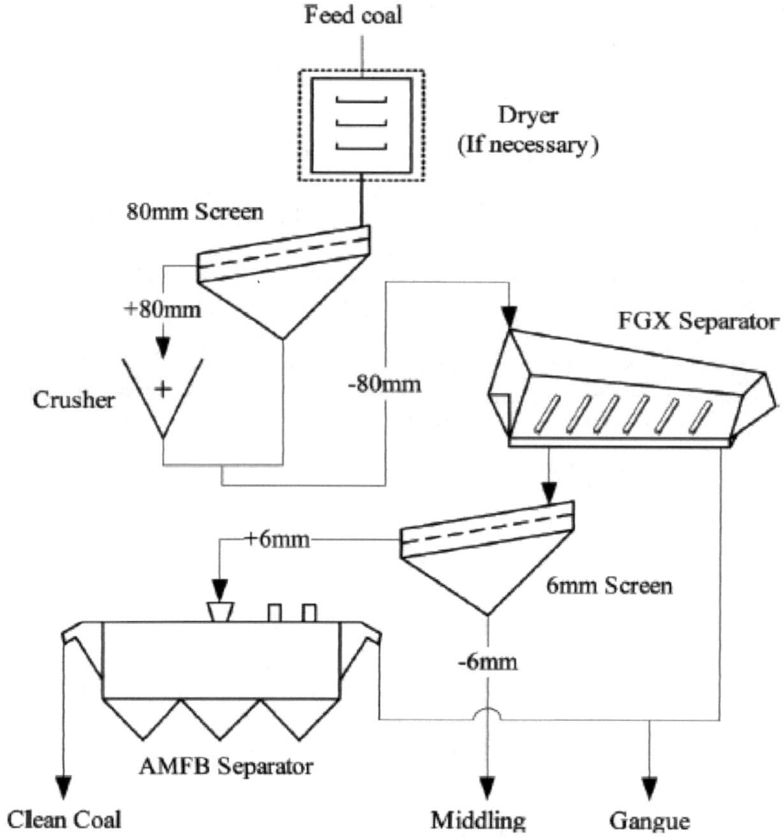

Figure 4. Flowsheet with full technological line based on FGX air table and AMFB separator [12].

Raw feed will enter into dryer to reduce moisture content (if it is necessary). Feed goes onto 80 mm screen where +80 mm particles from screen are crushed. Mixture of crushed product and particle size fraction -80 mm is directed to FGX separator. Coarse coal is separated into gangue and cleaned coal. After beneficiation coal is separated on 6 mm screen. Particle size fraction <6 mm is treated as middlings. Particles >6 mm are separated in AMFB separator (air fluidized bed separator). Gangues from AMFB and FGX are mixed together. Cleaned coal is obtained after beneficiation process conducted in AMFB Separator.

- There is a possibility of beneficiation of very small coal particles. Contact of these small particles with water can have negative effect for later use.
- Big simplification of the technological economy considering the smallest particles by eliminating them from water. It results in elimination of sludge thickening and dewatering of fine grain products.
- Elimination of drying and dewatering of final beneficiation products (roughly grained products).
- Dry separation can find wide applications in places of harsh climate like Mongolia, Russia, some parts of China etc., places with water shortages so it eliminates wet beneficiation methods.

The most important disadvantages are:

- Limited upper dimension of particle size fraction directed into beneficiation process. The maximum limit of particles directed to dry separation process is around 70(80) mm.
- Losses of carbonaceous substance in tailings and semi-products from enrichment processes. The best results are obtained by separation in devices working with aero suspension or with fluidized bed. The quality of tailings was better than in case of separation conducted in typical dry device (jig).
- Dry separation is dedicated for coals with low moisture content and for easily beneficiated coals. Coals harder to beneficiate, of higher ash content or of higher content of middlings are more difficult to separate by dry methods. Final results for separation this kind of coals is not satisfying. Coals with moisture content higher than 5% need to be dry before the enrichment process (increasing production costs).
- Due to the low settling coefficient it is necessary to classify raw feed into narrow particles fractions. It is necessary to develop the first step of classification by sieves etc.

References

[1] Blaschke S 1951 *Przeróbka Mechaniczna Węgla i Rud* (Państwowe Zakłady Wydawnictw Szkolnych: Warszawa) pp 309-23

[2] Laskowski T 1958 *Wzbogacanie Kopalin w Cieczach Ciężkich* (Wydawnictwo Górniczo-Hutnicze: Katowice) pp 211-12

[3] Baic I, Blaschke W and Szafarczyk J 2014 Dry Coal Cleaning Technology *Inżynieria Mineralna - Journal of the Polish Mineral Engineering Society* **34** pp 257-62

[4] Sobko W, Blaschke W and Baic I 2016 Constructional Improvements of FGX-1 Air Concentrating Table Aiming at Optimization of Operation *Inżynieria Mineralna - Journal of the Polish Mineral Engineering Society* **37** pp 37-46

[5] Krukowiecki W 1965 *Przeróbka Mechaniczna Kopalin* (PWN: Warszawa) pp 154-9

[6] Blaschke S 1972 *Przeróbka Mechaniczna Kopalin* (Wydawnictwo "Śląsk": Katowice) pp 314-21

[7] Laskowski T, Błaszczyński S and Ślusarek M 1979 *Wzbogacanie Kopalin w Cieczach Ciężkich* (Wydawnictwo „Śląsk": Katowice) ISBN 83-216-0011-5 pp 293-6

[8] Blaschke S 1982 *Przeróbka Mechaniczna Kopalin Część I* (Wydawnictwo „Śląsk": Katowice) ISBN 83-216-0272-X 446-61

[9] Praca zbiorowa 1982 *Poradnik Górnika Tom 5* (Wydawnictwo „Śląsk": Katowice) pp 1216-7

[10] Chen Q and Yang Y 2003 Coal Preparation *Development of Dry Beneficiation of Coal in China* **23** pp 3-12

[11] Zhang B, Akbari H, Yang F, Mohanty M K and Hirschi J 2011 Performance Optimization of the FGX Dry Separator for Cleaning High-Sulphur Coal *International Journal of Coal Preparation and Utilization* **31** pp 161-86

[12] Xuliang Y, Yuemin Z, Gongmin L, Zhenfu L, Zengqiang Ch and Chuncheng L 2012 Establishment and Evaluation of a United Dry Coal Beneficiation System *International Journal of Coal Preparation and Utilization* **32** pp 95-102

[13] Davaasuren J, Byoung-Gon K, Ju-Hyoung L, Davaatseren G and Bazarragchaa M 2016 Dry Coal Preparation of Fine Particles by KAT Process *XVIII International Coal Preparation Congress* (Saint Pettersburg) ISBN 978-3-319-40943-6 pp 1171-6

[14] Tangshan Kaiyuan Coal Preparation Technology Co. Ltd *Materiały promocyjne i katalogi* http://www.kyxuanmei.com/

[15] Mijał W 2016 *Praca magisterska* Gliwice

[16] Mijał W, Blaschke W and Baic I 2018 Dry Coal Beneficiation Methods in Poland *25th World Mining Congress* (Astana) – not published materials

The role of β-diketones in the extraction of cobalt (II) and nickel (II) ions

K Witt and E Radzyminska-Lenarcik

Faculty of Chemical Technology and Engineering, UTP University of Sciences and Technology, Seminaryjna 3, 85-326 Bydgoszcz, Poland

wittkatarzyna@utp.edu.pl

Abstract. β-diketones were used for Co(II) and Ni(II) separation from nitrate solutions by liquid-liquid extraction method. Pentane-2,4-dione (1), 3-methyl-pentane-2,4-dione (2) and 3-ethyl-pentane-2,4-dione (3), all in dichloromethane solution were extractants in this process. Potentiometric studies were used to determinate dissociation constants of investigated ligands. Stability constants of the created complexes were found by spectrophotometric methods and the increased in the order of 2 < 1 < 3. Extraction was carried out from three types of aqueous phase: one-component Co(II) or Ni(II), and two-component Co(II)-Ni(II). It was proved that pentane-2,4-dione (1) is the most effective extractant.

1. Introduction

Recovery of non-ferrous metals from ores and metal bearing wastes is based either on pyrogenic or hydrometallurgical technologies. Selection of an appropriate manufacturing process depends on useful metal content. In a typical hydrometallurgical process, among the basic technologies (i.e. leaching, phase separation, extraction of metal ions from aqueous solutions), separation of the ions is of particular interest.

Removal of Co(II) and Ni(II) ions from aqueous solution usually takes place using precipitation methods, e.g. oxidative precipitation and sulphide precipitation [1].

The separation of Co(II) and Ni(II) ions from ammonia solution can be likewise conducted by hollow fiber supported liquid membranes (HFSLMs) with Alamine 300 (a water insoluble tri-n-octyl amine), which was a more effective extractant for Co(II) than for Ni(II) [2].

Sadyrbaeva [3] used liquid membranes which contain trialkylbenzylammonium chloride and tri-n-octylamine in 1,2-dichloroethane for electrodialytic separation of Co(II) and Ni(II) from 3–4 M HCl solutions into dilute solutions of various acids. In these conditions, the rate of the membrane transport of $CoCl_4^{2-}$ anions was increased with increasing density, acidity, and Co(II) concentration in the feed solution and only slightly was dependent on the Ni(II) concentration.

Both solvent sublation technique as well ion flotation were effectively applied for separation of Ni(II) and Co(II) ions from their solution after acidic leaching of chemical power sources [4]. Moreover, a method of microwave-assisted hydrothermal leaching and chemical deposition was proposed for extraction and separation of Co(II)-Ni(II) ions from saprolite laterite ore by Zhao et al.

[5]. It was shown that the extraction percentage of Ni(II) and Co(II) from the ore pre-roasted at 300°C for 5 h were 89.19% and 61.89%, respectively.

Solvent extraction can be also used for recovery Co(II) and Ni(II) ions. A different group of compounds can play the role of extractants in these processes. Mubarok and Hanif [6] used a combination of Versatic 10 (neodecanoic acid) and Cyanex 272 (bis-(2,4,4-trimethylpentyl)phosphinic acid). Their results showed that Cyanex 272 can be effectively used to separate Co(II) from Ni(II) in the solution with a low nitrate concentration at the absences of Fe(III) and Mg(II). Also Cyanex 302 (bis(2,4,4-trimethylpentyl)-monothiophosphinic acid) in solvent impregnated resins was proposed as extractant by Mu et.al [7].

A mixture of DEHPA (di-(2-ethylhexyl)phosphoric acid) and Cyanex 272 was used to determinate extraction percentage of the Ni(II) and Co(II) from solution containing Mn(II), Mg(II), Ca(II) and Cu(II) ions [8]. The order for the separation of metal ions was: Cu(II) ≈ Mn(II) > Mg(II) > Co(II) and Cu(II) ≈ Mn(II) > Ca(II) > Co(II) ≈ Mg(II) using Cyanex 272 and DEHPA, respectively. The extraction of Ni(II) was lower than 50% [8].

Alkyl derivatives of imidazole were used for the extraction of Co(II) from nitrate aqueous solutions. The most effective extractants were hydrophobic derivatives containing more than 8 carbon atoms in the alkyl substituent [9-11].

1-Hexyl-2-methylimidazole was used to separate Co(II) from a mixture of Co(II)-Ni(II) at nearly neutral pH and the separation coefficient Co/Ni was 8.9 [12].

A non-equilibrium solvent extraction of Co(II)-Ni(II) from ammoniacal sulphate solution by distearyl acid phosphate (P303) in sulphonated kerosine used for fully separating Co(II) from Ni(II) at pH above 9.5 [13].

Also β-diketone derivatives are successfully used as extractants in solvent extraction. Ochromowicz et al. [14] and Kyuchoukov et al. [15] tested extraction of Cu(II) from ammoniacal solutions by LIX 54-100 (mixture of six isomers: 1-phenyldecane-1,3-diones, heptane-8,10-dione and 1,3-diphenylpropane-1,3-dione). Also Ni(II) can be extracted by LIX 54 at pH 9.0 [16].

In our previous papers we proved that the highest extraction percentage of Cu(II) takes place when 3-allyl-acetylacetone [17] or dibenzoylmethane [18] were used as extratants.

The aim of this work was to investigate complexation and extraction properties of β-diketones: pentane-2,4-dione (**1**), 3-methyl-pentane-2,4-dione (**2**) and 3-ethyl-pentane-2,4-dione (**3**). The extraction processes were conducted in ammonia solutions at 25°C. Methylene chloride was used as a solvent.

2. Experimental

2.1. Reagents

Pentane-2,4-dione (**1**) was purchased from Avantor (Gliwice, Poland), while 3-methyl-pentane-2,4-dione (**2**) and 3-ethyl-pentane-2,4-dione (**3**) were purchased from Sigma Aldrich (Poznan, Poland) (figure 1).

The standard solution of Co(II) and Ni(II) nitrates were purchased from Sigma Aldrich (Poznan, Poland). The potassium nitrate, nitric acid, both of analytical reagent grade, were purchased from Avantor (Gliwice, Poland). The concentration of potassium nitrate was determined gravimetrically like potassium sulphate. Ammonia buffer was prepared from ammonia and ammonium nitrate (both analytical reagent grade, Avantor, Gliwice, Poland). Carbonate-free potassium hydroxide (the titrant, Avantor, Gliwice, Poland) was prepared and standardised against a standard potassium hydrogenphthalate solution. Methylene chloride and methanol (analytical reagent grade, Avantor, Gliwice, Poland) were used as received. The pH-meter was calibrated using commercial buffer solutions (Avantor, Gliwice, Poland) having a pH of 7.00 ± 0.05, 9.21 ± 0.05 and 10.00 ± 0.05.

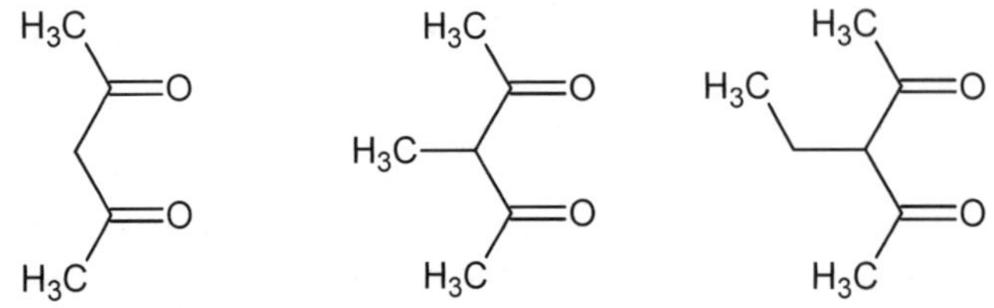

3-methyl- 3-ethyl-
pentane-2,4-dione pentane-2,4-dione pentane-2,4-dione

1 **2** **3**

Figure 1. Structure of the studied β-diketones.

2.2. Equipment
Potentiometric measurements were carried out using a pH-meter (pH meter, CX-731 Elmetron, Zabrze, Poland), with a combination pH electrode (ERH-126, Hydromet, Bytom, Poland). The atomic absorption spectrophotometer AAS 240FS (Agilent, Santa Clara, CA, USA) was used to determine Co(II) and Ni(II) concentration. A Cary 50 (Varian, Melbourne, Victoria, Australia) spectrophotometer was used for recording the absorption spectra of the Co(II) and Ni(II) complexes in the aqueous phase over the range of 200-800 nm.

2.3. Determination of dissociation constants (pK_a)
The dissociation constants (pK_a) of β-diketones (**1**-**3**) were determined by potentiometric titration. Measuring cell was thermostated at 25°C ± 0.5°C. Two water-methanol solutions were prepared. The first one (blank solution) contained nitric acid (1 mmol/dm^3), and a proper quantity of potassium nitrate to ensure a constant ionic strength (1 mol/dm^3). The second one contained nitric acid (1 mmol/dm^3), chosen β-diketone, and potassium nitrate (1 mol/dm^3). Measurements were taken for two β-diketone concentrations (60 mmol/dm^3 and 50 mmol/dm^3) in the second solution. A 50 cm^3 volume of the first solution was placed in the measuring cell, and the solution was titrated with standard solution of 5 mmol/dm^3 potassium hydroxide. Then, the second solution was titrated in the same way. Three series of measurements were made.

2.4. Determination of stability constants ($\log K_1$)
Stability constants ($\log K_1$) of complexes of β-diketones (**1**-**3**) with Co(II) and Ni(II) ions were determined based on absorption spectra and according to the previously described method [19].

2.5. Spectrophotometric studies of aqueous phase
UV-Vis spectra of the aqueous phase after extraction process were recorded in the wavelength range of 200 nm to 800 nm. Water solution of 0.5 mol/dm^3 potassium nitrate was used as a reference solution while testing the aqueous phase. The spectra of the samples were recorded with spectral slits of 1 nm and with 1 nm steps using curettes having 1 cm path lengths.

2.6. Extraction Procedure
The measurements were run at 25°C, at a fixed ionic strength (0.5 mol/dm^3) maintained in the aqueous phase with potassium nitrate. A constant pH was maintained using ammonia buffer (NH_3 + NH_4NO_3, 1:1).

Extraction was carried out from three types of aqueous phase:

- *Co(II) aqueous phase* consisting of: Co(II), NH_3 + NH_4NO_3, KNO_3

- *Ni(II) aqueous phase* consisting of: Ni(II), $NH_3 + NH_4NO_3$, KNO_3
- *Co(II)-Ni(II) aqueous phase* consisting of: Co(II), Ni(II), $NH_3 + NH_4NO_3$, KNO_3.

Before the extraction, concentrations of Co(II) and Ni(II) ions in the aqueous phases were held constant (10 mmol/dm^3) whereas the ligand (**1**-**3**) concentrations in organic phase (methylene chloride) were varied (from 20 to 50 mmol/dm^3). A volume of 5 cm^3 of the aqueous phase was placed in a graduated test tube and an equal volume of β-diketones (**1**-**3**) solution in the organic solvent was added. The test tubes were then shaken for 30 min. The equilibrium was established after a few minutes; however, never longer than 30 min. Afterwards, it was checked if any changes in the phase volumes had occurred, then the phases were separated and the pH of the aqueous phase was measured. The concentration of Co(II) and Ni(II) ions in the aqueous phase was determined by atomic absorption spectrophotometry.

3. Results and discussion

3.1. Dissociation constants of β-diketones

β-Diketones dissociate according to the following reaction:

$$\beta\text{-diketone} \rightleftharpoons \beta\text{-diketone}^- \tag{1}$$

Dissociation constants (pK_a) of investigated extractants (**1**-**3**) were determined by potentiometric method using equilibrium constant of the above reaction:

$$K_a = \frac{[H_3O^+][\beta - diketone^-]}{[\beta - diketone]} \tag{2}$$

where: (H_3O^+), (β-diketone$^-$), and (β-diketone) are equilibrium concentrations of the oxonium ion, the extractant anion and its undissociated form, respectively.
In table 1, the pK_a values of studied β-diketones are summarised.

Table 1. The dissociation constants pK_a of studied β-diketones in aqueous-methanol (1:1) solution (I=1 mol/dm^3 KNO_3) at 25°C.

	β-diketone	pK_a	ref.
1	pentane-2,4-dione	9.65	[20]
2	3-methyl-pentane-2,4-dione	9.66	[this work]
3	3-ethyl-pentane-2,4-dione	9.62	[this work]

The given values of the carry ±0.01 tolerance

Results in table 1 are similar and indicate that 3-ethyl-pentane-2,4-dione (**3**) has the strongest acid properties among the investigated extractants. Pentane-2,4-dione **1** and 3-methyl-pentane-2,4-dione **2** have similar acid-base properties, but they are stronger base than 3-ethyl-pentane-2,4-dione (**3**).

3.2. Stability constants of β-diketone complexes with cobalt(II) and nickel(II) ions

The absorption spectra were measured in the range of 200–400 nm wavelengths and at various β-diketone concentration (**1**-**3**) versus metal ions mole ratios (figure 2). In figure 1, the Ni(II)/(pentane-2,4-dione) complexes are formed in the studied systems. Vanishing of the band due to pentane-2,4-dione (**1**) ligand (280 nm) and the appearance of new bands (309 nm) together with the isosbestic points at 248 and 295 nm were also recorded.

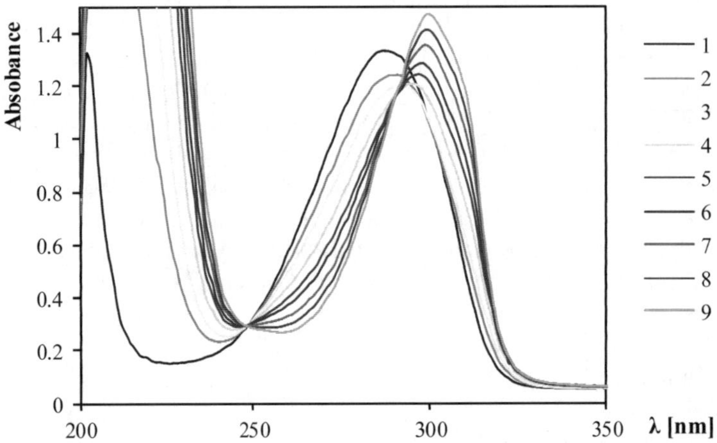

Figure 2. Absorption spectra of Ni(II) complexes with 1 in various pentane-2,4-dione concentration versus Ni(II) mole ratios: 1 : 0 (line 1); 1 : 0.125 (line 2); 1 : 0.25 (line 3); 1 : 0.375 (line 4); 1 : 0.5 (line 5); 1 : 0.625 (line 6); 1 : 0.75 (line 7); 1 : 0.875 (line 8); 1 : 1 (line 9).

Based on such spectra the values of stability constants of β-diketone complexes with Co(II) and Ni(II) ions were calculated and were summarised in table 2.

A methyl derivative of β-diketone (2) forms the most stable complexes. The methyl substituent located in the third position of pentane-2,4-dione increases the affinity of the metal for the ligand.

Table 2. The stability constants $log\ K_1$ of β-diketone complexes with cobalt(II) and nickel(II) ions in aqueous-methanol (1:1) solution.

β-diketone		$log\ K_1$	
		Co(II)	Ni(II)
1	pentane-2,4-dione	5.40	4.96
2	3-methyl-pentane-2,4-dione	5.56	4.99
3	3-ethyl-pentane-2,4-dione	5.03	4.97

The given values of the carry ±0.01 tolerance

The tendency of the methylene group to bind a proton to alkyl substituent is then noticeable. It is well known that increasing degree of enolisation, reduces the acidity of the proton in the enol form of compound. That makes it easier to detach hydrogen and bind the metal ion present in the solution [21]. Complexes of 3-methyl-penatane-2,4-dione (2) can be formed by binding a metal ion both with carbon from the methyl group and with oxygen from the carbonyl group.

Stability of the 3-ethyl-pentane-2,4-dione (3) complexes is lower in comparision with the same complexes with 3-methyl-pentane-2,4-dione (2) and with unsubstituted pentane-2,4-dione (1), hence the increase in the order of 2 < 1 < 3. The presence of an ethyl substituent (longer than methyl group) causes a steric hindrance and impedes the binding of metal by the molecule.

Table 2 shows that stability constants of Co(II) complexes are higher that constants of Ni(II) complexes. The reason for this phenomenon may be that Co(II) complexes have the ability to change the shape of a coordination polyhedron from an octahedron to a tetrahedron [9-11]:

$$(M(H_2O)_6)^{2+} + nL \leftrightarrow (M(H_2O)_{6-n}L_n)^{2+} + nH_2O \tag{3}$$

$$(M(H_2O)_6)^{2+} + nL \leftrightarrow (M(H_2O)_{4-n}L_n)^{2+} + (n+2)H_2O \tag{4}$$

where L denotes the extractant molecule.

The reactions described by equation (4) involving a change in the coordination number, are typical for Co(II). In such cases, configuration equilibria are established (equation 5) between ML_n complexes, which have different coordination sphere structures, for instance:

$$(M(H_2O)_2L_4)^{2+} \leftrightarrow (ML_4)^{2+} + 2H_2O \tag{5}$$

The occurrence of configuration equilibria will lead to higher values of its corresponding stability constant, K (table 2), because it is the sum of the stability constant of the tetrahedral complex, K_t, and the octahedral complex K_o ($K = K_t + K_o$) [10,11].

3.3. Spectrophotometric studies

The following figures show exemplary spectra of the Ni(II) aqueous phase after extraction using extractants (pentane-2,4-dione (**1**), 3-methyl-pentane-2,4-dione (**2**) and 3-ethyl-pentane-2,4-dione (**3**)) in different concentrations thereof in organic phase.

Above spectra clearly show that extraction process of investigated metal ions from aqueous phase is closely related to the concentration of extractant in organic phase and is the most effective when pentane-2,4-dione (**1**) is used (figure 3). Concentration of Ni(II) ions in aqueous phase decreases together with an increase in the concentration of pentane-2,4-dione (**1**) in organic phase (figure 3A). These changes are caused by the complexation reaction with pentane-2,4-dione (**1**), which takes place at the interface of aqueous and organic phase. At this moment in organic phase complexes of Co(II) and Ni(II) ions with pentane-2,4-dione (**1**) are formed.

The extractants: 3-methyl-pentane-2,4-dione (**2**) (figure 3B) and 3-ethyl-pentane-2,4-dione (**3**) (figure 3C), are significantly less effective in comparison with unsubstituted pentane-2,4-dione (**1**), even at their high concentration in organic phase. This phenomenon may be caused by the formation of steric hindrance in the molecules.

Similar results were received during extraction from Co(II) and Co(II)-Ni(II) mixture in the aqueous phase.

IOP Conf. Series: Materials Science and Engineering **427** (2018) 012004 doi:10.1088/1757-899X/427/1/012004

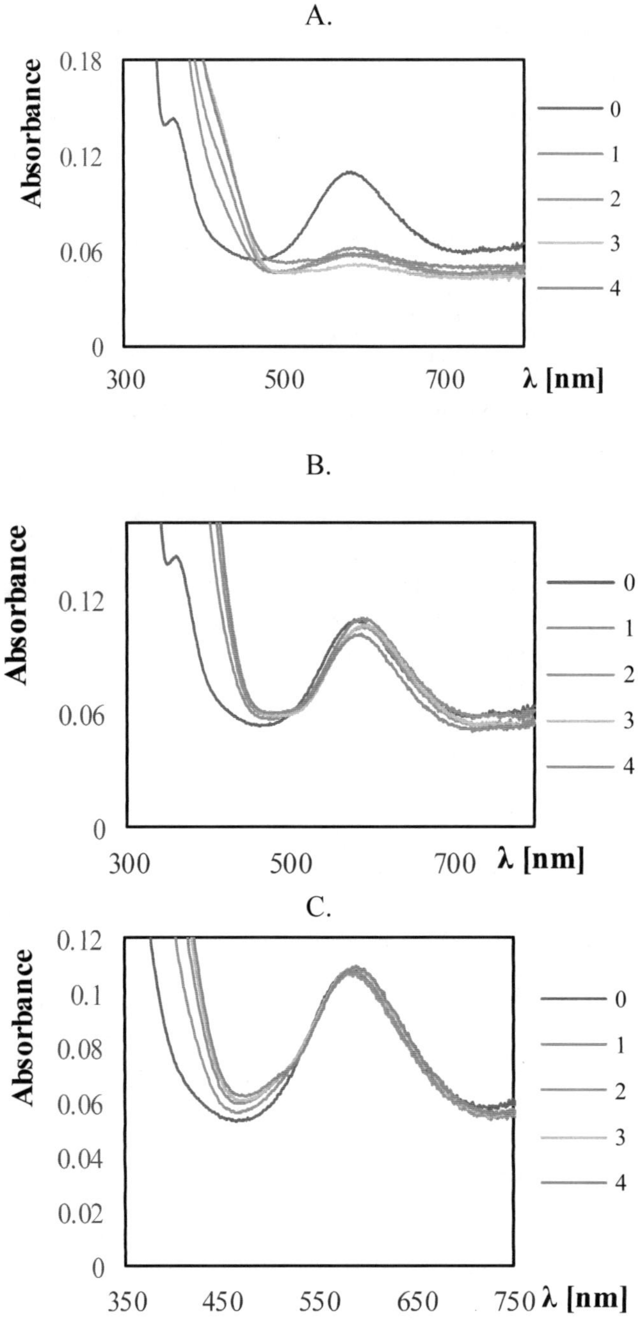

Figure 3. Absorption spectra of Ni(II) aqueous phase after extraction using A. pentane-2,4-dione, B. 3-methyl-pentane-2,4-dione and C. 3-ethyl-pentane-2,4-dione depending on extractant concentration: 0 mmol/dm^3 (line 0), 20 mmol/dm^3 (line 1), 30 mmol/dm^3 (line 2), 40 mmol/dm^3 (line 3), 50 mmol/dm^3 (line 4).

3.4. Extraction processes
Amount of metal ions extracted from aqueous phase (E%) is described by equation 6:

$$E\% = \frac{C_0 - C_E}{C_0} \cdot 100\% \qquad (6)$$

where: C_0 – initial concentration of metal ions in aqueous phase, C_E – concentration of metal ions in aqueous phase after extraction.

Figure 4 shows dependences between the extraction percentage of Co(II) and Ni(II) ions for one-component aqueous phase depending on pH after extraction equilibrium was achieved.

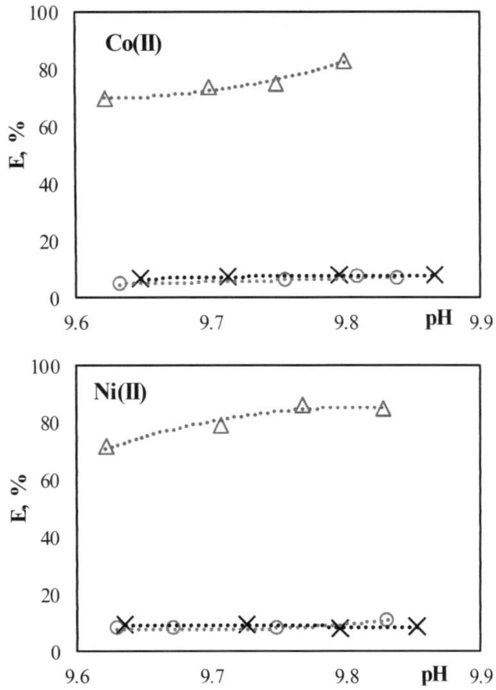

Figure 4. Extraction percentage of Co(II) and Ni(II) ions from one-component solutions after extraction using: Δ pentane-2,4-dione, o 3-methyl-pentane-2,4-dione, and x 3-ethyl-pentane-2,4-dione, depending on pH.

Results presented in figure 4 confirmed that both extraction percentages of Co(II) and Ni(II) had the highest values when pentane-2,4-dione (**1**) was used as an extractant and they were increased with the increase in pH of aqueous phase.

The pentane-2,4-dione (**1**) was the most efficient extractant for two-component Co(II)-Ni(II) aqueous phase (figure 5). Extraction percentage increased with the increase in pH of aqueous phase.

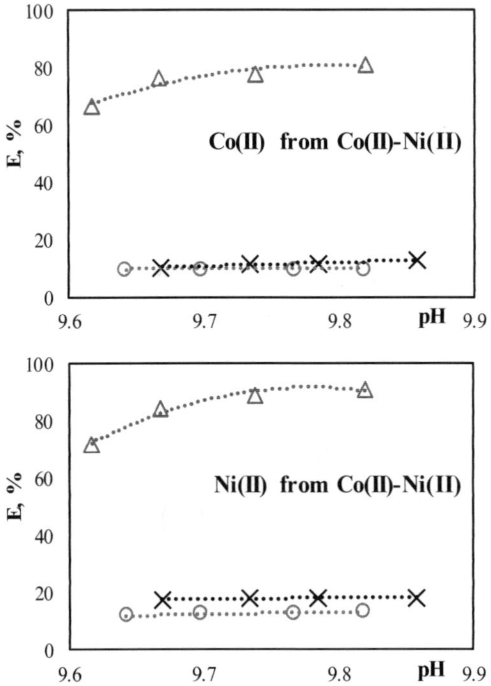

Figure 5. Extraction percentage of Co(II) and Ni(II) ions from two-component Co(II)-Ni(II) solution after extraction using: Δ pentane-2,4-dione, ○ 3-methyl-pentane-2,4-dione, and x 3-ethyl-pentane-2,4-dione, depending on pH.

Absence of substitution in the position 3 and high basic properties (table 1) caused **1** to be the best extractant in comparison with **2** and **3**.

Table 3. The values of extraction percentage depending of extractant concentration in organic phase.

extractant	concentration of extractant in organic phase (mmol/dm^3)	E%			
		Co(II) aq. phase	Ni(II) aq. phase	Co(II)-Ni(II) aq. phase/Co(II) extraction	Co(II)-Ni(II) aq. phase/Ni(II) extraction
1	20.0	69.58	71.33	66.31	71.33
	30.0	73.74	79.18	76.26	84.41
	40.0	74.64	84.46	77.64	88.70
	50.0	82.85	86.41	80.74	90.75
2	20.0	4.75	7.79	9.73	11.74
	30.0	6.00	7.83	9.70	12.30
	40.0	7.00	7.86	9.86	12.77
	50.0	7.13	10.73	9.86	13.20
3	20.0	6.56	8.01	10.17	17.43
	30.0	7.09	8.69	11.31	17.64
	40.0	7.94	9.18	11.57	18.05
	50.0	7.94	9.41	12.65	18.06

The given values of the carry ±0.01 tolerance

Table 3 presents data of extraction percentage vs. concentration of extractant in organic phase made for three different aqueous phases: one-component Co(II) or Ni(II) and two-component Co(II)-Ni(II) solution.

Results summarised in table 3 proved that extraction percentage of both metal ions depends on extractant concentration in the organic phase. The highest percentage value was achieved for 50.0 mmol/dm^3 of each extractant.

Extraction percentage of Ni(II) ions was higher that Co(II), either in one-component or two-component solutions, but separation coefficient Ni(II)/Co(II) decreased. Separation coefficient for Co(II)-Ni(II) in two-component solution was equal to 1.12, 1.34, and 1.43 using extractant **1**, **2**, **3**, respectively.

Extraction of metal ions from two-component aqueous phases using 2 and 3 is more efficient than from one-component solution.

Although the stability of Co(II) complexes is higher than that of Ni(II) complexes (table 2), table 3 shows that Ni(II) ions in comparison with Co(II) ions can be removed more effectively from aqueous solution using all investigated β-diketones as extractants. These differences are the result of differences in the process of the complex formation of Co(II) or Ni(II) with extractant molecules.

4. Conclusion

A process of Co(II) and NI(II) extraction was studied from one-component and two-component solutions using pentane-2,4-dione (**1**), 3-methyl-pentane-2,4-dione (**2**), and 3-ethyl-pentane-2,4-dione (**3**) in dichloromethane. The pentane-2,4-dione (**1**) was the most efficient extractant in all studied aqueous phases, maybe because there is no substituent in position 3 in its molecule.

Extraction depends on concentration of extractant in organic phase and pH of aqueous phase. The highest percentage values were achieved for 50.0 mmol/dm^3 of each extractant and pH around 9.8. In these conditions for one-component solutions the extraction percentage of Co(II) and Ni(II) were equal to 83% and 86%, respectively. For two-component solutions the extraction percentage were equal to 81% and 91% for Co(II) and Ni(II), respectively.

The Ni(II) ions can be removed from aqueous solution more effectively than Co(II) ions, although the stability constants of Co(II) complexes with investigated β-diketones are higher than the stability constants of Ni(II) complexes with the same extractants.

Acknowledges

The new developments presented above were carried out within the 2007-2013 Innovative Economy Operational Programme, Sub-action 1.3.2., Support of the protection of industrial property generated by scientific entities as result of R&D works within project no. UDA-POIG.01.03.02-04-077/12-01, financed by the European Regional Development Fund (ERDF) (85% of co-financing) and from a designated subsidy (15% of co-financing).

References

[1] Flett D S 2004 *Chem. Sustainable Develop.* **12** p 81
[2] Leepipatpiboon N, Pancharoen U and Ramakul P 2013 *Korean J. Chem. Eng.* **30(1)** p 194
[3] Sadyrbaeva T Zh 2013 *Russian Journal of Applied Chemistry* **86(2)** p 186
[4] Sobianowska K, Grzesiak A, Wierzbicka W and Sobianowska-Turek A 2016 *Chem. Environ. Biotechnol.* **19** p 13
[5] Zhao Y, Gao J-M, Yue Y, Peng B, Que Z-Q, Guo M and Zhang M 2013 *Int. J. Min. Met. Mater.* **20(7)** p 612
[6] Mubarok M Z and Hanif L I 2016 *Procedia Chem.* **19** p 743
[7] Mu F T, Jia Q, Tian Y M and Shang Q K 2008 *Adsorption* **14** p 31
[8] Santanillal A J M, Tenório J A S and Espinosa D C R 2014 *Annual Meeting Supplemental Proceedings TMS (The Minerals, Metals & Materials Society)* p 1157

IOP Conf. Series: Materials Science and Engineering **427** (2018) 012004 doi:10.1088/1757-899X/427/1/012004

[9] Radzyminska-Lenarcik E and Witt K 2015 *Sep. Sci. Technol.* **50(5)** p 676
[10] Radzymińska-Lenarcik E 2008 *Polish J. Chem. Technol.* **10** p 73
[11] Lenarcik B and Ojczenasz P 2004 *Sep. Sci. Technol.* **39** p 199
[12] Radzyminska-Lenarcik E and Wasilewska A 2015 *Challenges of Modern Technology* **6(3)** p 20
[13] Wang K-Y, Niu C-W, Qian D, Liu J-H, Chen X-Y and Lai D-Y 2001 *J. Cent. South Univ. Technol.* **8(1)** p 50
[14] Ochromowicz K, Jeziorek M and Wejman K 2014 *Physicochem. Probl. Miner. Process.* **50(1)** p 327
[15] Kyuchoukov G, Bogacki M B and Szymanowski J 1998 *Ind. Eng. Chem. Res.* **37(10)** p 4084
[16] Alguacil F J and Cobo A 1998 *Hydrometallurgy* **48** p 291
[17] Radzyminska-Lenarcik E and Witt K 2018 *Sep. Sci. Technol.* **53** p 1223
[18] Witt K and Radzyminska-Lenarcik E 2017 *E3S Web of Conferences* **18** 01016
[19] Lis S, Pawlicki G, Staniszewski B, Urbaniak W and Witt K 2011 *Chem. Pap.* **65(2)** p 221
[20] Witt K, Radzyminska-Lenarcik E and Urbaniak W 2016 *Sep. Sci. Technol.* **51(15-16)** p 2620
[21] Borduin W G, Guter G A and Hammond G S 1959 *J. Am. Chem. Soc.* **81** p 4682

IOP Conf. Series: Materials Science and Engineering **427** (2018) 012005 doi:10.1088/1757-899X/427/1/012005

Copper recovery from model chloride solution using polymer inclusion membranes with 1-decyl-2,4-dimethylimidazle

E Radzyminska-Lenarcik and K Witt

Faculty of Chemical Technology and Engineering, UTP University of Sciences and Technology, Seminaryjna 3, PL-85326 Bydgoszcz, Poland

elaradz@utp.edu.pl

Abstract. The recovery of Cu(II) ions from chloride solutions by the transport of a binary Cu(II)-Zn(II) and Cu(II)-Cd(II), and a ternary Cu(II)-Zn(II)-Cd(II) mixture ions through inclusion membranes doped with 1-decyl-2,4-dimethylimidazole (DDMI) was studied ($c_{M(II)}$ = 0.001 M; pH 5.5). The efficiency of metal ions separation by DDMI increased as: Cu(II) > Zn(II) > Cd(II). Selectivity coefficients Cu(II)/Zn(II) and Cu(II)/Cd(II) were equal 1.5 and 13.5, respectively for binary solutions and 1.6 and 27.7 for ternary solution. The step limiting the rate of transport of metal ions by PIM doped with DDMI may be membrane diffusion coefficients of cation complexes with DDMI. The recovery factor of Cu(II), Zn(II), and Cd(II) ions during transport across PIMs from an equimolar mixture of cations after 24 hrs was above 96%, 60% and 3.5% for Cu(II), Zn(II) and Cd(II), respectively. PIM containing DDMI are stable for 120 hrs.

1. Introduction

Due to their wide application, copper, zinc and cadmium are classified as strategic metals [1]. Such a wide industrial use of these metals causes the generation of metal-bearing wastes in the form of sewage, dust, sludge, waste heap, etc. Due to its threat to the environment and human health, all waste should be managed or cleaned of toxic metals [2]. An increasing demand for metal production has led to researching more efficient and economical methods for purification required by the industry.

Membrane technology has become an important alternative to normal processes employed for wastewater treatments, separation and recovery of target metals.

Polymer inclusion membranes (PIMs) represent an attractive liquid-to-liquid extraction for the selective removal and concentration of nonferrous metal ions (Zn(II), Co(II), Ni(II), Cd(II), Cu(II) and Pb(II)) from aqueous solutions. The transport of such metal ions across PIMs can be described as the simultaneous extraction and back-extraction operations combined in a single stage [3,4]. The cations transport and the possibility of their selective separation are determined by the way the cations react with the carriers and by the structure and stability of the complexes being formed [3-6].

IOP Conf. Series: Materials Science and Engineering **427** (2018) 012005 doi:10.1088/1757-899X/427/1/012005

Table 1. Example of nonferrous metal ions transport across PIMs doped imidazole derivatives from chloride or nitrate aqueous solutions reported in the literature.

Carriers	Target solutions	Solution	J_0, mol/m^2·s	RF, %	Ref.
imidazole azocrown ethers	Zn(II), Cu(II), Pb(II)	NO_3^-	1.03-0.29·10^{-7}	< 20.0$_{Zn(II)}$	[12,13]
imidazole azothiacrown ethers	Zn(II), Pb(II), Ni(II), Co(II), Cd(II), Cu(II)	NO_3^-	0.27-1.13·10^{-7}	< 30.0$_{Zn(II)}$	[14]
1-alkylimidazole	Cu(II), Zn(II), Co(II), Ni(II)	NO_3^-	1.19-2.02·10^{-6}	-	[15]
1-alkylimidazole	Cu(II), Zn(II), Co(II), Ni(II)	Cl^-	1.7-2.15·10^{-6}	-	[16]
1-alkyl-2-methylimidazole	Cu(II), Zn(II), Co(II), Ni(II)	NO_3^-	2.42-0.2·10^{-6}	-	[17]
1-hexyl-2-methylimidazole	Cu(II), Zn(II), Co(II), Ni(II)	NO_3^-	0.72·10^{-6}	53.5$_{Cu(II)}$	[18]
1-hexyl-4-methylimidazole	Zn(II), Cd(II), Ni(II), Co(II)	Cl^-	3.64-0.16·10^{-6}	96.0$_{Zn(II)}$	[19]
1-heptyl-2-methylimidazole	Cu(II), Zn(II), Co(II), Ni(II)	NO_3^-	3.86-0.21·10^{-6}	85.0$_{Cu(II)}$	[20]
1-decyl-2-methylimidazole	Cu(II), Zn(II), Co(II), Ni(II)	NO_3^-	1.23-0.14·10^{-6}	92.0$_{Cu(II)}$	[21]
1-decyl-4-methylimidazole	Zn(II), Cd(II), Ni(II), Co(II)	Cl^-	3.72-0.15·10^{-6}	< 95.0$_{Zn(II)}$	[22]

In a majority of studies, commercial extractants which exhibit low selectivity separation are used as ion carriers in transport through the liquid membranes [7-11]. Also imidazole and its derivatives, have been used for the membrane separation of metals such as Co(II), Ni(II), Cu(II), Zn(II), Fe(III), and Cd(II). Several works on their application in transport of nonferrous metal ions across PIMs have been collected in table 1.

The data compiled in table 1 show that alkyl imidazole derivatives more efficiently separate metal ions compared to crown ethers, as well as that the alkyl substituent placed in the 2 or 4 position in alkylimidazole molecules increases the possibility of effective separation of metal ions.

In this work, the authors present the results of investigation of the competitive transport of nonferrous metal ions as copper(II), zinc(II), and cadmium(II) across polymer inclusion membranes doped with 1- decyl-2,4-dimethylimidazole (DDMI) from dilute chloride solutions. The results were compared with the previously published results of liquid-liquid extraction of a mixture of Cu(II), Zn(II), and Cd(II) ions [23].

2. Experimental

2.1. Reagents

Analytical grade chemical reagents: copper (II), zinc(II) and cadmium(II) chlorides, hydrochloric acid, tetramethylammonium hydroxide were purchased from POCh (Gliwice, Poland). All aqueous solutions were prepared using analytical reagent grade chemicals and deionized water (conductivity –

0.10 μS/cm). Solutions of Cu(II), Zn(II) and Cd(II) were prepared by dissolving appropriate amounts of metal chlorides in water.

Analytical grade organic reagents, i.e. cellulose triacetate (CTA), o-nitrophenyl pentyl ether (o-NPPE) and dichloromethane (all from Fluka) were used without further purification. 1-Decyl-2,4-dimethylimidazole (DDMI) (figure 1) was synthesized by the alkylation of 2,4-dimethylimidazole (Sigma-Aldrich Company, Poland) in yields amounting to 63% according to the procedure described in paper [24].

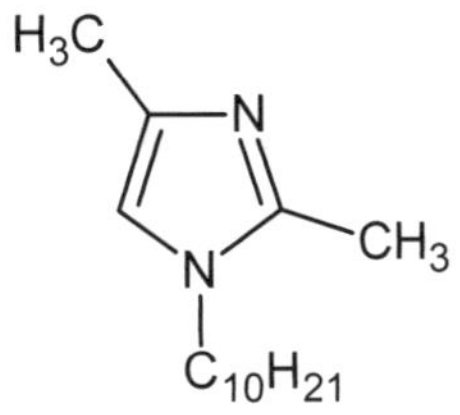

b.p. 104-106 °C/8mmHg

$pK_a = 8.62$

Figure 1. Structure of the carrier 1-decyl-2,4-dimethylimidazole (DDMI), and its boiling point and acid dissociation constant (pK_a)

2.2. The preparation of PIMs

The polymer membranes were prepared according to the procedure reported in the previous paper [15-22]. Organic solutions consisting of support (CTA), carrier (DDMI) and plasticizer (NPPE) in dichloromethane were prepared. A portion of this solution was poured into a membrane mould comprised of a 6.0 cm glass ring attached to a glass plate with CTA – dichloromethane glue. The organic solvent was allowed to evaporate overnight and the resulting membrane was separated from the glass plate by immersion in cold water. The thickness of the PIM samples was measured using a digital micrometer (Panametrics® Magna-Mike® 8500 (San Diego, CA, USA)) with an accuracy of 0.0001 mm.

The thickness of a membrane was measured 10 times for each case and shown as the average value of these measurements, with the standard deviation below 1%. The thickness of the membrane before and after transport was the same. The average PIM thickness varied in the range was 28-32 μm. Each experiential point was repeated 4 times, i.e. membrane formed by immobilization, thickness measured and transport parameters calculated. Experimental reproducibility was high with standard deviation below 1% of the measured values.

2.3. Transport studies

Transport experiments were carried out in a permeation cell described in an earlier paper [15-22]. The membrane film (surface area 4.9 cm^2) was tightly clamped between two cell compartments. Both the source phase and the receiving aqueous phase (45 cm^3 each) were mechanically stirred at 600 rpm. Metal chlorides were used in the source phase, whereas the receiving phase was deionized water. The PIM transport experiments were carried out at 20 ± 0.2°C. Small samples of the aqueous receiving phase were taken periodically from the sampling port equipped with a syringe and analysed by atomic absorption spectroscopy (AAS 240FS Spectrometer, Agilent, Santa Clara, CA, USA) to determine Cu(II), Zn(II), and Cd(II) concentrations. The pH of the source phase equaling 5.5 was kept constant using tetra-methyl-ammonium hydroxide.

The kinetics of transport across the PIMs was described as a first-order process with respect to the metal-ion concentration [25] expressed by equation (1):

$$\ln\left(\frac{c}{c_0}\right) = -kt \tag{1}$$

where: c is the metal ions concentration (M) in the source aqueous phase at a given time, c_0 is the initial metal ions concentration in the source phase, k is rate constant values, and t is the time of transport (s).

In order to calculate the value of k, ln(c/c0) versus time was plotted. The rate constant values for two independent transport experiments were averaged and the standard deviation was calculated. The correlation between ln(c/c0) and time was linear, which was confirmed by the high correlation coefficient (R^2) ranging from 0.9906 to 0.9974. The permeability coefficient (P) was calculated according to equation (2):

$$P = -\frac{V}{A}k \tag{2}$$

where: V is the volume of the aqueous source phase (m^3), and A is an effective membrane area (m^2). The initial flux (J_0) is equal to:

$$J_0 = P \cdot c_0 \tag{3}$$

The selectivity ratio (S) was defined as the ratio of initial fluxes for M_1 and M_2 metal ions, respectively:

$$S = J_{0,M1} / J_{0,M2} \tag{4}$$

In order to describe the efficiency of metal removal from the source phase, the recovery coefficient (RF) was calculated:

$$RF = \frac{c_0 - c}{c_0} \cdot 100\% \tag{5}$$

The reported values correspond to the average values of three replicates, with deviations within 5%.

3. Results and discussion

3.1. Membrane characterization

An AFM image of PIM doped with DDMI in three- dimensional forms in a 5.0 x 5.0 μm format is shown in figure 2.

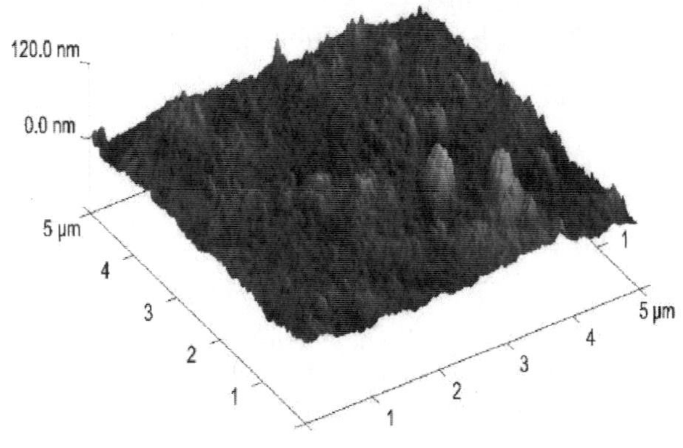

Figure 2. 3D atomic force microscopy for PIM with DDMI, scan area 5x5 µm.

Calculated by atomic force microscopy (AMF) the porosity and roughness of PIM doped with DDMI was 23.7 and 5.8, respectively. Effective pore sizes for membrane was 0.065 µm, respectively.

3.2. Separation of Cu(II), Zn(II) and Cd(II) ions from their equimolar solution across PIMs

Findings on the transport of non-ferrous metal ions from chloride solutions, across PIMs containing DDMI as ion carrier are discussed below. The concentration of the ion carrier in the membrane was 1.0 M (relative to the plasticizer) since earlier studies [15-22] indicate that the concentration is optimum for that group of carriers. The studies were carried out from single- or multi-component solutions, containing metals each at a concentration of 0.001 M. The initial flux values and selectivity coefficients for the metal ion transport across the PIM's are shown in table 2.

Table 2. Initial fluxes (J_0), selectivity order and selectivity coefficients (S) for the competitive transport of Cu(II), Zn(II), and Cd(II) ions across PIMs doped DDMI.

solutions	metal ion	J_0, µmol/m^{2}s	$S_{Cu(II)/M(II)}$
I	Cu(II)	10.82	-
II	Cu(II) Zn(II)	8.56 5.84	Cu(II) > Zn(II) 1.5
III	Cu(II) Cd(II)	8.91 0.63	Cu(II) > Zn(II) 13.5
IV	Cu(II) Zn(II) Cd(II)	7.21 4.57 0.26	Cu(II) > Zn(II) > Cd(II) 1.6 27.7

It follows from the data shown in table 2 that the initial flux value for Cu(II) ion transport from a unary solution is higher than that from multi-component solutions. For a ternary mixture, the initial fluxes of metal ions transport across PIMs containing DDMI decrease in the following order: Cu(II) > Zn(II) > Cd(II). Selectivity coefficients (S) Cu(II)/Zn(II) and Cu(II)/Cd(II) were equal 1.5 and 13.5, respectively for binary solutions and 1.6 and 27.7 for ternary solution.

Decyl substituents at position 1 of the imidazole ring distinctly affect the hydrophobic properties of the molecule. An additional two methyl substituents at position 2 and 4 of the imidazole ring has been found to increase basicity by one order of magnitude, compared with that of 1-decylimidazole (pK_a = 7.39 [26]).

IOP Conf. Series: Materials Science and Engineering **427** (2018) 012005 doi:10.1088/1757-899X/427/1/012005

Ion carriers embedded in the membrane interact with metal ions in the diffusion-controlled water phase to form complexes. These form electrically neutral ion pairs with anions that penetrate the membrane in the co-transport:

$$M^{2+} + L + 2A^{-} \leftrightarrow (ML)A_2 \quad (6)$$

where M^{2+} is the metal ion, L is the carrier, and A^{-} is the co-transported anion (Cl^{-}).

Table 3. Comparison of the stability constants β_n of Cu(II), Zn(II), and Cd(II) complexes with DDMI, at 25°C, ionic strength 0.5 M (KNO_3) [23].

metal ion	log β_1	log β_2	log β_3	log β_4
Zn(II)	2.17	4.48	6.39	9.87
Cu(II)	3.95	7.18	9.42	11.69
Cd(II)	2.53	4.21	5.68	7.56

The stability constants, β_n of the studied metal complexes with DDMI are collected in table 3.

The presence of the substituent at position 2 or 4 of the imidazole ring hinders the formation of each metal complex (steric effect). The stability constants of Cu(II), Zn(II) and Cd(II) complexes increase in the following order: Cd(II) > Zn(II) > Cu(II) (table 3).

In aqueous solutions, cations of Zn(II), Cd(II) and Cu(II) exist in the form of octahedral aqua complexes $(M(H_2O)_6)^{2+}$. Owing to the effect of ligands (extractants), octahedral aqua complexes of certain cations tend to change their coordination number (c.n.) from 6 to 4 and they change their coordination sphere into tetrahedral flat or, depending on the structure of their d-electron layer. The process is illustrated by the equation.

$$(M(H_2O)_{6-n+1}L_{n-1})^{2+} + L \leftrightarrow (M(H_2O)_{4-n}L_n)^{2+} + (n+2)H_2O \quad (7)$$

The steric effect of the substituent at position 2 decreases the stability constants of octahedral complexes of all the metals studied though it does not hinder the formation of tetrahedral species [23]. The formation of tetrahedral complexes with 1-alkylimidazoles has been proven for the Zn(II) and Cu(II) ions [27,28]. The Cu(II) ions were transported across PIMs most effectively. Perhaps the reason for this is the coordination sphere Cu(II), which is the most flexible, i.e. easier to deform [29]. It seems (compare tables 2-3) that the initial flux values correspond to those of the stability constants of the complexes.

3.3. Recovery of metal

The recovery factor (RF) of Cu(II) transport across PIMs from chloride solutions of Cu(II), Zn(II), and Cd(II) into deionized water is presented in fig. 3.

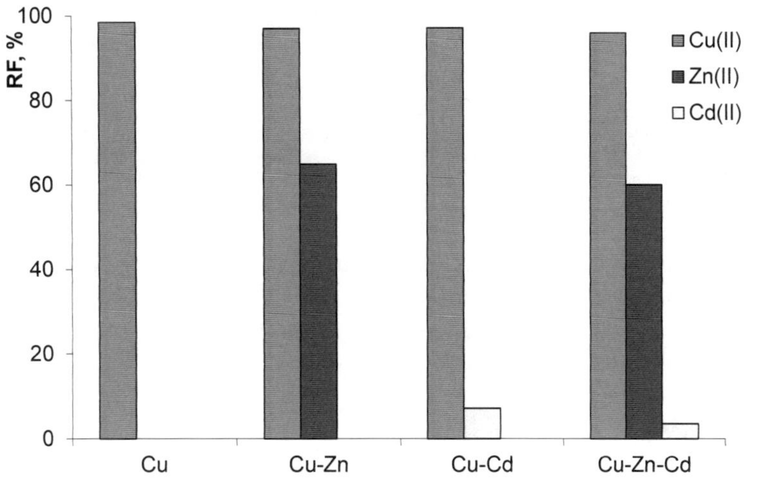

Figure 3. Metal ion recovery factor (RF) during transport across PIMs doped with 1.0 M DDMI.

The highest recovery factors (RF) were found for Cu(II) ions for unary solution. The RF values of Cu(II) ions from binary Cu(II)-Zn(II) and Cu(II)-Cd(II) solutions are almost the same. In the case of a ternary Cu(II)-Zn(II)-Cd(II) mixture, the RF for Cu(II), Zn(II) and Cd(II) were 96, 60, and 3,5 percent, respectively.

The lowest RF values were obtained for Cd(II) ions, which are the slowest transported by this type of membrane. Practically, Cd(II) ions remain in the feeding phase.

3.4. Membrane diffusion coefficients of Cu(II), Zn(II), and Cd(II) complexes with DDMI

In figure 4, the correlation graphs $(M^{2+})_0$-$(M^{2+})_t$ versus time of metal ions transport across PIM doped with DDMI is presented.

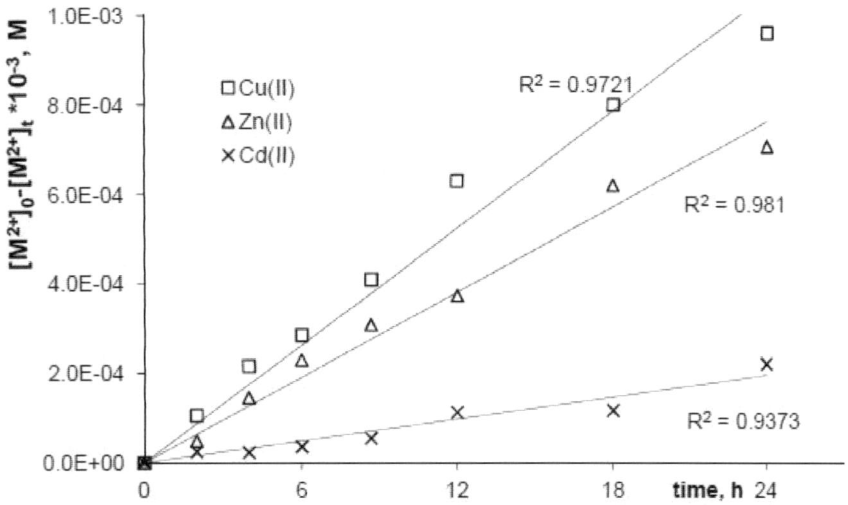

Figure 4. Relation of $(M^{2+})_0$-$(M^{2+})_t$ plotted vs. time for Cu(II), Zn(II), and Cd(II) transport across PIMs doped with 1.0 mol/dm^3 DDMI.

IOP Conf. Series: Materials Science and Engineering **427** (2018) 012005 doi:10.1088/1757-899X/427/1/012005

The diffusion coefficient of M(II) was calculated from the equation:

$$D_o = d_o/\Delta_o \tag{8}$$

where: d_o is the thickness of the membrane (0.003 cm) and Δ_o could be evaluated by plotting $(M^{2+})_0$-$(M^{2+})_t$ vs. time.

The corrected (normalized) membrane diffusion coefficient $D_{o,n}$ [30], which considers the morphological features inside the membrane (ε - porosity and τ - tortuosity), was calculated from equation:

$$D_{o,n} = D_o \cdot (\varepsilon / \tau) \tag{9}$$

where: ε - porosity and τ – tortuosity.

Obtained values of diffusion coefficients are presented in table 4.

Values of diffusion coefficient determined in this study are comparable with presented in literature data for different membranes are in the range from 10^{-12} to 10^{-6} cm^2/s and show that limiting step of the process is the transfer of metal complex across membrane barrier. The value of the diffusion coefficient of M(II)-carrier species of $4.64 \cdot 10^{-11}$ - $1.06 \cdot 10^{-9}$ cm^2/s is smaller than the value of $1.5 \cdot 10^{-7}$ cm^2/s reported for the lead complex with the D_2EHPA in PIM's reported by Salazar-Alvarez et al. [30].

The values of normalized diffusion coefficients (considering membrane porosity and tortuosity) of M(II)-carrier complexes, obtained in the process of transport across PIM's containing DDMI from the solution Cu(II)-Zn(II)-Cd(II) are in the range $1.90 \cdot 10^{-12}$ - $1.03 \cdot 10^{-10}$ cm^2/s. Thus, the rate of transport of non-ferrous metal ions across PIM's doped with DDMI is determined by the diffusion rate of the complexes M(II)-carrier across the membrane.

Table 4. Diffusion coefficients normalized for competitive transport of Cu(II), Zn(II), and Cd(II), ions through PIM with DDMI.

metal ion	Δ_o, s/m	D_o, s/m	$D_{o,n}$, s/m
Zn(II)	$10^{6.45}$	$1.06 \cdot 10^{-09}$	$1.03 \cdot 10^{-10}$
Cu(II)	$10^{5.53}$	$8.85 \cdot 10^{-09}$	$3.63 \cdot 10^{-09}$
Cd(II)	$10^{7.81}$	$4.64 \cdot 10^{-11}$	$1.90 \cdot 10^{-11}$

Table 5. Comparison of solvent and membrane extraction of Cu(II), Zn(II), and Cd(II) complexes using DDMI as extractant/carrier.

solvent extraction [23]		membrane extraction	
selectivity order			
Cu(II)>Zn(II)>Cd(II)		Cu(II)>Zn(II)>Cd(II)	
selectivity coefficient			
$S_{Cu(II)/Zn(II)}$	$S_{Cu(II)/Cd(II)}$	$S_{Cu(II)/Zn(II)}$	$S_{Cu(II)/Cd(II)}$
1.4	1.6	1.5	27.7
recovery factor of Cu(II), %			
94		96	
solvent			
dichloromethane		-	
reaction conditions			
nitrate, ionic strength 0.5 mol/dm^3 (KNO_3)		chloride, pH=5.5	

3.5. Comparison of transport across PIMs with solvent extraction (SX)
Previously, the possibility of Cu(II) ion separation from an equimolar Cu-Cd-Zn tertiary mixture by solvent extraction was investigated [23]. The separation of Cu(II) ions from the mixture of Cu(II)-Cd(II)-Zn(II) ions by solvent and membrane extraction is compared in table 5. In both processes, DDMI was used as the extractant or ion carrier.

It follows from the data shown in table 5 that the selectivity order in both extraction processes was the same. For PIMs selectivity coefficients $S_{Cu(II)/Zn(II)}$ and $S_{Cu(II)/Cd(II)}$ increased. Separation of Cu(II)/Zn(II) in both processes was almost the same, while the Cu(II)/Cd(II) separation is definitely better in the case of membrane extraction.

In addition, no toxic solvents are used in the membrane separation.

4. Conclusion
Copper(II) cations can be effectively removed from aqueous chloride solutions in the hydrometallurgical process of polymer inclusion membrane transport with 1-decyl-2,4-dimethylimidazole as ion carriers. With the use of this carrier the competitive transport of metal ions across the PIMs the order is: Cu(II) > Zn(II) > Cd(II). It seems that the initial flux values correspond to those of the stability constants of the complexes. The rate of transport of non-ferrous metal ions is determined by the diffusion rate of the complexes M(II)-carrier across the membrane. The selectivity coefficients of Cu(II)/Zn(II) and Cu(II)/Cd(II) were equal 1.5 and 13.5, respectively for binary solutions and 1.6 and 27.7 for ternary solution. Owing to the high steric effect, 1-decyl-2,4-dimethylimidazole fairly well differentiates the properties of its complexes with Cu(II), Zn(II), and Cd(II). The highest recovery factor (after 24 h) was obtained for Cu(II). In the case of a ternary solution, the RF for Cu(II), Zn(II), and Cd(II) were 96, 60, and 3,5 percent, respectively. The lowest RF values were obtained for Cd(II) ions, which are the slowest transported by this type of membrane.

IOP Conf. Series: Materials Science and Engineering **427** (2018) 012005 doi:10.1088/1757-899X/427/1/012005

References

[1] Cote G 2000 Hydrometallurgy of Strategic Metals *Solv. Extr. Ion Exch.* **18** pp 703-27

[2] Rao S R 2006 *Resource, recovery and recycling from metallurgical wastes* (Elsevier: Amsterdam)

[3] Nghiem L D, Mornane P, Potter I D, Perera J M, Cattrall R W and Kolev S D 2006 Extraction and transport of metal ions and small organic compounds using polymer inclusion membranes (PIMs) *J. Membr. Sci.* **281** pp 7-41

[4] Kislik V S (eds) 2010 *Liquid membranes. Principles and application in chemical separation and wastewater treatment* (Elsevier: UK)

[5] Almeida M I G S, Cattrall R W and Kolev S D 2012 Recent trends in extraction and transport of metal ions using polymer inclusion membranes (PIMs) *J. Membr. Sci.* **415-416** pp 9-23

[6] Abdul-Halim N-S, Whitten P G and Nghiem L D 2013 Characterising poly(vinyl chloride)/Aliquat 336 polymer inclusion membranes: Evidence of phase separation and its role in metal extraction, *Sep. Purif. Technol.* **119** pp 14-8

[7] Sgarlata C, Arena G, Longo E, Zhang D, Yang Y and Bartsch R A 2008 Heavy metal separation with polymer inclusion membranes *J. Membr. Sci.* **323** pp 444-51

[8] Hosseini S S, Bringas E, Tan N R, Ortiz I, Ghahramani M and Shahmirzadi M A A 2016 Recent progress in development of high performance polymeric membranes and materials for metal plating wastewater treatment: A review *J. Water Proc. Eng.* **9** pp 78-110

[9] San Roman M F, Bringas, Ibanez E R and Ortiz I 2010 Liquid membrane technology: fundamentals and review of its applications *J. Chem. Technol. Biotechnol.* **85** pp 2-10

[10] Araki T and Tsukube H 2000 *Liquid Membranes: Chemical Applications* (CRC Press Inc. Boca Raton: Florida)

[11] Kozlowska J, Kozłowski C A and Koziol J J 2007 Transport of Zn(II), Cd(II), and Pb(II) across CTA plasticized membranes containing organophosphorous acids as an ion carriers *Sep. Purif. Technol.* **57** pp 211-9

[12] Ulewicz M, Sadowska K and Biernat J F 2007 Selective transport of Pb(II) across polymer inclusion membrane using imidazole azocrown ethers as carriers *Physicochem. Probl. Miner. Process.* **41** pp 133-43

[13] Luboch E, Wagner-Wysiecka E, Fainerman-Melnikova M, Lindoy L F and Biernat J F 2006 Pyrrole Azocrown Ethers. Synthesis, Complexation, Selective Lead Transport and Ion-Selective Membrane Electrode Studies, *Supramol. Chem.* **18** pp 593-601

[14] Ulewicz M, Sadowska K and Biernat J F 2007 Facilitated transport of Zn(II), Cd(II) and Pb(II) across polymer inclusion membrane doped with imidazole azocrown ethers, *Desalination* **214** pp 352-64

[15] Radzyminska-Lenarcik E and Ulewicz M 2012 Selective transport of Cu(II) across a polymer inclusion membrane with 1-alkylimidazole from nitrate solutions *Sep. Sci. Technol.* **47** pp 1113-8

[16] Radzyminska-Lenarcik E and Ulewicz M 2014 The use of 1-alkylimidazoles for selective separation of zinc ions in the transport process across a polymer inclusion membrane *Physicochem. Probl. Mineral Process.* **50** pp 131-42

[17] Radzyminska-Lenarcik E and Ulewicz M 2017 Application of polymer and supported membranes with 1-alkyl- 2-methylimidazoles for separation of some transition metal ions *Des. Water Treat.* **64** pp 425-31

[18] Ulewicz M and Radzyminska-Lenarcik E 2012 Supported liquid (SLM) and polymer inclusion (PIM) membranes pertraction of copper(II) from aqueous nitrate solutions by 1-hexyl-2-methylimidazole *Sep. Sci. Technol.* **47** pp 1383-9

[19] Ulewicz M and Radzyminska-Lenarcik E 2015 Application of polymer membranes doped with 1-hexyl-4-methylimidazole for pertraction of zinc(II) ions, *Physicochem. Probl. Mineral Process.* **51** pp 447-60

[20] Radzyminska-Lenarcik E and Ulewicz M 2015 The use of the steric effect of the carrier

molecule in the polymer inclusion membranes for the separation of cobalt(II), nickel(II), copper(II), and zinc(II) ions *Pol. J. Chem. Technol.* **17** pp 51-6

[21] Ulewicz M and Radzyminska-Lenarcik E 2012 Application of supported and polymer membrane with 1-decyl-2-methylimidazole for separation of transition metal ions *Physicochem. Probl. Min. Proces.* **48** pp 91-102

[22] Ulewicz M and Radzyminska-Lenarcik E 2014 Application of Polymer and Supported Membranes with 1-Decyl-4-Methylimidazole for Pertraction of Transition Metal Ions *Sep. Sci. Technol.* **49** pp 1713-21

[23] Radzyminska-Lenarcik E and Witt K 2017 Studies on the separation of some transition metals using trialkylimidazole as selective extractant *E3S Web of Conferences* **18**, 01017, MEC2017, DOI: 10.1051/e3sconf/20171801017

[24] Pernak J, Krysinski J and Skrzypczak A 1987 Bakterizide wirkung von iminiumverbindungen *A. Tens. Surf. Det.* **24** pp 276-86

[25] Danesi P R 1984-1985 Separation of metal species by supported liquid membranes *Sep. Sci. Technol.* **19** pp 857-94

[26] Lenarcik B and Ojczenasz P 2002 The influence of the size and position of the alkyl groups in alkylimidazole molecules on their acid – base properties *J. Heterocyc. Chem.* **39** pp 287-90

[27] Lenarcik B and Kierzkowska A 2004 The influence of alkyl length on stability constants of Zn(II) complexes with 1-alkylimidazoles in aqueous solutions and their partition between aqueous phase and organic solvent *Solv. Ext. Ion Exch.* **22** pp 449-71

[28] Radzyminska-Lenarcik E 2007 The influence of the alkyl chain length on extraction equilibrium of Cu(II) complexes with 1-alkylimidazole in aqueous solution/organic solvent system *Solv. Extr. Ion Exch.* **25** pp 53-64

[29] Salazar-Alvarez G, Bautista-Flores A N, San Miguel E R, Muhammed M and De Gyves J 2005 Transport characterization of a PIM system used for the extraction of Pb(II) using D2EHPA as carrier *J. Membr. Sci.* **250** pp 247-57

[30] Gâzo J, Bersuker I B, Garaj J, Kabešová M, Kohout J, Langfelderowá H, Melník M, Serator M and Valach F 1976 Plasticity of the coordination sphere of copper(II) complexes, its manifestation and causes *Coord. Chem. Rev.* **19** pp 253-97

IOP Conf. Series: Materials Science and Engineering **427** (2018) 012006 doi:10.1088/1757-899X/427/1/012006

Influence of dextrin on beneficiation of components from copper flotation concentrate

A Bakalarz[1], M Duchnowska[1] and R Kubik[2,3]

[1] Wrocław University of Science and Technology, Wybrzeże Wyspiańskiego Street 27, 50-370 Wrocław, Poland

[2] Polish Center for Technology Development (Wroclaw Research Centre EIT+ Ltd.), Stabłowicka 147, 54-066 Wrocław, Poland

[3] University of Silesia, Faculty of Earth Sciences, Będzińska 60, 41-200, Sosnowiec, Poland

alicja.bakalarz@pwr.edu.pl

Abstract. Depressants are one of the most important reagents in flotation process. They are used to improve upgrading selectivity. Natural depressants, especially starch and its derivatives, are one of the most significant polysaccharides used in polymetallic sulphide ores flotation. In the paper influence of dextrin on flotation of copper, organic carbon and mineral components from rougher concentrate was investigated. It has been shown, that dextrin is a good depressant for organic carbon and clay minerals with micas as long as the appropriate collector for sulphides are added. Moreover, significantly better flotation of all copper sulphides and pyrite with marcasite was observed, in comparison with flotation test using collector and frother only. Using dextrin as a single reagent does not affect beneficiation of examined concentrate. However, dextrin had not depressed galena particles. It can suggest that various factors can influence on good flotation of galena, e.g. associations with other sulphides, mineral composition and structure of examined feed.

1. Introduction

The largest European sedimentary copper ore, located at Legnica-Glogow Copper Basin (SW Poland), is processed by three Concentrator Plants which are the part of KGHM Polska Miedz SA [1]. The Concentration Plants process around 33 million tonnes of ore per year [2]. Ore consist of three lithological rock types: sandstone, dolomite and shale. Each of rock type differs mainly in chemical and mineral compositions, physical properties and floatability. The main copper sulphides are chalcocite, bornite, chalcopyrite and covellite. The ore also contains sulphides of other metals like lead (galena), zinc (sphalerite) and iron (pyrite and marcasite) [3,4].

In the case of the Fore-Sudetic deposit located in Poland, the shale ore is main carrier of organic matter (especially organic carbon). It is the main reason for technological problems in the metallurgical process [5]. However, this rock is quite rich in copper and other accompanying metals. One of the method to separate organic carbon from useful minerals is the application of depressant for organic matter and flotation of useful sulphides in the presence of appropriate collector with frother.

Depressants are used for increasing upgrading selectivity. This is one of the most important group of flotation reagents. In many processing plants the consumption of depressant is much higher than collector (even 20–30 times) [6]. Usually depressants are classified in two groups: organic and nonorganic. Sodium cyanide, zinc sulphate, sodium sulphide and dichromates have the great significance and the highest consumption amongst nonorganic depressants [7,8]. Sodium cyanide is used in depression of pyrite, arsenopyrite, sphalerite and pyrrhotite [9], zinc sulphate depresses zinc minerals and sodium sulphide – copper and iron sulphide minerals. Dichromates are applied to separate lead and copper minerals [8]. The most important organic depressants are: guar gum, starch and derivatives, especially dextrin. Guar gum is used in depression of talc and galena [10]. Starch is one of the most important polysaccharide used as depressing reagent in flotation process, especially for iron ores, flotation of muscovite and to separate molybdenite from copper sulphides [11]. Dextrin is received from starch by partial thermal degradation in acidic environment. It can be used to separate micas from sulphides, lead and iron minerals, like galena or pentlindite, from copper minerals, like chalcopyrite [12]. Dextrin and starch can be used in the depression of weathered silicates and carbonaceous matter and assist in depressing the nickel-bearing minerals [8]. According to Liu and Laskowski [13], chalcopyrite can be depressed by dextrin while galena can be floated with xanthate collector in acidic solution. However, in alkaline solutions chalcopyrite minerals are floated while galena is depressed [13].

The paper focuses on the analysis of influence of dextrin on the flotation of copper concentrate obtained after rougher flotation of sedimentary copper ore from Fore-Sudetic Monocline deposits. Three flotation tests with detailed chemical and mineral analysis of products were performed. All experiments were conducted in the presence of an aqueous solutions of industrial reagents. The upgrading selectivity of copper, organic carbon and mineral components was evaluated by mass balance and the plotting of separation curves. Flotation of copper, organic carbon and mineral components were analyzed and described.

2. Materials and methods

For the purpose of conducting the depression tests, a sample of shale-dolomitic copper ore was collected from the Rudna Concentrator Plant of KGHM Polska Miedz S.A. in Poland. Three rougher flotation experiments in the presence of collector and frother were conducted. A mixture of sodium ethyl and isobutyl xanthate with sodium O,O-diethyl dithiophosphate was used as the collector, while a mixture of polyglycol ethers was utilized as the frothing reagent. Both reagents were prepared directly before flotation test. From each flotation test one concentrate was collected after 30 minutes. These samples were used as a feed in depression tests. Chemical and mineralogical analyses of each rougher concentrate sample were conducted. The contents of copper and organic carbon were found to be about 7.4% and 8.8%, respectively. The content of all copper sulphides was about 8.9%, consisting of 4.5% chalcocite, 2.6% bornite, 0.9% chalcopyrite, and lower contents of covellite and enargite. Moreover, the concentrate samples contained about 1.4% pyrite and/or marcasite, 0.4% galena and 0.2% sphalerite. In table 1 the particle size distribution of rougher concentrate is presented.

Table 1. Particle size distribution of rougher concentrate.

Particle size, mm	Yield, %
0–0.020	37.4
0.020–0.040	27.7
0.040–0.071	22.8
0.071–0.100	7.9
+0.100	4.2

Three flotation tests of rougher concentrate samples were conducted. All conditions of performed tests are presented in table 2. In the course of each test one concentrate and tailings were collected.

IOP Conf. Series: Materials Science and Engineering **427** (2018) 012006 doi:10.1088/1757-899X/427/1/012006

Table 2. Conditions of performed flotation tests.

Flotation number	Dextrin dose, g/Mg	Collector dose, g/Mg	Frother dose, g/Mg	Comments
1	–	20	5	–
2	2 000	80		–
3	2 000	–		Before depression test, the feed was purified with distilled water to remove all collector

All experiments were conducted in a Denver D12 laboratory flotation machine using flotation cell with capacity of 1.5 dm^3. The density of the flotation suspension was 300 g/dm^3. In each test technological water from Rudna Concentrator Plant was used. Dextrin was used as an organic carbon depressant next to standard reagents applied as the collecting and frothing reagents. During each experiment, the air flow rate was kept constant at 20 dm^3/h. Flotation tests were conducted at stirrer speeds between 700 and 850 rpm. The concentrates were collected after 20 minutes. All flotation tests were conducted in natural pH (between 7.8–8.0). It should also be pointed out that dextrin solution was added before collector and frother in the second flotation. The suspension was mixing with depressant for 10 minutes, with collector 3 minutes and 1 minute with frother.

After the flotation all products were washed in distilled water to remove all soluble compounds. Clear samples were dried at 65 °C for 24 hours and weighed. After that, the representative samples from each product were collected to perform chemical and mineral analysis. Mineralogical investigation of each sample was performed using QEMSCAN® system with advanced procedure that was created and developed based on Polish copper ores and products of their processing [14]. Mineral library dedicated for this type of material have been previously used in another research related to the final concentrates and tailings from three different Concentration plants in Poland [15]. Data obtained for each sample contain such information as mineral composition and particle size distribution. The contents of copper and organic carbon in each flotation product were determined by using iodometric titration and spectrometric method, respectively. All obtained data were recalculated and balanced.

3. Results

The results of flotation were evaluated by the mass balance of components and products of separation, as well as by plotting recovery-recovery and grade-recovery separation curves (figures 1-5). Obtained results allow for comparison of the upgrading selectivities of copper, organic carbon and all mineral components. Moreover, in figure 6 the upgrading results of particle size in each flotation test are presented.

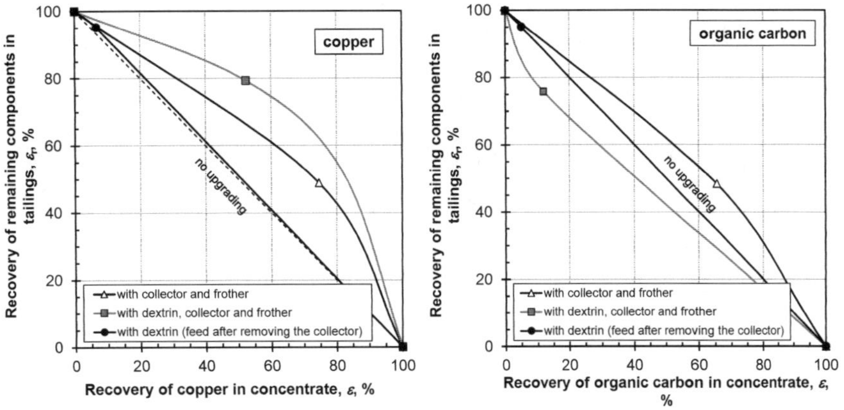

a b

Figure 1. Upgrading curves as a relationship between recovery of copper (a) and organic carbon (b) in the concentrate and recovery of remaining components in tailing for all performed flotation experiments.

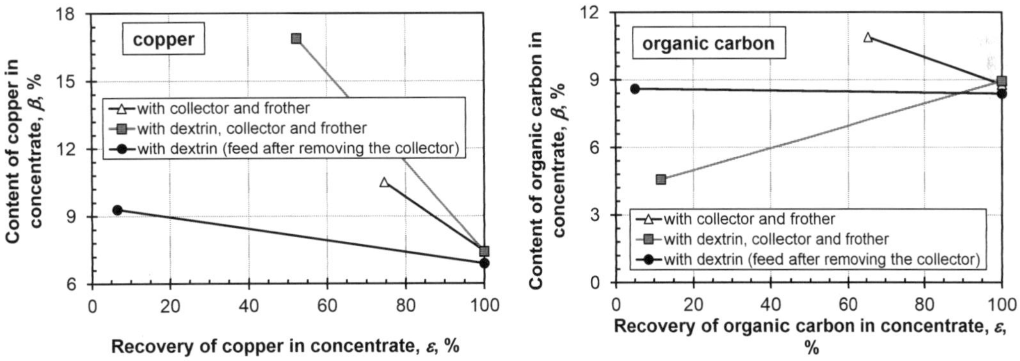

a b

Figure 2. Upgrading curves as a relationship between recovery of copper (a) and organic carbon (b) in the concentrate and content of this component in concentrate for all performed flotation experiments.

As it is presented in figures 1 and 2, the best results of upgrading of copper were obtained in the presence of dextrin, collector and frother as reagents. In this experiment the best quality of flotation concentrate was obtained: the content of copper equals to 17% with recovery of this metal at the level of 53%. The addition of organic depressant in the presence of sulphides collector significantly improves the copper selectivity (figures 1a and 2a) and decreases the flotation of organic carbon (figures 1b and 2b). The content of this component in concentrate was reduced from about 9% to 4.5% while the recovery of organic carbon in concentrate decreased from about 65% to 12%. Furthermore, addition of dextrin as a single reagent (without collector) results in the lack of upgrading of copper and organic carbon in the concentrate. It indicates that to depress the organic carbon the presence of dextrin with sulphide collector is essential.

In figures 3-5 the upgrading results of copper, lead and iron sulphides are presented. Additionally, the results for useless components: clay minerals with micas, carbonates and quartz are shown. As it is presented, the best upgrading results for each copper sulphide in the presence of dextrin, collector and frother are obtained. The best results were observed for chalcopyrite while the lower upgrading selectivity was obtained for covellite. For bornite and chalcocite quite similar results were achieved. It confirmed, that the addition of depressant reagent improves the upgrading of each copper sulphide, especially the easiest floating mineral – chalcopyrite. As it can be seen that, when only collector and

IOP Conf. Series: Materials Science and Engineering **427** (2018) 012006 doi:10.1088/1757-899X/427/1/012006

frother was used, the upgrading results for copper sulphides are quite comparable (similar curves courses). Application of dextrin as a single reagent results in the lack of copper sulphides flotation, with the exception of covellite. For this mineral low flotation level is observed. Similar results were achieved for galena and pyrite/marcasite flotation (figure 4). Moreover, the addition of dextrin (in the presence of collector and frother) does not reduce the flotation of the main lead-bearing mineral – galena. This sulphide floats similar in test with collector, frother and dextrin and also in the test using only collector with frother. It means that the dextrin does not depress galena. Furthermore, the addition of dextrin into flotation in the presence of collector significantly increases the upgrading selectivity of pyrite/marcasite.

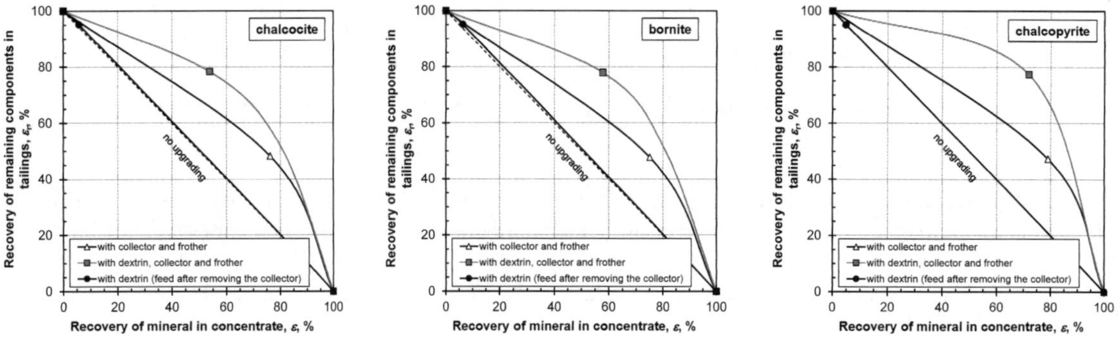

Figure 3. Upgrading curves as a relationship between recovery of chalcocite/bornite/chalcopyrite in the concentrate and recovery of remaining components in tailings for all performed flotation experiments.

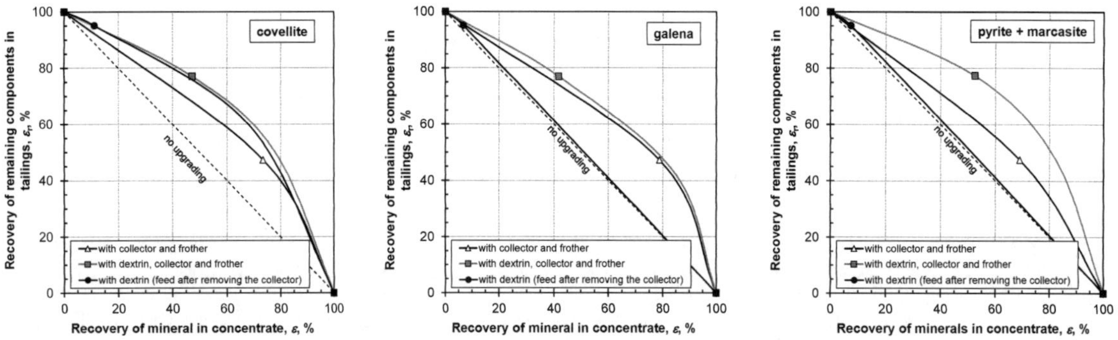

Figure 4. Upgrading curves as a relationship between recovery of covellite/galena/pyrite+marcasite in the concentrate and recovery of remaining components in tailings for all performed flotation experiments.

Application of dextrin with collector as a flotation reagents results in worse flotation of clay minerals with micas – the main organic matter carrier (figure 5). For the rest of analyzed components – rock-forming minerals (calcium and magnesium carbonates and quartz) a bit better flotation was achieved in the presence of depressant with collector. No effect of dextrin usage (without collector addition) during the flotation of all rock-forming components was observed.

IOP Conf. Series: Materials Science and Engineering **427** (2018) 012006 doi:10.1088/1757-899X/427/1/012006

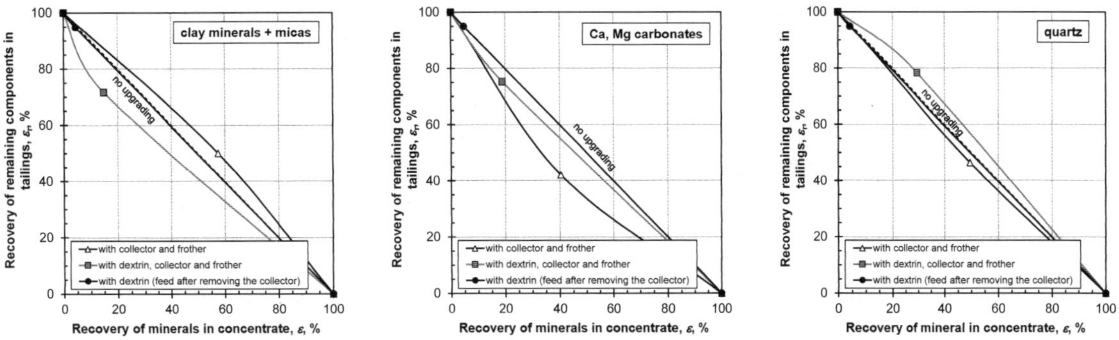

Figure 5. Upgrading curves as a relationship between recovery of clay minerals+micas/carbonates/quartz in the concentrate and recovery of remaining components in tailings for all performed flotation experiments.

The analysis of influence of dextrin on the particle size flotation was conducted. As it can be seen in figure 6, the presence of collector with frother as flotation reagents give similar flotation of each analyzed size fraction. The addition of dextrin give better flotation of all particles, with the exception of the smallest one (below 0.020 mm). This fraction markedly floats worse than the rest. Definitely it is connected with the depression of fine and very fine clay minerals, which are major carrier of organic matter. The most interesting results were obtained in the experiment with dextrin without collector addition. There is no upgrading into concentrate particles coarser than 0.020 mm. Moreover, a low flotation of these fractions into tailings are observed. Only for the smallest particles (below 0.020 mm) a weak flotation into concentrate was observed. Probably, this observation explains the reason of quite good flotation of covellite in this test. This mineral usually occurs as fine and very fine grains and forms a lot of outgrowths with other sulphides [16]. Small size of covellite effects in a weak flotation of this mineral in concentrate with the presence of dextrin and without using collector.

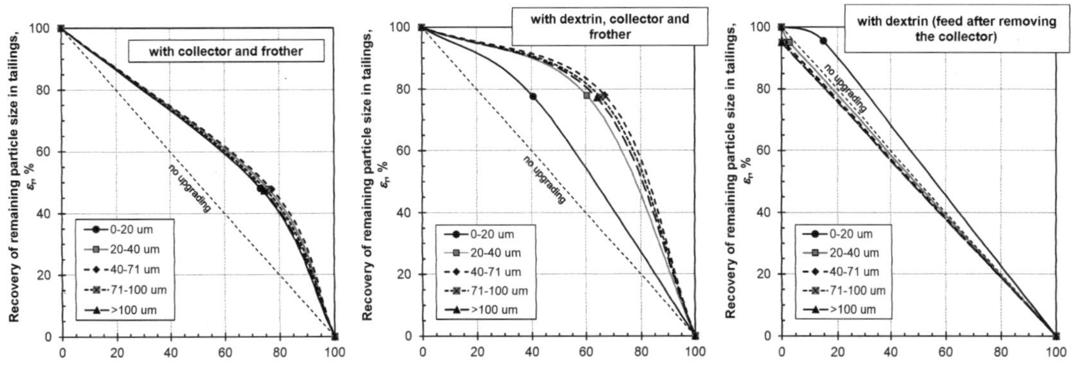

Figure 6. Upgrading curves as a relationship between recovery of particle size in the concentrate and recovery of remaining particle sizes in tailings for all performed flotation experiments.

4. Conclusions

Three flotation tests were conducted in the presence of dextrin and/or collector and frother. Chemical and mineralogical analyses of all upgrading products were performed.

According to presented results, dextrin is a good depressant for organic carbon and clay minerals with micas as long as the appropriate collector for sulphides is added. Significantly better upgrading of each copper sulphide and pyrite/marcasite is observed, in comparison with flotation test in which only collector and frother were used. Using dextrin as a single reagent does not affect beneficiation of examined feed. The addition of dextrin improves a bit flotation of useless minerals such as carbonates and quartz. In the presence of dextrin with sulphide collector, flotation of very fine particles reduces substantially. Certainly it is connected with depressing effect of dextrin on the clay minerals, which usually form fine and very fine grains. It is quite interesting that dextrin does not depress main lead-bearing mineral – galena. According to available worldwide literature, in the presence of polysaccharides (starch or dextrin) and xanthate collector at a neutral or slightly alkaline pH, copper sulphides (like chalcopyrite or chalcocite) are depressed while galena floats well [11,13,17,18]. Markedly different results were obtained in the paper and other research carried out in Authors department. There may be several reasons for the described phenomenon that can relate to mineralization, composition and/or structure of examined material. Received results can indicate galena particles that occur in strongly associations with copper sulphides and float with copper-bearing minerals in the presence of sulphide (usually xanthate) collector. Moreover the methodology can significantly influence on the flotation of galena. This research has potential for further investigation.

Acknowledgments
Financial support by Statutory Research grant (0402/0007/17) is greatly acknowledged.

References
[1] Yu-Zhuang Sun, Leischner, A and Göbbels, F-J 2000. Residuary toxic elements and pahs in sediments of the Zbiornik Gilow tailings pond and Zimmica stream from Lubin district, southwest Poland *Environmental Geochemistry and Health,* **22** pp 249–61
[2] Ore Enrichment, Website: http://kghm.com/en/our-business/processes/ore-enrichment, May 13, 2018
[3] Piestrzynski A, Banaszak A and Zaleska–Kuczmierczyk M 2007 *Okruszcowanie Monografia KGHM Polska Miedz S.A.* pp 67–9
[4] Kucha H 2007 Mineralogia kruszcowa i geochemia ciala rudnego zloza Lubin-Sieroszowice *Biuletyn Panstwowego Instytutu Geologicznego* **423** pp 77–94
[5] Konieczny A, Pawlos W, Krzeminska M, Kaleta R, and Kurzydlo P 2013 Evaluation of organic carbon separation from copper ore by pre-flotation *Physicochemical Problems of Mineral Processing,* **49(1)** pp 189–201
[6] Bulatovic S M and Wyslouzil D M 1994 Selection and evaluation of different depressants systems for flotation of complex sulphide ores *Minerals Engineering* **8** pp 63–76
[7] Pearce M J 2005 An overview of the use of chemical reagents in mineral processing *Minerals Engineering* **18** pp 139–49
[8] Cytec 2002 *Mining Chemicals Handbook, Revised Edition,* Cytec Industries Inc.
[9] Kostovic M and Vucinic D 2014 The influence of cyanide salts and ferrous sulphate on pyrite flotation *Physicochemical Problems of Mineral Processing* **52** pp 609–19
[10] Zhao K, Yan W, Wang X, Hui B, Gu G and Wang H 2017 The flotation separation of pyrite from pyrophyllite using oxidized guar gum as depressant *International Journal of Mineral Processing* **161** pp 78–82
[11] Drzymala J, Kapusniak J and Tomasik P 2003 Removal of lead minerals from copper industrial flotation concentrates by xanthate flotation in the presence of dextrin *International Journal of Mineral Processing* **70** pp 147–55
[12] Lopez-Valdivieso A, Celedón Cervantes T, Song, S and Laskowski J S 2004 Dextrin as a non-toxic depressant for pyrite in flotation with xanthates as collector *Minerals Engineering* **17(9-10)** pp 1001–6
[13] Liu Q, Laskowski J S, Li Y and Wang D F 1994 Synergetic effect of mineral surface constituents in dextrin adsorption *International Journal of Mineral Processing* **42** pp 251–66
[14] Kubik R 2018 Creation and development of QEMSCAN® analysis procedure based on Polish

copper ores investigation *18th International Multidisciplinary Scientific Geoconference SGEM 2018 Conference Proceedings* **18** Issue **1.1.** pp 59–66

[15] Kubik R, Bakalarz A and Duchnowska M 2018 Characteristics of the processing products from Polish copper ores in terms of the quality of concentrates and wastes composition, based on mineralogical analyzes *18th International Multidisciplinary Scientific Geoconference SGEM 2018 Conference Proceedings* **18**, Issue **1.4**. pp 19–26

[16] Banaszak A and Banas M 1996 *Geologia, cz. II Monografia KGHM Polska Miedz SA,*

[17] Liu Q and Laskowski J S 1989 The role of metal hydroxides at mineral surfaces in dextrin adsorption. II: Chalcopyrite–galena separations in the presence of dextrin *International Journal of Mineral Processing* **27** pp 147–55

[18] Bolin N J and Laskowski J 1991 Polysaccharides in flotation of sulfides: Part II. Copper/lead separation with dextrin and sodium hydroxide *International Journal of Mineral Processing* **33** pp 235–41

IOP Conf. Series: Materials Science and Engineering **427** (2018) 012007 doi:10.1088/1757-899X/427/1/012007

Influence of foam height on salt flotation of copper-bearing shale

T Ratajczak

Wroclaw University of Science and Technology, Faculty of Geoengineering, Mining and Geology, Wybrzeze Wyspianskiego 27, 50-370 Wroclaw, Poland

tomasz.ratajczak@pwr.edu.pl

Abstract. Influence of height of foam on salt flotation of copper-bearing shale, having different Cu and organic carbon contents, was investigated. It was found that the yield of flotation of copper-bearing shale in the presence of NaCl is directly dependent on the content of organic carbon and height of foam produced during flotation. It has been shown that the height of flotation froth *h* depends on type of shale, and its variation can be described by formula $h/h_{max}=1/(t+1)^n$, where *t* is the time of flotation and *n* is a constant.

1. Introduction

Flotation of mineral particles in aqueous salt solutions without additional chemical reagents is referred to as salt flotation [1-2]. Inorganic salts may act as foaming agents [1-7], and their effectiveness in flotation depends on the type and concentration of salt [2, 8].

Copper-bearing shale from the sedimentary deposits of the Legnica-Głogów Copper District is characterized by a different lithological structure, mainly due to its diverse chemical and mineralogical composition. For this reason, shales from the LGOM area are classified as clayey, dolomitic, clayey-dolomitic with varying content of copper (Cu) and organic carbon (Corg). The most valuable is the shale with increased content of organic carbon due to the higher content of copper. An important factor of the shale is Corg content, because its high content adversely affects the copper metallurgical processes. For the separation of carbonaceous substances from copper ore preflotation is used, which consists in flotation of ore in the presence of only froth agents. It is indicated that differentiated organic carbon content in shale can affect their floatability in the presence of foaming agents [11].

The purpose of paper, based on the results of Kurkiewicz [9] as well as Kurkiewicz and Ratajczak [10], was to determine the effect of foam height on salt flotation of copper-bearing shale having different content of copper and organic carbon.

2. Flotation of copper-bearing shale in presence of NaCl

The tests were performed using copper-bearing materials denoted as shale L, Ps and Pd [9, 10, 11]. They originated from LGOM [11]. Their Cu and C_{org} were determined in the Mineral Processing Laboratory WGGG PWR content is given in table 1. The size of particle was below 0.1 mm.

Table 1. Characteristics of shale [10, 11].

Shale symbol	Type of shale	Cu,%	C_{org},%
L	loamy-dolomite shale high-carbon and medium-copper	1.13	7.45
Pd	dolomite-loamy shale medium-carbon and high-copper	0.83	5.34
Ps	dolomite shale low-carbon and low-copper	0.58	0.57

Flotation of copper-bearing shale was carried out in a Mechanobr flotation machine with cell having capacity of 0.25 dm^3 and an air flow of 50 dm^3/h. The tests were performed in aqueous NaCl solutions at concentrations of 0.25M; 0.50M; 1.00M; and 2.00M. The flotation time was 30 minutes. The height of the flotation froth was measured at the beginning of collecting each flotation product, i.e. 1; 4; 10; 15 and 30 minutes of flotation. A more detailed research methodology is given in the work of Kurkiewicz and Ratajczak [10] and Kurkiewicz [9]. Figure 1 presents a summary of the results of flotation of investigated copper-bearing shales. Figures 1a-c refer to flotation yield, described by the 1st order kinetics in the work of Kurkiewicz and Ratajczak [10], while figures 1a'-c' the height of foam during flotation of the L, Pd, Ps shales.

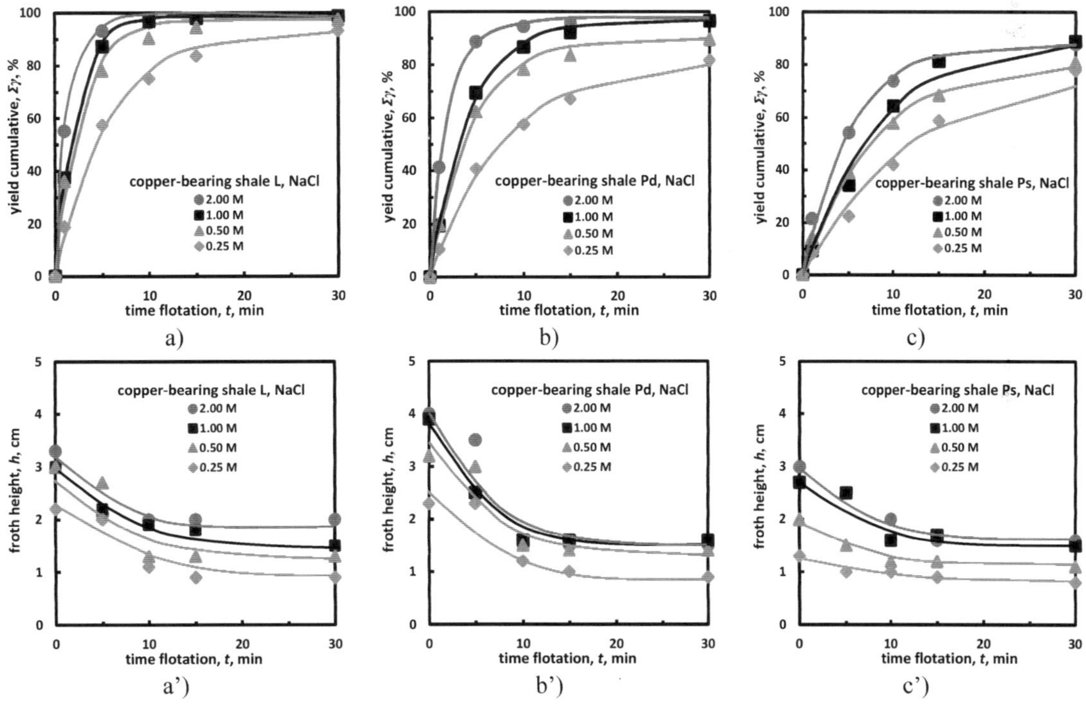

Figure 1. a-c) yield of flotation of copper-bearing shale [10] as a function of flotation time, a'-c') height of foam versus flotation time.

Figure 1 shows that copper-bearing shale L, Pd, Ps flotation depends on concentration of NaCl solution and type of shale. Additionally figure 2 presents a dependence of the maximum yield as a function of froth height measured during flotation (t = 30 min). Larger yields and higher froth were obtained for flotation of L shale, which was characterized by a higher content of organic carbon, in comparison to shale Ps (table 1).

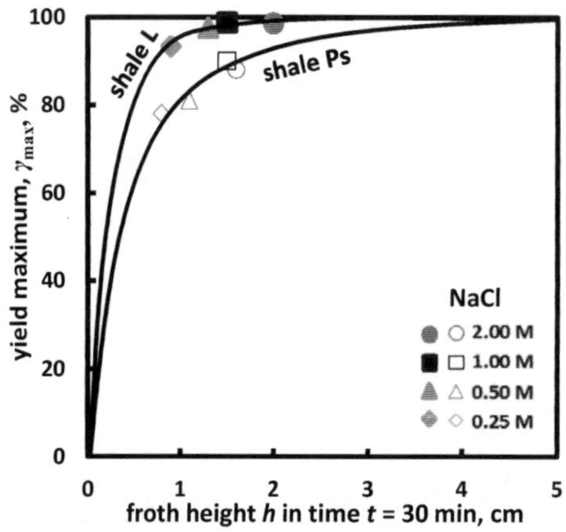

Figure 2. Dependence of shale flotation maximum yield on height of froth in the presence of NaCl.

3. Evaluation of copper-bearing shale flotation performed in the presence of NaCl

In order to plot curves which characterize the influence of flotation time on froth height for the studied copper-bearings shale, an approximation of results collected in fig. 1 a'-c' was carried out. Figure 3 shows a dependence of the height ratio of foam (h) and its maximum height (h_{max}) on time (t) of shale flotation. The waveforms realation, which is shown in figure 3, can be approximated with an exponential or polynomial curve, but the latter would lead to a multiparameter function, which is not beneficial when flotation results are used for comparison.

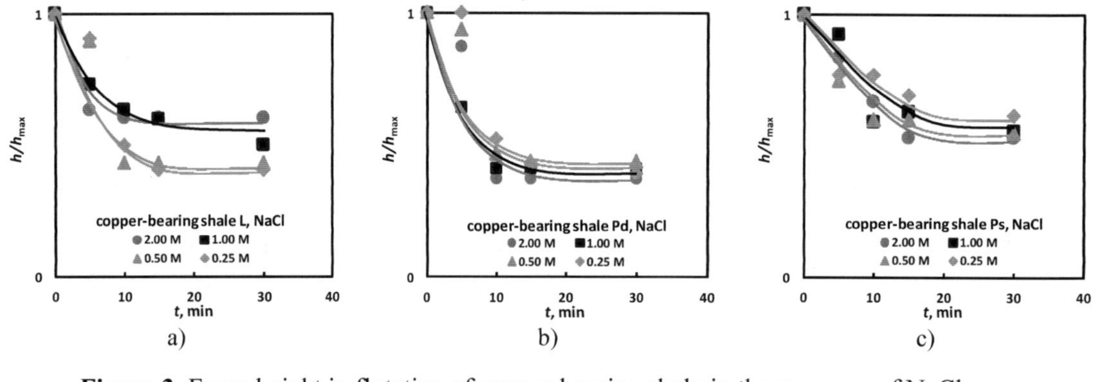

Figure 3. Foam height in flotation of copper-bearing shale in the presence of NaCl.

Figure 4 shows the relation between height of froth and time of flotation of the copper-bearing shale samples in NaCl solutions approximated with equation (1):

$$h/h_{max} = 1/(t+1)^n, \tag{1}$$

where: h - height of foam during salt flotation of copper-bearing shale, cm; h_{max} - maximum height of foam during salt flotation of copper-bearing shale, cm; t - flotation time, min; n - foam height indicator.

Figure 4 shows that the height of froth depends on the type of shale, and its height, characterized by index n, is directly dependent on the organic carbon content in the shale (table 1), and thus hydrophobicity of shale [10, 14]. However, the dependence of n and C_{org} has a minimum at the carbon content of approximately 5%. After that the value of the index describing the height of foam n begins to increase, which may indicate the parabolic nature of dependence of n and C_{org}. Research in this area should be continued, because basing on 3 experimental points it is difficult to clearly know the course of the dependence.

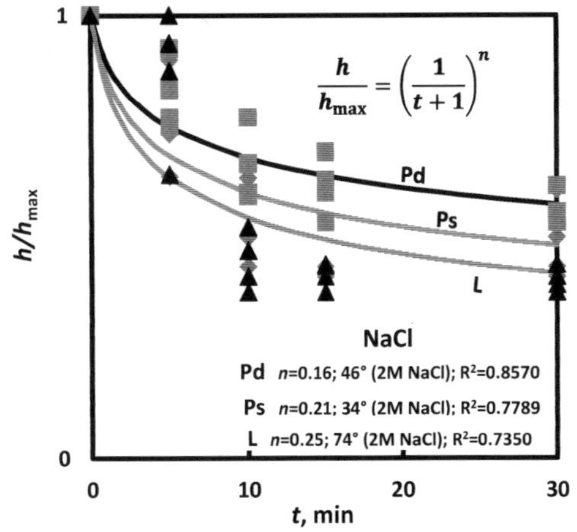

Figure 4. Froth height limit curves for flotation of copper-bearing shales in the presence of NaCl.

4. Summary

As a result of the conducted research, the influence of froth height on flotation yield of copper-bearing shale in presence of NaCl was presented. It was noted that height on froth flotation increases with increasing salt concentration and is highest at the beginning of each flotation of studied shale. During further flotation (1-15 min), froth height decreases and becomes constant for about 15 minutes. The initial increase in froth height may indicate that the particles of shale rapidly move to the surface of the froth, which increases the volume (froth height), and the gradual reduction in the particles concentration can cause a decrease of the froth height. In addition, it was shown that for L shale characterized by a higher content of organic carbon in relation to the content of Corg in shale Ps, the outflow and the flotation foam is higher (fig. 1a-a', c-c'; fig. 2). It was also found that the height of the froth during copper-bearing shale flotation depends on concentration of salt and type of copper-bearing shale, and its course can be described by equation $h/h_{max}=1/(t+1)^n$. The presented research results may be a basis for further work on the effectiveness of flotation measured by the recovery of Cu or C_{org} content in relation to changes in the froth height over time.

Acknowledgements

This work was partially financed by the Polish Statutory Research Grant 0401/0129/17.

References

[1] Laskowski J 1965 *J. Colliery Guardian* **211** pp 361-6
[2] Ratajczak T and Drzymała J 2003 *Flotacja solna* (Oficyna Wydawnicza Politechniki Wrocławskiej: Wrocław)
[3] Pugh R J, Weissenborn P and Paulson O 1997 *Int. J. Miner. Process.* **51** pp 125-38
[4] Grabowski B and Drzymała J 2008 *Annales Universitatis Mariae Curie-Skłodowska, Lublin-Polonia, Sectio AA*, Vol. LXIII **6** pp 68-72
[5] Lipniarski M, Ratajczak T and Drzymała J 2015 *III Polski Kongres Górniczy, Mineralurgia i wykorzystanie surowców mineralnych* (WGGG PWr: Wrocław) pp 35–9
[6] Smólska M and Ratajczak T 2017 *Łupek miedzionośny III* (WGGG PWr: Wrocław) pp 97-102
[7] Witan J and Ratajczak T 2017 *Łupek miedzionośny III* (WGGG PWr: Wrocław) pp 138-45
[8] Ratajczak T 2017 *Mineral Engineering Conference MEC2017* EDP Sciences art. 01028 pp 1-3

IOP Conf. Series: Materials Science and Engineering **427** (2018) 012007 doi:10.1088/1757-899X/427/1/012007

[9] Kurkiewicz S 2017 *Flotometryczna hydrofobowość łupka miedzionośnego w obecności soli* Praca dyplomowa, niepublikowana (WGGG PWr: Wrocław)
[10] Kurkiewicz S and Ratajczak T 2017 *Łupek miedzionośny III* (WGGG PWr: Wrocław) pp 103-9
[11] Drzymała J, Karwowski P, Borowski K, Pązik P and Kowalczuk P B 2017 *Łupek miedzionośny III* (WGGG PWr: Wrocław) doi: 10.5277/lupek1702 pp 52-58
[12] Ratajczak T, Drzymała J and Kowalczuk P B 2016 *Mineral Engineering Conference MEC2016* EDP Sciences art. 01033 pp 1-5
[13] Drzymala J, Ratajczak T and Kowalczuk P B 2017 *Physicochem. Probl. Miner. Process.* **53(2)**, pp 983–95
[14] Kubiak B and Drzymała J 2017 *Łupek miedzionośny III* (WGGG PWr: Wrocław) pp 64-68

IOP Conf. Series: Materials Science and Engineering **427** (2018) 012008 doi:10.1088/1757-899X/427/1/012008

Analysis of distributions of various coal types properties by means of statistical methods

T Niedoba, P Pięta and A Surowiak

AGH University of Science and Technology, Faculty of Mining and Geoengineering, Department of Environmental Engineering and Mineral Processing, Mickiewicza 30, 30-059 Krakow, Poland

tniedoba@agh.edu.pl

Abstract. Coal can occur in various natural forms and its quality depends on its various properties. Polish coal mines extract and process various coal types, according to the Polish nomenclature. The paper presents an attempt of determining whether is possible to statistically differ the typical coal types occurring in Polish hard coal mines by means of traditional statistical tests. To this purpose three coal types, namely 31, 34.2 and 35 were selected from three Polish mines located in Upper Silesia. After appropriate preparation of the raw coals selected coal properties were examined, including combustion heat, ash contents, sulfur contents, volatile parts contents and analytical moisture. Then, by means of several statistical tests it was verified whether it is possible to prove statistically that these coal types differ significantly not knowing the type of distribution they represent. To this purpose such tests as Kruskal-Wallis test, test of series, Wilcoxon test and median test were applied. The results were statistically evaluated and the conclusions were made. The presented methodology may serve to compare not only coal types but also other raw materials.

1. Introduction

Coal as a fuel is still crucial one in Poland. Despite the environmental aspects and European Union policy its production is still very important for the country. However, its quality is also very important taking into consideration both environmental and economic aspects. Polish hard coal is located mainly in Upper Silesian Industrial Region, where several mines are situated. They produce various sorts of hard coal. There is a question – is it possible to easily recognize the type of coal basing only on its selected properties. The Authors already tried to answer this question but this time the analysis was conducted by means of traditional non-parametric statistical tests, without assumption of known form of distribution function. The whole calculating procedure is described in further part of the paper.

2. Preparation of data

Three types of coal are considered in the paper, namely 31, 34.2 and 35 according to Polish nomenclature. On the basis of the results of their basic properties investigations such parameters as combustion heat, ash contents, sulfur contents, volatile parts contents and analytical moisture were examined. These properties are treated as random variables of unknown distribution functions. By means of selected statistical methods the distributions of the same property for various coal types were compared in purpose of determining their similarity. To achieve this goal the separation of coal feeds of all considered coal types into particle size fractions 0.50-1.00; 1-00-3.15, 3.15-6.30, 6.30-8.00,

8.00-10.00, 10.00-12.50, 12.50-14.00, 14.00-16.00, 16.00-20.00 was performed. The measurements of the researched properties were done and the obtained results are presented in tables 1-3 for coal types 31, 34.2 and 35, respectively. Remark: data for combustion heat are given in J, but the calculations were made for coal.

Table 1. Juxtaposition of the results for coal, type 31.

Size *d*, mm	Combustion heat, J	Ash, %	Sulfur, %	Volatile parts, %	Analytical moisture W_a
0.50-1.00	21218.70	31.59	1.31	26.02	3.32
1.00-3.15	20506.95	30.92	1.45	26.35	2.78
3.15-6.30	20929.81	31.71	1.42	27.58	2.87
6.30-8.00	19770.07	33.53	1.23	25.43	2.78
8.00-10.00	18526.59	34.20	0.99	25.90	2.77
10.00-12.50	19430.94	32.88	1.23	26.02	2.94
12.50-14.00	18258.63	34.56	1.10	25.59	2.70
14.00-16.00	18802.92	34.74	0.98	25.78	2.79
16.00-20.00	19929.17	33.17	0.97	25.62	2.98

Table 2. Juxtaposition of the results for coal, type 34.2.

Size *d*, mm	Combustion heat, J	Ash, %	Sulfur, %	Volatile parts, %	Analytical moisture W_a
0.50-1.00	23169.75	28.75	0.89	25.19	0.66
1.00-3.15	23077.64	28.98	0.66	24.30	1.05
3.15-6.30	22792.94	27.49	0.64	24.74	0.96
6.30-8.00	23278.61	31.91	0.58	24.55	0.88
8.00-10.00	22591.97	31.05	0.50	23.92	1.12
10.00-12.50	23211.62	31.74	0.30	24.60	1.05
12.50-14.00	22324.02	29.44	0.46	25.65	0.97
14.00-16.00	26456.39	28.09	0.66	22.63	0.74
16.00-20.00	19313.71	28.78	0.34	22.25	1.01

Table 3. Juxtaposition of the results for coal, type 35.

Size *d*, mm	Combustion heat, J	Ash, %	Sulfur, %	Volatile parts, %	Analytical moisture W_a
0.50-1.00	22922.73	31.17	0.60	17.56	1.08
1.00-3.15	22051.88	33.29	0.57	17.55	1.02
3.15-6.30	22139.80	32.41	0.58	17.16	1.37
6.30-8.00	21490.84	34.09	0.58	14.40	1.23
8.00-10.00	22085.37	34.88	0.61	17.54	1.37
10.00-12.50	21788.11	34.44	0.50	17.31	1.36
12.50-14.00	21507.59	34.33	0.57	17.81	1.40
14.00-16.00	21930.46	33.89	0.52	17.39	1.32
16.00-20.00	21411.30	34.77	0.69	17.25	1.39

3. Methodology and results

Let mark the random variable related to j^{th} property in i^{th} coal type ($i = 1, 2, 3; j = 1, 2, 3, 4, 5$) as X_{ij} and distribution function of X_{ij} as $F_{ij}(x)$. First, the verification of H_0 hypothesis was performed that the distribution function of jth coal property is identical for all considered coal types. That means that

$$H_0: F_{1j}(x) = F_{2j}(x) = F_{3j}(x) \tag{1}$$

To verify this hypothesis the Kruskal-Wallis χ^2 statistics was applied, which represents asymptotic χ^2 distribution function of two degrees of freedom [1-4]. The χ^2 statistics is in form:

$$\chi^2 = \frac{12}{n(n+1)} \sum_{i=1}^{3} \frac{R_i^2}{n_i} - 3(n+1) \tag{2}$$

where: n_i – quantity of the sample for i^{th} coal type (i = 1, 2, 3); $n = n_1+n_2+n_3$; R_i – sum of ranges of the results of ith sample (i = 1, 2, 3).

The values of the χ^2 test for individual coal properties are presented in the table 4.

Table 4. Values of χ^2 test.

Coal property	χ^2
Combustion heat	20.174
Ash	17.620
Sulfur	20.056
Volatile parts	25.580
Analytic moisture	25.056

The critical area for Kruskal-Wallis test is range $(\chi^2_\alpha(2), +\infty)$ [2-4]. For the significance level α=0.05 the value of χ^2_α=5.991. On the basis of the results presented in table 4 the values of the test for all coal properties belong to the critical area, so the H_0 hypothesis should be rejected. This means that the distributions of the investigated coal properties depend on the coal type.

Next, the distributions of the researched coal properties were compared in pairs. So, the H_0 hypotheses were in form:

$$H_0: F_{i,j}(x) = F_{k,j}(x) \tag{3}$$

where: i, k = 1, 2 ,3; j = 1, 2, ..., 5; i≠k

To this purpose the test of series and the Wilcoxon's test were applied [2-4]. The results of both selected samples were put in growing order. In the test of series the test statistics is number of series *K*, where a series is each maximum sub-series including only elements of the same sample [2]. The critical area for test of series is a range $<2, k_\alpha(n_1, n_2)>$, where $k_\alpha(n_1, n_2)$ is the value read from the statistical tables for series distribution and significance level α.

In Wilcoxon test U statistics is the tested value which is equal to the sum of inversions of the elements of one of the samples according to the elements of the second sample (equation (4).

$$U = u_1 + u_2 + \cdots + u_k \tag{4}$$

where u_i is the number of elements of first sample being not bigger than the ith element of second sample. The critical area of Wilcoxon test is sum of ranges $<0, u_\alpha(n_1, n_2)> \vee <n_1n_2 - u_\alpha(n_1, n_2), n_1n_2>$., where $u_\alpha(n_1, n_2)$ is a value read from statistical tables for significance level α.

The results of the test of series and Wilcoxon test are presented in tables 5 and 6.

Table 5. Test of series – values of *K*.

	(31, 34.2)	(31, 35)	(34.2, 35)
Combustion heat	4	2	5
Ash	6	10	4
Sulfur	2	2	9
Volatile parts	4	2	2
Analytic moisture	2	2	6

Table 6. Wilcoxon test – values of *U*.

	(31, 34.2)	(31, 35)	(34.2, 35)
Combustion heat	75	81	13
Ash	7	56	79
Sulfur	81	81	49
Volatile parts	0	0	77
Analytic moisture	2	81	81

The critical area for test of series (significance level α=0.05) is range <2, 6> while for Wilcoxon test is the sum of ranges <0, 11> and <70, 81>. On the basis of the results of both test it can be said that only in case of the random variable describing ash contents for coal types 31 and 35 as well sulfur contents for coal types 34.2 and 35 the H_0 hypotheses of identical distributions can be accepted. In case of the random variable describing combustion heat for coal types 34.2 and 35 the test of series rejected the hypothesis of distributions equality while Wilcoxon test did oppositely. In other cases, both tests rejected H_0 hypotheses. In these three cases the median test was applied additionally with the same form of H_0 hypothesis (distributions are identical) [3,4]. To this purpose the division of researched material into particle size and particle density fractions was done. In this way 72-element samples were obtained [1]. For each considered coal pair and their properties the median value was determined from all results considering certain property.

To verify the hypothesis of equality between distributions the χ^2 test was applied

$$\chi^2 = \sum_{i=1}^{2}\sum_{j=1}^{2}\frac{\left(n_{ij}-np_{ij}\right)^2}{np_{ij}} \tag{5}$$

where: n_{1i} – number of results being bigger than the median value for i^{th} coal type (i=1, 2); n_{2i} – number of results being smaller than the median value for i^{th} coal type (i=1, 2); $n=n_1+n_2$ – number of results being part of both considered samples; $p_{ij} = p_i \cdot p_j = \frac{n_{i1}+n_{i2}}{n} \cdot \frac{n_{1j}+n_{2j}}{n}$.

The results of median test are presented in table 7.

Table 7. Median test – results of χ^2 test.

	median	χ^2
Ash (31, 35)	29.44	0.0007
Sulfur (34.2, 35)	0.52	0.0279
Combustion heat (34.2, 35)	5448	0.2500

The χ^2 test is used in median test where the number of degrees of freedom is equal to 1. The critical area for this test is a range $\left(\chi^2_\alpha, +\infty\right)$, where χ^2_α is value read from χ^2 distribution statistical tables for significance level α. For α=0.05 the value of $\chi^2_\alpha = 3.841$. Because in all three cases the inequality $\chi^2 < \chi^2_\alpha$ occurs then the hypothesis about equality between distributions can be accepted.

4. Conclusions

As it was shown application of traditional non-parametric statistical tests to determine whether there are significant differences between the analyzed coal types is not always efficient. This means that either these differences are not so big, either the selection of the coal properties was not ideal. However, after applying several tests to this aspect it occurred that eventually the differences with combustion heat, ash contents, sulfur contents and analytic moisture taken into consideration can be treated as statistically important.

In works [1, 5, 6, 7, 8], in which the observational tunnels method was applied to identify the coal type on the basis of the measurements of their properties was proved that the sufficient coal features to identify the coal type are moisture, volatile parts contents and sulfur contents. The methods being applied in this particular work also proved that such properties as moisture, sulfur contents and volatile parts contents significantly differentiate individual coal types.

Acknowledgment

The paper is an effect of the statutory project no. *11.11.100.276.*

References

[1] Niedoba T 2013 *Multidimensional characteristics of random variables in description of grained materials and their separation processes* (Kraków: Wydawnictwo Instytutu Gospodarki Surowcami Mineralnymi i Energią PAN) [in Polish]
[2] Krysicki W, Bartos J, Dyczka W, Królikowska K and Wasilewski M 2012 *Probability theory and mathematical statistics in examples 2* (Warszawa: PWN) [in Polish]
[3] Dobosz M 2001 *Statistical analysis of investigation results* (Warszawa: Akademicka Oficyna Wydawnicza EXIT) [in Polish]
[4] Stanisz A 2007 *Easy course of statistics 3* (Kraków: Statsoft) [in Polish]
[5] Jamróz D and Niedoba T 2014 *Physicochemical Problems of Mineral Processing* **50** pp 185-203
[6] Jamróz D and Niedoba T 2013 *Archives of Mining Sciences* **58(4)** pp 1317-31
[7] Niedoba T, Pięta P and Surowiak A 2016 *E3S Web of Conferences* **8 01036** pp 1-10
[8] Jamróz D 2017 *Information Visualization* **16(4)** pp 346-60

IOP Conf. Series: Materials Science and Engineering **427** (2018) 012009 doi:10.1088/1757-899X/427/1/012009

Studies on Polish copper ore beneficiation in Jameson cell

P Pięta[1], T Niedoba[1], A Surowiak[1], O Şahbaz[2], C Karagüzel2 and Ö Canieren[2]

[1]AGH - University of Science and Technology, Faculty of Mining and Geoengineering, Department of Environmental Engineering and Mineral Processing, Cracow, Poland

[2]Dumlupinar University, Department of Mining Engineering, Kutahya, Turkey

ppieta@agh.edu.pl

Abstract. The paper presents the preliminary studies concern the results of Polish copper ore beneficiation in Jameson cell. To investigate the simultaneous influence of flotation machines type, grain size (0-25, 25-45, 45-75 μm), type (isobutyl and ethyl xanthate aqueous solution) and dosage (100, 150 g/Mg) of collector, flotation time on the efficiency of the process, the multi-level experiment has to be performed for carbonate lithological type of copper. Compering results of flotation test it can be concluded that the best results were obtained for 0-25 μm with the use of isobutyl xanthate aqueous solution in a dosage equal to 100 g/Mg. Tests of significance (MANOVA, ANOVA) showed a high influence of factors on the difference between an efficiency of separation in the most cases. This paper shows an attempt to create a model for the copper ore enrichment process at Jameson cell. The equations were described using a logarithmic substitution. The model evaluation was made based on mean squared error MSE and coefficient of determination R^2.

1. Introduction

Copper prices on world markets and the necessity of obtaining copper ore from more and more hard to process and poorer parts of deposits mean that the efficiency of the processing is constantly being improved. In the case of exploitation of such minerals, flotation remains the main process with goal of separating metal minerals from gangue. At the same time, it is not easy to carry out because many factors affect the result. Among them, it can be listed the entire range of properties of the feed, parameters and conditions of separation as well as the characteristics of the devices [1, 2, 3, 4].

Polish copper ore deposits are of sedimentary type, which are characterized by different mineralogical-petrographic and physicochemical features. Additionally, these deposits have complex chemical composition and a fine-grain mineralization. The crucial property is an occurrence of three lithological types (carbonate, shale and sandstone) conditioning a high variability in susceptibility to enrichment processes [3, 4, 5]. In the work [6] a big variation in copper content in the deposit was described. However, the correlation between the content of copper in the deposit and beneficiation effect is not significant.

To obtain a high-quality copper concentrate a multi-level flotation is applied in Polish enrichment plants. The most important condition for its application is a high level of minerals liberation of different sizes [7]. Numerous works indicate that the best flotation results are achieved for a feed with particle size below 0.02 mm or even less. The problem of deep grinding appears and it is unavoidable for Polish copper deposit [8]. It is important to consider the feed properties in an individual way, which was justified by presence of diversified lithological types. Considering a sulphide mineral size distribution, it was stated that copper minerals for sandstone type are 0.1-0.5 mm, carbonate around 0.14-5 mm and for shale type copper minerals are equal to 0.007-0.01 mm. In particular mines, these values differ among themselves. In the work enrichment results of copper ore from Polkowice-Sieroszowice mine were presented. It is characterized by the biggest share of dolomite lithological type (about 62%) characterizing by mineral size below 100 μm.

Considering the size of Polish copper mineral particles, which determine high efficiency of flotation, it is necessary to ensure the appropriate size of air bubble particles. This guarantees a high probability of creating solid-air bubble aggregates. Flotation efficiency of the mechanical cells for fine particles (<20 μm) is believed to be low due to relatively low collision probability of fine particles and the high levels of turbulence, respectively. The Jameson cell have a greater potential for recovery in this regard compared to mechanical cells. The reason is a possibilities of produce finer bubbles (e.g., 200-700 μm) in the Jameson cell [9, 10]. It can be stated that there are certain indications that the second type of device should be used.

2. Jameson cell

At the beginning, the Jameson flotation cell was used for coal slimes in Australia [11] but since then, it has been successfully applied for coal and copper beneficiation, as well processing of molybdenite and lead ores. The Jameson cell is suitable for processing of fine particles (below 100 μm) due to fine bubbles (200–1000 μm) generated in the cell [9]. The flotation of coarse particles is unsatisfactory, which can be explained by low energy of an air bubble, which is not able to keep mineral particle on its surface or turbulences during a floating [9, 12]. Common features of Jameson cell include small tank residence times, high concentrate production rates and a high shear rate to produce very fine bubbles. However, this machines operates with not compressed air input and the vacuum developed within the downcomer produces interactions unlike conventional pneumatic devices [13, 14].

The Jameson cell has two main parts, the downcomer (providing a primary contact of particles and bubbles) and the separation tank (elementary phenomena of creating the bubble-particle aggregate and a floating it to the foam). Pulp is pumped into the downcomer through a nozzle, creating a high-pressure jet and air entrainment till the bubble formation with a plunging jet is achieved. The pulp zone is where secondary contact of bubbles and particles occurs, and air bubbles disengage from the pulp. The froth zone is where entrained materials are removed from the froth by froth drainage and/or froth washing [9, 15].

3. Aims and experimental description

The main purpose of the work was to examine the effectiveness of the flotation process of hardly enriched copper ores in Jameson cell. A multi-parameter analysis was carried out basing on the results of the flotation tests and the process models were generated. The sample of copper ore was collected in Polkowice Enrichment Plant in Poland. The carbonate lithological type was separated from the samples, subjected to further preparatory work and finally it was tested in Jameson cell to investigate the simultaneous influence of particle size, type and dosage of collector, flotation time on the efficiency of the flotation. The multi-level experiment was performed in Dumlupinar University in Kütahya, Turkey. For carbonate lithological type of copper ore three size fractions (45-75, 25-45, 0-25 μm) were separated. During the flotation tests two types of collectors (isobutyl and ethyl xanthate aqueous solution) in two dosages (100, 150 g/Mg) were used. The single experiment assumed testing of feed in a cycle with a cleaning flotation. Samples of the froth concentrates were collected after the following time: 1, 2, 4, 6, 9, 12, 17, 22, 30 [min]. The solid ratio was equal to 1.96%. Nasfroth in dosage 50 g/Mg was used as a frother during tests. Samples were examined by means of X-ray fluorescence to determine a copper content in flotation product. The evaluation of the process was carried out in accordance with a copper content and copper recovery in concentrate and tailings.

4. Calculation methods

To assess flotation efficiency evaluation indicators were calculated (1-4). They are described in table 1 [1, 3, 12]. The results were compiled with factor parameters, which was a basis to carry out a statistic analysis and a modelling.

Table 1. Evaluation process indicators.

Copper recovery in concentrate ε_i	Copper recovery in tailings η_i	Copper content in concentrate β_i	Copper content in tailings υ_i
$\varepsilon_i = \frac{\beta_i}{\alpha}\sum_{j=1}^{i}\gamma_j$ (1)	$\eta_i = \frac{\vartheta_i}{\alpha}\sum_{j=i}^{n}\gamma_j$ (2)	$\beta_i = \frac{\sum_{j=1}^{i}\gamma_j\lambda_j}{\sum_{j=1}^{i}\gamma_j}$ (3)	$\vartheta_i = \frac{\sum_{j=i}^{n}\gamma_j\lambda_j}{\sum_{j=i}^{n}\gamma_j}$ (4)
α – a copper content in the feed,		λ_i – *content of copper in a product,*	
$\gamma_i = \frac{m_i}{\sum_{i=1}^{n} m_i}$ – *a yield of a product,*		m_i – an amount of a product,	
$\sum_{j=1}^{i}\gamma_j = 100 - \sum_{j=1}^{i-1}\gamma_j$			

The family of variance analysis methods is widely use and universal tools to examine complex phenomenon, which depend on many different factors. Those methods enable to check an existence and a strength of the influence of parameters, like particle size, types and dosages of collector over the final results of flotation process. Based on mean sums of square effects and errors it is possible to verify the accuracy of experiments performance for appropriate number of repetition count. Before the use of ANOVA/MANOVA some basic assumptions have to be made (representativeness of the dataset and sampling, the normality of distributions of the dependent variables, the homogeneity of the intergroup variance). The verification of normality assumption can be based on the Kolmogorov-Smirnov, Shapiro-Wilk, Lilliefors, D'Agostino-Pearson or Q-Q plot test. In case of homogeneity of variance, it is possible to use following tests: Neyman and Pearson, Bartlett, Cochran, Scheffy, Levene, Brown and Forsythe ones. The choice depends on type of data [10, 16, 17].

ANOVA allows to examine an influence of one parameter (e.g. collector type) on one dependable variable (e.g. copper recovery in tailings). It can be use in multifactorial scheme, so this approach enables studying the influence of different parameters on flotation results describing only by one evaluation index. Despite the fact those methods are still one-dimensional contrary to MANOVA, which is multidimensional variance analysis. It makes possible to examine an influence of one or more parameters on efficiency of flotation in terms of all chosen indicators (β, υ, ε, η). Based on that information it can be concluded that comparison of parameters and indicators values must be done in different manner which depends on sort of methods. In ANOVA case operations are proceeded on population mean values but MANOVA is more extensive and it uses the multivariate vector of dependent variables. A detailed description of calculation methodology is presented in works [10, 18, 19, 20] for both approaches.

Information obtained from the application of ANOVA/MANOVA is useful to determine flotation mathematical models. It provides an opportunity to eliminate some insignificant factors for final results of flotation test. A multiple regression allows to determine quantitative relationship between parameters (x_1, x_2, x_3, x_4,) and response variable (y_i). Formula (5) describes the general regression equations, which form is the basis for modelling.

$$Y_i = b_0 + b_1 \cdot x_1 + b_2 \cdot x_2 + b_3 \cdot x_3 + b_4 \cdot x_4 + e_i \quad (5)$$

where: y_i – process evaluation indicators,
x_1 – collector type,

x_2 – collector dosage,
x_3 – size fraction,
x_4 – flotation time,
b_i – model parameter,
e_i – standard estimation error.

In order to assess the accuracy of matching the regression functions to the experimental data, a mean squared error and a coefficient of determination were used. They are described by formula (6) and (7) [10, 18, 20, 21]:

$$MSE = \sqrt{\frac{1}{n-k}\sum(y_{ei} - y_{ti})^2} \quad (6)$$

$$R^2 = \frac{\sum_{i=1}^{n}(y_{ei}-\bar{y})}{\sum_{i=1}^{n}(y_{ti}-\bar{y})} \quad (7)$$

where: y_{ei} – an empirical value of parameter,
y_{ti} – a theoretical value of parameter,
$\acute{y}$ – an arithmetic mean of empirical values of Y -random variable,
n – number of cases,
k – number of variables in the model.

5. Results and discussion

In order to carry out an experimental plan it was necessary to collect copper ore in Poland and prepare samples to flotation test in Dumlupinar University, where Jameson cell was available to use. Cooperation of DPU and AGH University results in performance of studies. The final results were presented in the figure 1.

It can be concluded that the beneficiation of copper ore for size fractions below 75 µm is hard to perform in both Jameson cell and conventional Denver flotation machine. Table 1 presents the results of flotation tests for narrow size fractions (0-25 µm, 25-45 µm, 45-75 µm). Considering the copper recovery and copper content in concentrates it can be stated that in order to obtain the best results during the separation process in Jameson cell the process should be conducted for 0-25 µm particle size. In case of 25-45 µm size fraction evaluation indicators values decline. The copper recovery decreases on average about 20% according to collector type and dosage. For particle size 45-75 µm the process is unprofitable. Differences between each size fractions are not too big, but it is a factor, which affect significantly the final results.

During the flotation an isobutyl xanthate aqueous solution (Z) and an ethyl xanthate aqueous solution (E) were used as collectors. Reagents were added in two dosages which were equal to 100 g and 150 g per 1 Mg of feed, respectively. For finest size fraction, 0-25 µm the best results were obtained for the isobutyl xanthate aqueous solution in dosage 100 g per 1 Mg of feed. The copper recovery was similar for an ethyl xanthate aqueous solution in dosage 150 g per 1 Mg of feed, but the quality of flotation concentrate decreased significantly. It can be stated very clearly that an isobutyl xanthate aqueous solution caused an increase in copper recovery in concentrates during processing of 25-45 µm particle size fraction. This reagent seems to be better for finest size fractions. In the last case (45-75 µm) a little bit better results were obtained when an ethyl xanthate aqueous solution was used for the separation, but the particle size distribution is inadequate for the Jameson cell work conditions.

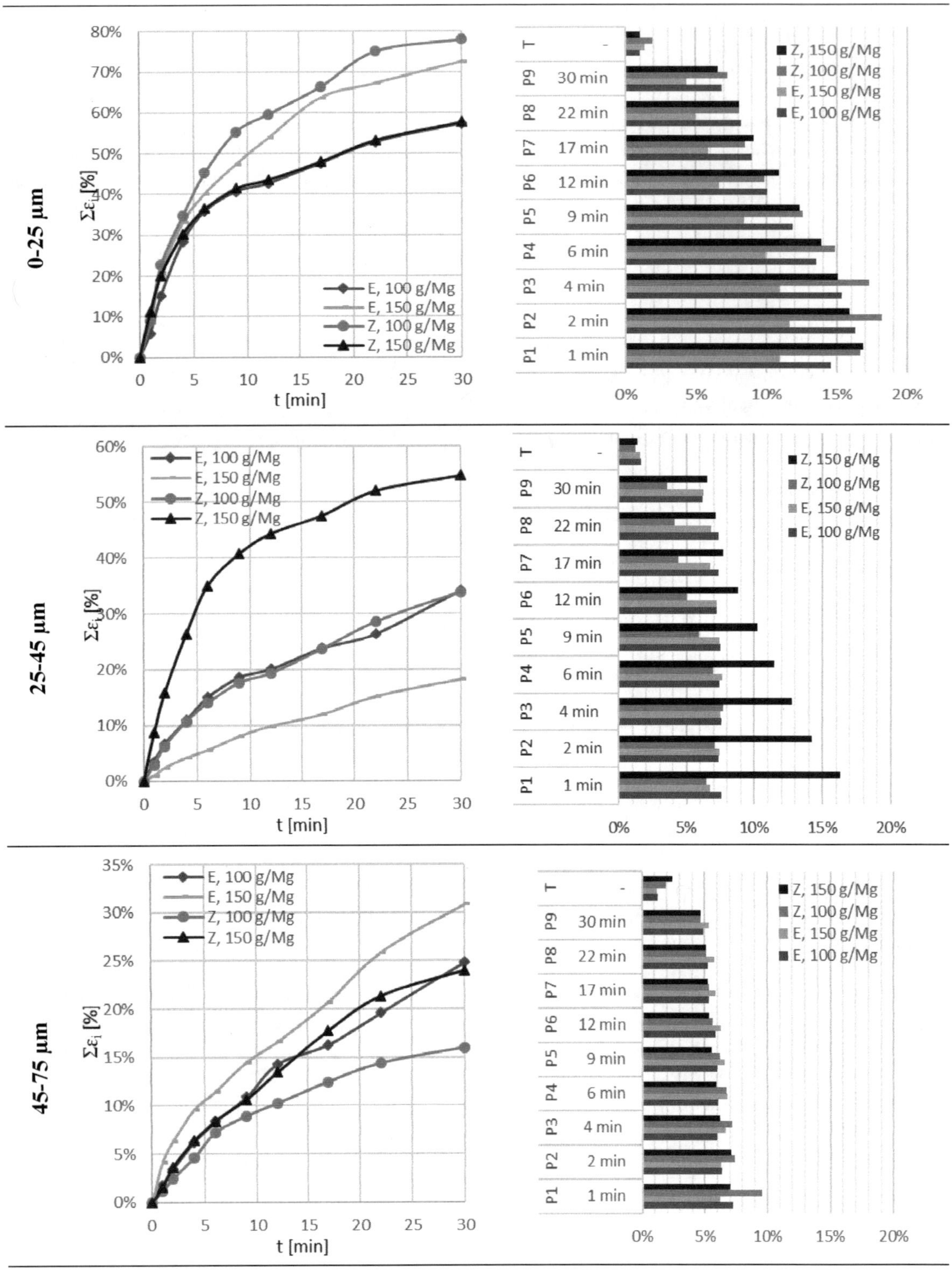

Figure 1. Final results of copper ore beneficiation in Jameson cell.

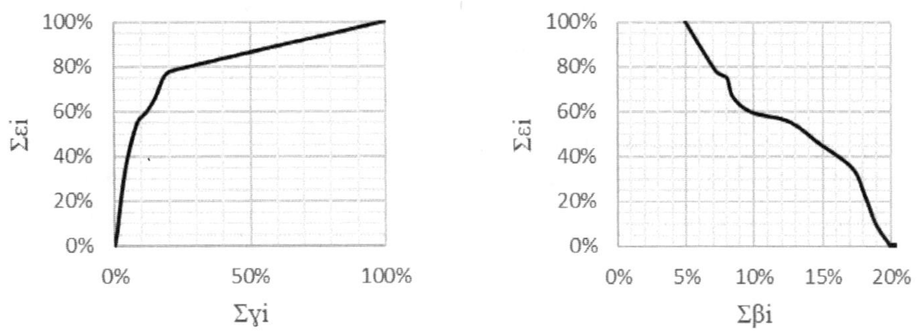

a) The Mayer upgrading curve b) The Halbich upgrading curve

Figure 2. Beneficiations curves for a dolomite copper ore flotation test in Jameson cell (0-25 μm, Z, 100 g/Mg).

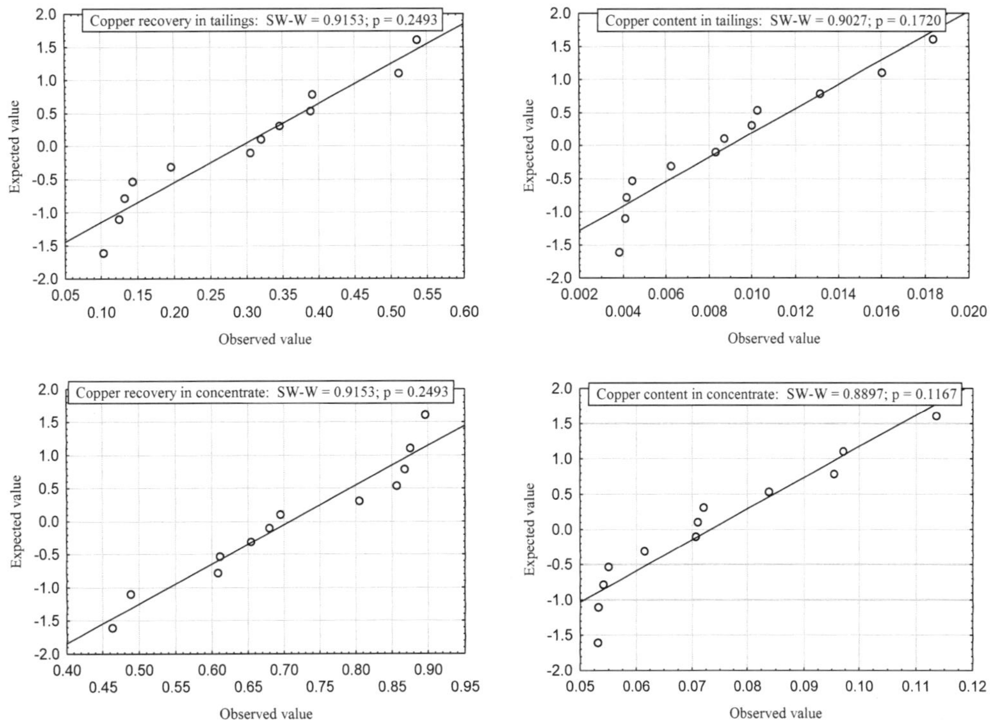

Figure 3. Normal probability plots for examined variables with results for Shapiro-Wilk test (SW-W).

For the selected case of dolomite copper ore enrichment upgrading curves were drawn (figure 2). Their shapes confirm an occurrence of some problems with beneficiation. Especially Halbich upgrading curve indicates this conclusion.

Further analysis was based on a variance analysis. A large number of factors differentiating the effects of flotation enrichment affected the high degree of variability of the copper content and its recovery in the concentrate and tailings, and as a result, the normality assumptions were not met. Therefore, it was decided to eliminate the variable "time", which obviously and significantly affects the final effects of the enrichment process. After this all the assumptions were met for the ANOVA and MANOVA application (figure 3).

The assumption of variance homogeneity was not met for all tested variables. However, after application of an elementary logarithmic substitution instead of original data the results of Levene's test confirmed the possibility of the implementation of the planned analyses.

Result of MANOVA was shown in the table 2. It can be concluded that simultaneous effect of parameters (time, size fraction, collector type and dosage) is important for variability in test results of flotation process in Jameson cell. The conclusion was drawn on the basis of the multidimensional significance (Wilks' Lambda test).

In order to check a single effect of parameters on the efficiency of the process ANOVA tests were performed for each factor. Taking into account a copper content in concentrates and tailings it can be concluded that factor "dosage of collector" was not significant for their change. In case of copper recovery in both products the parameters "type of collector" occurred to be unambiguous. Additionally, it can be stated that the variables are not correlated. The values of Pearson correlation coefficient were equal to 0.

Table 2. MANOVA results.

Independent variables	Wilks' Lambda λ	F	p
Particle size	0.182901	234.2	0.00
Flotation time	0.033369	141.5	0.00
Collector type	0.853637	30.0	0.00
Collector dosage	0.879644	23.9	0.00

Table 3. ANOVA results.

Independent variables	Copper content in concentrate		Copper content in tailings		Copper recovery in concentrate		Copper recovery in tailings	
	F	p	F	p	F	p	F	p
Particle size	602.44	0.00	235.76	0.00	774.82	0.00	774.82	0.00
Flotation time	30.69	0.00	134,47	0.00	556.01	0.00	556.01	0.00
Collector type	74.48	0.00	7.73	0.01	2.32	0.13	2.32	0.13
Collector dosage	0.04	0.84	0.04	0.84	8.36	0.01	8.36	0.01

Models for flotation process in Jameson cell were described by semi-logarithmic functions which are presented below as formulas (8), (9), (10) and (11). There some factors are omitted according to ANOVA results. The parameter "collector dosage" was not important from statistical point of view in cases of copper content in product. The similar situation concerned copper recovery in concentrate and tailings in connection with the influence of "collector type" on efficiency of flotation process.

$$\beta = 0.08706 - 0.00012 \cdot ln\, x_1 - 0.0005 \cdot ln\, x_3 - 0.01074 \cdot ln\, x_4 \pm 0.01924 \quad (8)$$

$$\vartheta = 0.02297 - 0.00001 \cdot ln\, x_1 - 0.00006 \cdot ln\, x_3 - 0.00404 \cdot ln\, x_4 \pm 0.00332 \quad (9)$$

$$\varepsilon = -0.16139 + 0.04378 \cdot ln\, x_2 - 0.00227 \cdot ln\, x_3 + 0.18835 \cdot ln\, x_4 \pm 0.09082 \quad (10)$$

$$\eta = 1{,}16139 - 0.04378 \cdot ln\, x_2 + 0.00227 \cdot ln\, x_3 - 0.18835 \cdot ln\, x_4 \pm 0.09082 \quad (11)$$

where: β – copper content in concentrate,
υ – copper content in tailings,
ε – copper recovery in concentrate,
η – copper recovery in tailings,
x_1 – collector type,
x_2 – collector dosage,
x_3 – size fraction,
x_4 – flotation time.

In all cases, the clouds of experimental points are wide (figure 4) which directly affects the great difficulties in the modelling process. It is not possible to obtain a model that will describe the phenomenon with 100% accuracy.

IOP Conf. Series: Materials Science and Engineering **427** (2018) 012009 doi:10.1088/1757-899X/427/1/012009

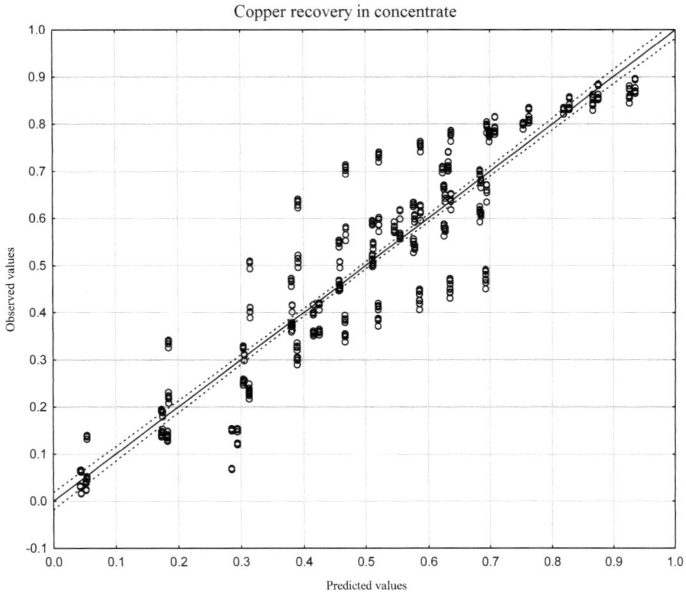

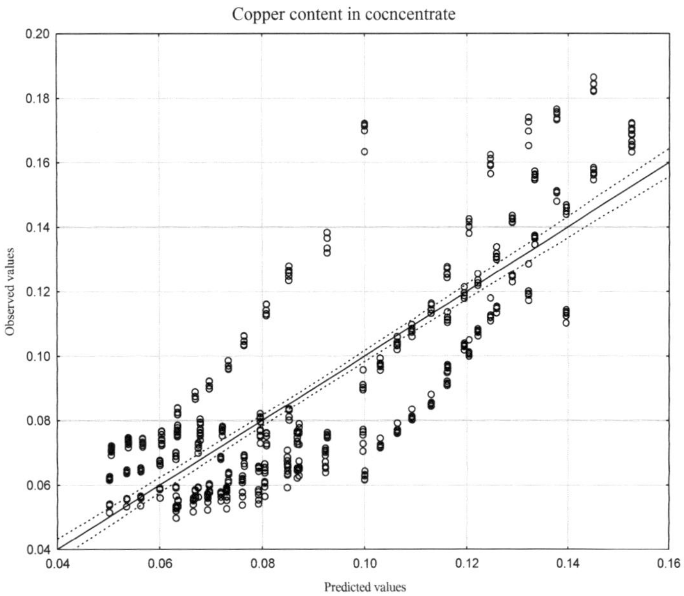

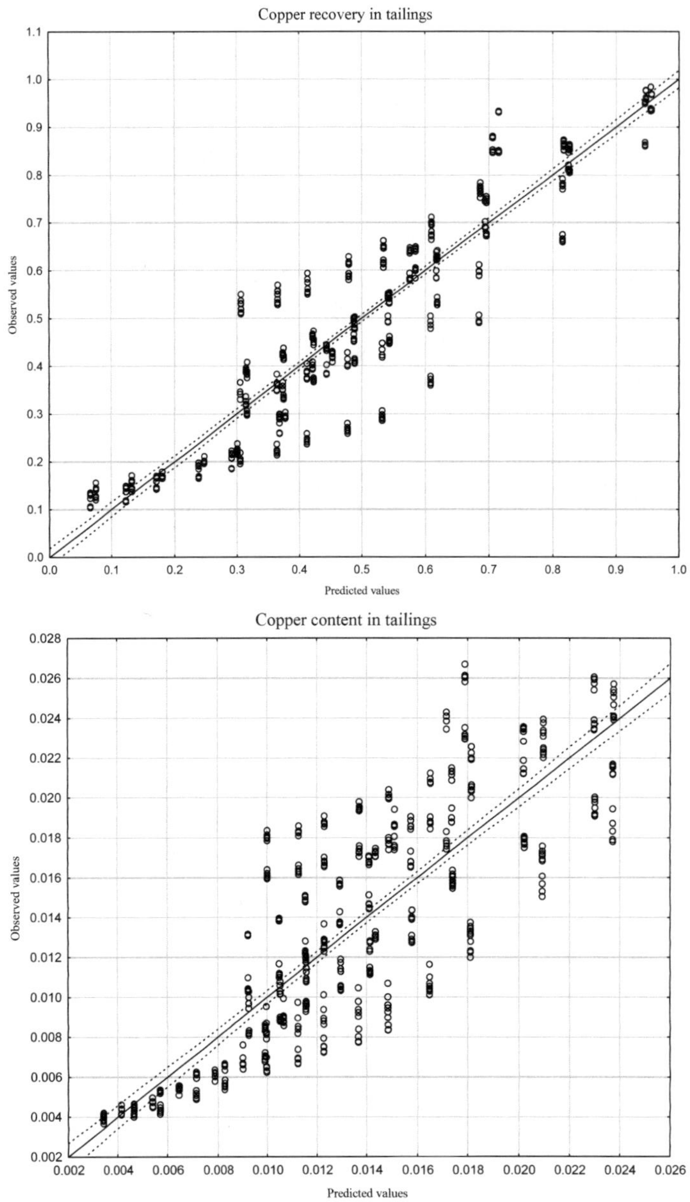

Figure 4. Comparison of experimental points and overall models for variables

The match of models and process evaluation indicators are presented in the table 4. In the overall approach, some difficulties can be noticed for accurately describing the phenomenon of flotation through created functions. Both values of the MSE and R^2 confirm the occurring deviations. The selected factors have a significant impact on the final result of copper ore flotation in the Jameson cell and therefore it was a reason to generate additional models of copper recovery and content in products for size fractions. This is a very important parameter affecting the enrichment of raw materials at this machine. In a similar approach, as in the general models, equations (12-23) were created describing the effects of copper ore enrichment.

$$\beta_{<25} = 0.4527 - 0.00023 \cdot \ln x_1 - 0.0627 \cdot \ln x_2 - 0.01957 \cdot \ln x_4 \pm 0.0104 \quad (12)$$

$$\vartheta_{<25} = 0.03457 - 0.00001 \cdot \ln x_1 - 0.0031 \cdot \ln x_2 - 0.00546 \cdot \ln x_4 \pm 0.00193 \quad (13)$$

$$\varepsilon_{<25} = 0.35914 - 0.00016 \cdot ln\, x_1 - 0.02853 \cdot ln\, x_2 + 0.21797 \cdot ln\, x_4 \pm 0.07118 \quad (14)$$

$$\eta_{<25} = 0.64086 + 0.00016 \cdot ln\, x_1 + 0.02853 \cdot ln\, x_2 - 0.21797 \cdot ln\, x_4 \pm 0.07118 \quad (15)$$

$$\beta_{25-45} = -0.18266 - 0.00014 \cdot ln\, x_1 + 0.05663 \cdot ln\, x_2 - 0.00755 \cdot ln\, x_4 \pm 0.01646 \quad (16)$$

$$\vartheta_{25-45} = 0.01382 + 0.00005 \cdot ln\, x_1 + 0.00227 \cdot ln\, x_2 - 0.00385 \cdot ln\, x_4 \pm 0.00123 \quad (17)$$

$$\varepsilon_{25-45} = -0.38388 - 0.00116 \cdot ln\, x_1 + 0.08875 \cdot ln\, x_2 + 0.18589 \cdot ln\, x_4 \pm 0.05889 \quad (18)$$

$$\eta_{25-45} = 1.38388 + 0.00116 \cdot ln\, x_1 - 0.08875 \cdot ln\, x_2 - 0.18589 \cdot ln\, x_4 \pm 0.05889 \quad (19)$$

$$\beta_{45-75} = 0.05386 + 0.000004 \cdot ln\, x_1 + 0.00389 \cdot ln\, x_2 - 0.00509 \cdot ln\, x_4 \pm 0.0584 \quad (20)$$

$$\upsilon_{45-75} = 0.0128 - 0.00006 \cdot ln\, x_1 + 0.00124 \cdot ln\, x_2 - 0.00281 \cdot ln\, x_4 \pm 0.00102 \quad (21)$$

$$\varepsilon_{45-75} = -0.2345 + 0.00106 \cdot ln\, x_1 + 0.07117 \cdot ln\, x_2 + 0.16119 \cdot ln\, x_4 \pm 0.04465 \quad (22)$$

$$\eta_{45-75} = 1.2345 - 0.00106 \cdot ln\, x_1 - 0.07117 \cdot ln\, x_2 - 0.16119 \cdot ln\, x_4 \pm 0.04465 \quad (23)$$

For each of the equations a matching assessment was based on the coefficient of determination R^2 and the mean squared error MSE (table 4). Considering the division into size fractions, the modelling results can be much better. The greatest difficulties appear in the description of copper content in concentrate for size fractions 25-45 μm and 45-75 μm. For the remaining flotation indicators, the accuracy of the description was above 90%. Only models of copper recovery and copper content in products deriving from flotation of the 0-25 μm size fraction are entirely characterized by a better fitting. This may suggest the general models will more accurately reflect the final results of the enrichment of this size fraction.

Table 4. The assessment of model fit to experimental data.

Model	MSE [%]			
	All cases	0-25	25-45	45-75
Copper recovery in products	9.05	7.04	5.92	4.42
Copper content in concentrate	1.92	1.03	1.63	0.64
Coper content in tailings	0.79	0.19	2.16	0.1
Model	R^2 value [%]			
	All cases	0-25	25-45	45-75
Copper recovery in products	86.71	91.63	92.52	94.41
Copper content in concentrate	68.33	87.66	48.49	47.45
Coper content in tailings	70.66	90.53	90.53	95.14

6. Conclusions

The Jameson Cell is a co-current flotation device, in which a feed slurry is pumped through a fixed diameter orifice. Ultrafine bubbles are produced in the downcomer because of the high shearing provided by the jet action of the feed slurry. As a result of the high air-fraction generated within the downcomer, a superior bubble-particle collision environment is produced which provides greater flotation rate constant values for the hydrophobic species. However, the research on various raw materials proved (e.g. coal, copper or zinc ore) that the efficient flotation for fine feed particles is

IOP Conf. Series: Materials Science and Engineering **427** (2018) 012009 doi:10.1088/1757-899X/427/1/012009

possible when the probability of formatting permanent aggregates is high. In the case of Polish copper ore, the highest efficiency of the process can be indicated for 0-25 μm of the size fraction with an application of the isobutyl xanthate aqueous solution in a dosage of 100 g/Mg of copper ore. In other cases, there is clear deterioration.

The separation of useful minerals from gangue in the Jameson cell depends on many different factors. Simultaneous analysis of all of them is not possible, but it is worth to use some multidimensional statistical methods for such complex operations. The analysis was performed for some parameters basing on ANOVA and MANOVA tests. It is necessary to pay attention to the fact that the use of variance analysis and the modelling require to introduce some data changes, such as eliminating of the time variable and inserting of a logarithmic transformation. The reason for that was the large variation of the results of flotation tests which was related to the factor effect as well as the measurement error which is not possible to eliminate.

The effect of four important operating parameters on separation performance main indicators, which are copper content and recovery in flotation products, was studied in detail with a goal to understand the process better. The results of the variation analysis showed that the factor "dosage of collector" was not significant for the copper content in products. At the same time a "type of collector" parameter was unambiguous for copper recovery. Furthermore, the models describing the copper content in products are linear functions of the collector type, particle size and separation time. On the other hand, the copper recovery functions depend on three variables: the collector dosage, the particle size of the feed and a separation time. In their entirety models are not fully describing the flotation indicators. When the problem is considered for each size fraction separately, fittings of models are better. In case of 0-25 μm size fraction values of MSE and R^2 confirm that the generated mathematical functions characterize a copper ore flotation in the best way. The matching is maintained at the level 90% with a mean square error equal to 7% for copper recovery equation, 1% for copper content in concentrate and 0.2% for copper content in tailings.

References

[1] Drzymała J 2007 *Mineral processing : foundations of theory and practice of minerallurgy* Oficyna Wydawnicza Politechniki Wrocławskiej [in Polish]

[2] Foszcz D 2013 *Principles for determining optimal results of enrichment of multi-component copper ores* IGSMiE PAN [in Polish]

[3] Potulska A 2008 *An influence of fine grinding operations on flotation effectiveness of domestic copper ore* PhD thesis PWr WGGG Wroclaw [in Polish]

[4] Zarudzka E 2010 *Influence of acid leaching of sulfid ores* PhD thesis PWr WGGG Wroclaw [in Polish]

[5] Piestrzyński A 1996 *Poland Monography KGHM Polska Miedź S.A. CBPM Cuprum* 200-323

[6] Wieniawski A and Skorupska B 2016 *E3S Web of Conference MEC2016* 8 01064

[7] Foszcz D, Trybalski K, Tumidajski T and Pawlos W 2009 *Cuprum: scientific and technical journal of ore mining* **2** pp 47-59

[8] Łuszczkiewicz A and Wieniewski A 2006 *Mining and Geoengineering* **3(1)** pp 181-96

[9] Sahbaz O, Uçar A and Öteyaka B 2013 *Minerals Engineering* **41** pp 79-85

[10] Stanisz A 2007 *An affordable statistic course using STATISTICA PL on examples of medicine* Poland. StatSoft [in Polish]

[11] Harbort G J, Manlapig E V and DeBono S K 2002 *Trans. IMM (Section C: Mineral Process. Extr. Metall). In: 111/Proceedings of the Australas. Inst. Min. Metall.* **307** C_1-C_2

[12] Gonjito F C, Fornasiero D and Ralston J 2007 *Can. J. Chem. Eng.* **85** 739-47.

[13] Harbort G J, De Bono S, Carr D and Lawson V 2003 *Minerals Engineering* **16** pp 1091-101

[14] Lawson V 2016 *Enhancing the Performance of Cu-Mo Circuits Using the Jameson Cell 12th IMPC*

[15] Tasdemir A, Tasdemir T and Oteyaka B 2007 *Minerals Engineering* **20** pp 1331-6

[16] Napier Munn T J 2014 *Julius Kruttschnitt Mineral Research Centre* 0980362245
[17] Pillai R N S and Bagavathi V 2010 *Practical Statistics* India Chand and company
[18] Jajuga K 1993 *Statistical multidimensional analysis Poland* PWN [in Polish]
[19] Niedoba T, Surowiak A and Pięta P 2016 *J. Pol. Miner. Eng. Soc.* **17(1)** pp 15-22
[20] Maxwell S E and Delaney H D 2014 *Lawrence Erlbaum Associates*
[21] Pease J D, Curry D and Young M F 2006 *Minerals Engineering* **19** pp 831-40

IOP Conf. Series: Materials Science and Engineering **427** (2018) 012010 doi:10.1088/1757-899X/427/1/012010

Determination of flakiness index (FI) and shape index (SI) using open source digital image analysis software

Oleksik K, Ostrowski K and Mrocheń D

AGH University of Science and Technology, Al. A. Mickiewicza 30, 30-059, Kraków, Poland

koleksik@agh.edu.pl

Abstract. The aim of the article was to verify the possibility of using open source digital image analysis software to determine the particle shape. Devices using laser diffraction or dynamic image analysis could be used to determine size and shape of particles. However, the laboratory devices for digital image analysis are relatively expensive. For this reason, it was decided to use open source digital image analysis software Fiji Is just ImageJ and mobile phone with a 12 Mpx camera. Coarse aggregates of basalt and granite were used in the study. Flakiness index (FI) and shape index (SI) were determined using two methods: in accordance with the European Standard and with the use of digital image analysis (reference method). Based on the obtained results it was found that there is an appropriate relationship between the European Standard method and the reference method.

1. Introduction

High quality of aggregate is very important in industries such as road construction and architectural engineering. The size and shape of aggregate depends on processing methods (mainly grinding method) used during the production [1].

Aggregate for road construction and architectural engineering should have regular, cubical shape, since the ratio of length to width of particles has essential impact on their strength. The shape characteristics of aggregate impacts the asphalt performance [2]. The durability, workability, tensile strength, shear resistance, optimum binder content of the mixture and fatigue response of asphalt concrete depends on the shape of coarse aggregate [3]. Grading and shape of aggregate highly affects the strength of concrete [4, 5, 6, 7].

To test geometrical properties of aggregates two European Standard are used: EN 933-4:2012 [8] to determine the flakiness index (FI) and EN 933-4:2008 [9] to determine shape index (SI). However, also digital image analysis is used to assess the aggregate quality [10]. For this purpose, high resolution digital cameras [11] or stereomicroscopes could be used [12]. The latest devices for measuring the size and/or shape of particles are using laser diffraction or dynamic image analysis. A very valuable advantage of this type of devices is high measurement accuracy, speed and automaticity of measurement. The measurement can be carried out on a dry or wet sample. There are many manufacturers of such devices such as:

- Fritsch (Germany) – Analysette 28 ImageSizer (20 μm – 20 mm), Analysette 22 NanoTec (0.01 – 2100 μm) [13, 14)

- Kamika (Poland) – AWK 3D analyzer (200 µm – 31,5 mm), mini 3D analyzer (0.5 µm – 2000 µm), IPS UA analyzer (0.5 – 2000 µm) [15, 16]
- Malvern Panalytical (the Netherlands) – Morphologi G3 analyser (0.01 – 3500 µm) [17]
- Retsch Technology (Germany) – CAMSIZER X2 (0.8 µm – 8 mm), CAMSIZER P4 (20 µm to 30 mm) [18, 19, 20]

2. Materials and methods

Coarse aggregate of basalt and granite were used in the study. At the first stage the material was crushed in a jaw crusher. Afterwards the sieve analysis using square meshes of size 4 mm and 8 mm was carried out (figure 1) in accordance with the European Standard [21]. The particle size fraction 4 – 8 mm was then sieved into three narrow fractions:

- 4 – 5 mm,
- 6.3 – 8 mm,
- 8 – 6.3 mm.

A 100-gram representative samples of each type of coarse aggregate were selected for further analysis.

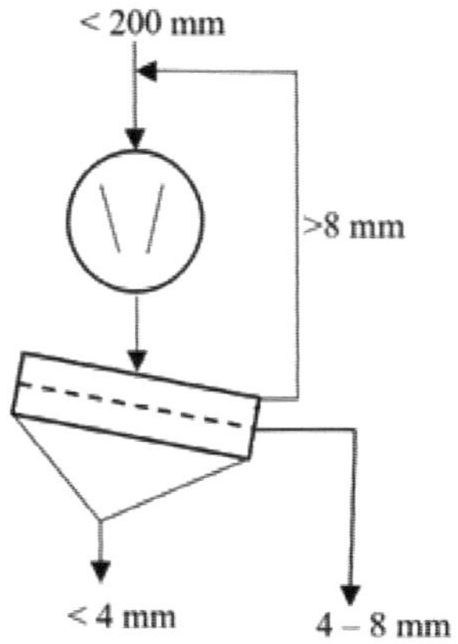

Figure 1. Laboratory-scale comminution circuit.

In accordance with the European Standard [8, 9] to test for geometrical properties of aggregates two indexes were determined:

- FI:

$$FI = \frac{M_2}{M_1} * 100 \tag{1}$$

where:

M_1 – the sum of the masses of the particles in each of the particle size fraction d_i/D_i [g]

M_2 – the sum of masses of the particles in each particle size fraction passing the corresponding bar sieve (figure 2) of slot $D_i/2$ [g]

Particle size fractions (d_i/D_i) is a fraction of an aggregate passing the larger (D_i) of two sieves and retained on the smaller (d_i). The overall flakiness index (FI) shall be recorded to the nearest whole number.

IOP Conf. Series: Materials Science and Engineering **427** (2018) 012010 doi:10.1088/1757-899X/427/1/012010

Figure 2. Laboratory bar sieve.

- SI:

$$SI = \frac{M_2}{M_1} * 100 \quad (2)$$

where:

M_1 – the mass of test portion [g]

M_2 – the mass of the non-cubical particles [g]

Individual particles in a sample of coarse aggregate are classified on the basis of the ratio of their length L to thickness E using a particle slide gauge (figure 3) where necessary. The shape index is calculated as the mass of non-cubical particles (particles with ratio of dimensions L/E more than 3) expressed as a percentage of total dry mass of particles tested.

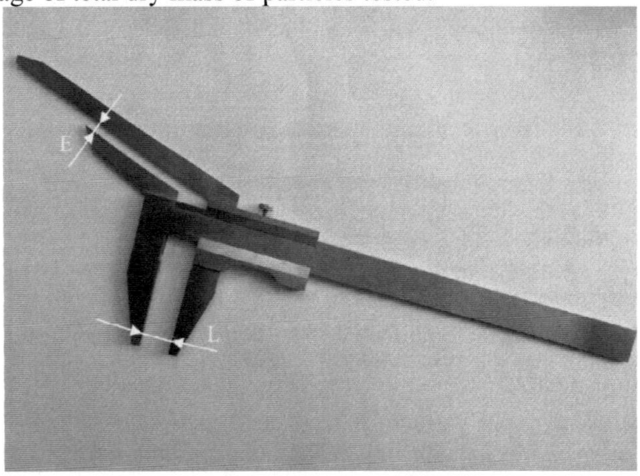

Figure 3. Particle slide gauge.

For digital image analysis high resolution photos with special filter were taken in two projection: from top and front view. A mobile phone with a 12 Mpx camera was used to take pictures. The figure 4 and 5 show examples of the taken photos.

IOP Conf. Series: Materials Science and Engineering **427** (2018) 012010 doi:10.1088/1757-899X/427/1/012010

Figure 4. Top view of regular basalt particles (6.3 – 8 mm).

Figure 5. Front view of regular basalt particles (6.3 –8 mm).

Open source software Fiji Is just ImageJ [22] was used to measure the length and width of particles. In order to determine the flakiness index (FI) and shape index (SI) using the digital image analysis the following assumptions were made:

- length (L) of the particle is the higher value of L from top and front view of particle (figure 6 and 7)
- width (W) of the particle is the lower value of W from top and front view of particle (figure 6 and 7)
- a single particle should pass the corresponding bar sieve of slot $D_i/2$ if the value of width (W) is lower than $D_i/2$.
- a single particle is non-cubical when L/W ratio is higher than 3

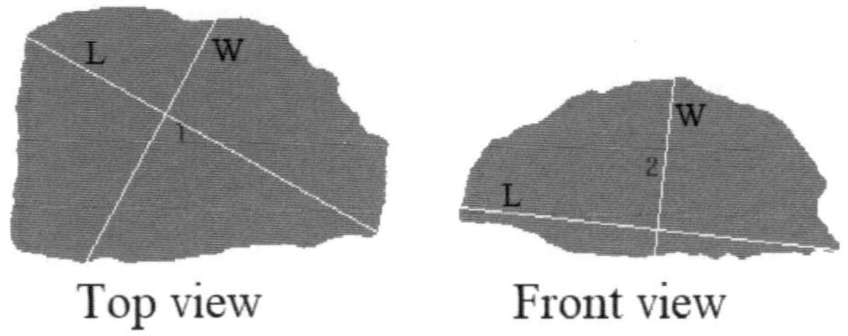

Figure 6. Top and front view of regular basalt particle (6.3 – 8 mm).

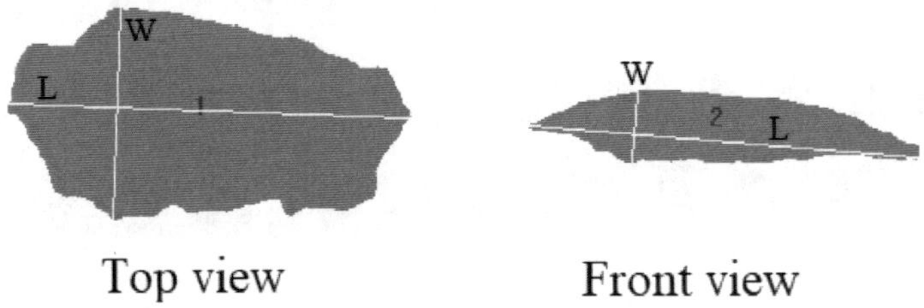

Figure 7. Top and front view of irregular basalt particle (6.3 – 8 mm).

3. Analysis of results

For basalt and granite coarse aggregate (4 – 8mm) FI was determined using two methods: in accordance with the European Standard (table 1) and with using of the digital image analysis (table 2). SI of basalt and granite coarse aggregate (4 – 8 mm) was determined using two methods: in accordance with the European Standard (table 3) and with using of digital image analysis (table 4.)

Table 1. Flakiness index (FI) based on European Standard.

Type of coarse aggregate	Particle size fraction	Mass M_1 [g]	Mass M_2 [g]	Flakiness Index (FI) [%]
Basalt	4 – 8	100	31.7	32
Granite	4 – 8	100	24.1	24

Table 2. Flakiness index (FI) based on digital image analysis.

Type of coarse aggregate	Particle size fraction	Mass M_1 [g]	Mass M_2 [g]	Flakiness Index (FI) [%]
Basalt	4 – 8	100	27.56	28
Granite	4 – 8	100	21.72	22

Table 3. Shape index (SI) based on European Standard.

Type of coarse aggregate	Particle size fraction	Mass M_1 [g]	Mass M_2 [g]	Shape Index (SI) [%]
Basalt	4 – 8	100	30.20	30
Granite	4 – 8	100	28.02	28

Table 4. Shape index (SI) based on digital image analysis.

Type of coarse aggregate	Particle size fraction	Mass M_1 [g]	Mass M_2 [g]	Shape Index (SI) [%]
Basalt	4 – 8	100	25.91	26
Granite	4 – 8	100	26.47	26

When analyzing the results it can be noticed that for both measurement methods the values of FI and SI are higher for basalt than for granite coarse aggregate. It indicates that granite coarse aggregate contains more regular particles than basalt coarse aggregate after same comminution method.

For granite coarse aggregate the difference in the FI and shape index SI value determined in accordance with the European Standard and with the use of the digital image analysis is 2 %. For basalt coarse aggregate the difference is higher. FI and SI values determined in accordance with the European Standard are 4% greater, than the value determined with the digital image analysis. The reason for differences in results for both methods is the occurrence of low width particles with a curved surface (figure 8.). These particles are classified as irregular according to the European Standard. However, based on digital image analysis, some of them are considered as regular particles. Most particles of this type contain basalt coarse aggregate grade 4 – 5 mm. In order to get more accurate results using digital image analysis three projection: from top, front and side view should be used, especially with 4 – 5 mm grade.

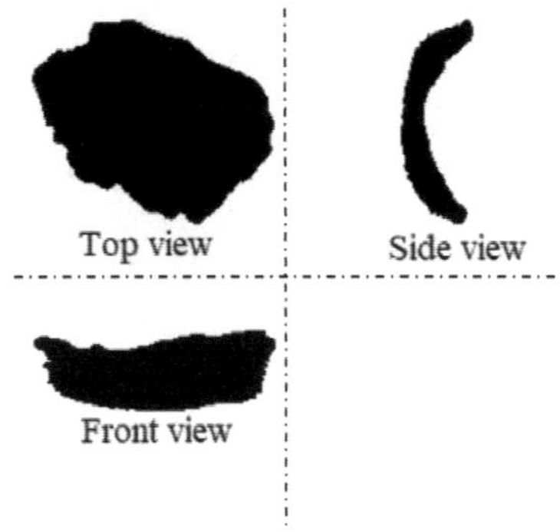

Figure 8. Top, front and side view of the irregular basalt particle (4 – 5 mm).

4. Summary

The results of tests of geometrical properties of aggregates proved appropriate relationship between the European Standard method and the reference method (digital image analysis). The presented method of digital image analysis is a proper tool to determine the FI and SI at low cost. In addition, open source digital image analysis software Fiji Is just ImageJ allows to determine the size (area, diameter, perimeter, width, length) and shape descriptors (circularity, aspect ratio, roundness, solidity) of particles [23, 24].

Experimental work allows to draw the following conclusions:

- With same comminution procedure, the lower values of FI and SI were achieved for granite coarse aggregate,
- The differences between values of FI and SI achieved with the European Standard method and the reference method (digital image analysis) are 2 % for granite coarse aggregate and 4 % for basalt coarse aggregate,
- Results obtained from the photos taken in two projections (top and front view) are sufficient if the coarse aggregate does not contain low width particles with a curved surface.

References

[1] Gawenda T 2015 Innowacyjne technologie produkcji kruszyw o ziarnach foremnych *Mining Science – Mineral Aggregates***22(1)** pp 45-59

[2] Singh D, Zaman M and Commuri S 2013 Effect of production and sample preparation methods on aggregate shape parameters. *Inter. J. of Pavement Engr.* **14(2)** pp 154-75

[3] Arasan S, Yenera E, Hattatoglu F, Hinislioglua S and Akbuluta S 2011 Correlation between shape of aggregate and mechanical properties of asphalt concrete *Road Materials and*

Pavement Design **12(2)** pp 239-62.

[4] Muhit I B, Haque S and Rabiul Alam Md 2013 Influence of Crushed Coarse Aggregates on Properties of Concrete *AJCEA,* Vol. **1**, No. **5**, 103-6

[5] Neville A M 2000 Właściwości betonu, *Polski Cement*, Kraków

[6] The European Guidelines for Self-Compacting Concrete. Specification, Production and Use. *SCC European Project Group*, 2005

[7] Ostrowski K 2017 The influence of coarse aggregate shape on the properties of high-performance, self-compacting concrete *Technical Transaction, Civil Engr.* **5** pp 25-33

[8] EN 933-4:2012 Tests for geometrical properties of aggregates – Part 3: Determination of particle shape – Flakiness index

[9] EN 933-4:2008 Tests for geometrical properties of aggregates – Part 4: Determination of particle shape – Shape Index

[10] Carter R M, Yan Y 2005 Measurement or particle shape using digital imaging techniques *JPCS* **15** pp 177-82

[11] Gu X, Tran Y and Hong L 2014 Quantification of coarse aggregate shape in concrete *Front. Struct. Civ. Eng.* **8(3)** pp 308–21

[12] Xie X, Lu G, Liu P, Wang D, Fan Q and Oeser M 2017 Evaluation of morphological characteristics of fine aggregate in asphalt pavement *Constr. Build. Mater.* **139** pp 1-8

[13] Product leaflet Particle sizer Analysette 28 ImageSizer, https://www.fritsch-international.com [access: 28.06.2018]

[14] Product leaflet Laser particle sizer analysette 22 NanoTec, https://www.fritsch-international.com [access: 28.06.2018]

[15] Kamiński S and Kamińska D 2015 Analizator AWK 3D, urządzenie do pomiaru uziarnienia i kształtu cząstek stałych w powietrzu oraz jego zastosowanie do przemysłowych pomiarów on-line granulacji nawozów sztucznych *Konferencja Granulacja 2015,* Nałęczów

[16] Kamiński S and Kamińska D 2009 Pomiar granulacji surowców w mineralurgii przy użyciu nowoczesnych elektronicznych urządzeń pomiarowych *Górnictwo i Geoinżynieria* **33/4** pp 35-39

[17] Szymura P and Malinowska-Limanówska M 2015 Zastosowanie analizatora wielkości i kształtu cząstek morphologi g3 do badań teksturalnych osadów *Badania Fizjograficzne Seria A - Geografia Fizyczna* **63** pp 139-51

[18] Product leaflet, Particle Analyzer CAMSIZER® P4, Soulutions in Particle Sizing, www.retsch-technology.com [access: 28.06.2018]

[19] Product leaflet, Product leaflet Particle Analyzer CAMSIZER® X2, Soulutions in Particle Sizing, www.retsch-technology.com [access: 28.06.2018]

[20] Lv H, Wang Z and Li J 2017 Experimental study of planar shock wave interactions with dense packed sand wall *Int. J. Multiph. Flow.* **89** pp 255-65

[21] EN 933-4:2008 Tests for geometrical properties of aggregates – Part 1: Tests for geometrical properties of aggregates. Determination of particle size distribution. Sieving method

[22] http://fiji.sc/ [access: 28.06.2018]

[23] Ostrowski K and Oleksik K 2018 Comparative analysis of the coarse aggregate shapes used to manufacturing high performance self-compacting concrete *Civil Engr.* **7** pp 61-72 [in press]

[24] Ferreira T and Rasband W 2012 ImageJ User Guide, IJ 1.46r, Revised Edition, online access: 28.06.2018 https://imagej.nih.gov/ij/docs/guide/user-guide.pdf

IOP Conf. Series: Materials Science and Engineering **427** (2018) 012011 doi:10.1088/1757-899X/427/1/012011

Application of kinetic, incentive, limits and arrheniusan curves for flotation characterization

J Drzymala

Wroclaw University of Science and Technology, Faculty of Geoengineering, Mining and Geology, Wybrzeze Wyspianskiego 27, 50-370 Wroclaw, Poland

jan.drzymala@pwr.edu.pl

Abstract. Separation curves such as kinetic (recovery R versus process time t), limits (kinetic constant k versus maximum recovery R_{max}), incentive (R vs incentive parameter i), and arrheniusan (ln k vs 1/energy - for a regent concentration ($i=c$) being $kT \cdot \ln(c/c_o)$), were characterized and discussed. Their usefulness for flotation was described and relations between the curves were presented. All the separation curves provide valuable parameters. The arrheniusan curves are exceptionally useful because they deliver activation energy, expressed in the terms of energy, a parameter indicating how well a separation system responds to a change of an incentive parameter. The incentive parameter can be temperature, reagent concentration, pH, particles size and many others.

1. Introduction

Flotation is one of many separation processes capable to separate useful components of ores and raw materials from unwanted elements [1]. The process of flotation is very unique because, in addition to numerous parameters, both thermodynamics and kinetics play there an important role [2, 3, 4]. Therefore, a meaningful characterization of flotation requires taking onto account many parameters simultaneously. In such a case, graphical methods of flotation characterization become important because two or more parameters can be evaluated [5]. The goal of this work is to present a system of graphical representation of flotation results based on the so-called kinetic [1], limits [6, 7], arrheniusan [8] and incentive curves, that can be used for characterization of separation processes.

2. Kinetic curves

Flotation can be performed as a batch as well as a continuous process. In both cases the kinetics (process rate dR/dt or kinetic constant k) and thermodynamics (ultimate recovery R_{max}) determine the outcome of the process (figure 1). For a given flotation process both parameters can be determine if the relationship between flotation recovery (one of two or many components) ($R=\varepsilon$) or yield (all components) ($R=\gamma$) and flotation time is available. This relation, known as the flotation kinetic curve, provides useful information about the process. The kinetic curves can be found in numerous papers.

IOP Conf. Series: Materials Science and Engineering **427** (2018) 012011 doi:10.1088/1757-899X/427/1/012011

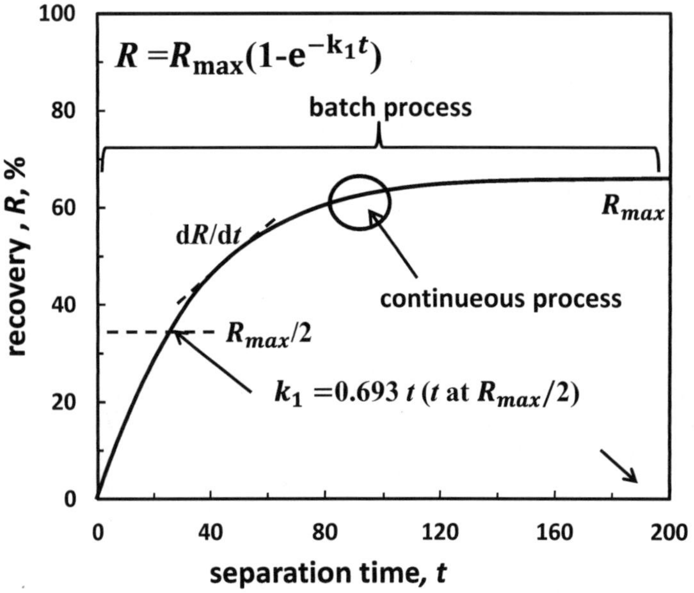

Figure 1. The kinetic separation curve. Here separation process follows the first order kinetics with kinetic constant $k = k_1$. Other properties of the curve are given.

By an appropriate manipulation of flotation data available as the kinetic curve, both kinetics of the process in the form of either kinetic constant k (or constants) or specific rate [6] and the thermodynamics of the process in the form of ultimate (maximum) yield γ_{max} (for all components) or recovery ε_{max} (for a selected component), or shortly R for both yield and recovery, is obtained. The calculation of k and R_{max} is very easy when the process follows the first order kinetic law as is shown in figure 1.

3. Incentive curves

Frequently, flotation process is run by using different levels of the investigated parameters such as temperature, reagent concentration, pulp density, particle size distribution etc. In such a case the data from the kinetic curve (figure 1) are usually used for plotting another relation, that can be called the incentive curve. The incentive curve relates recovery R (either recovery of all components called yield γ or recovery of a selected component ε for a given flotation time) and the numerical value of the incentive parameter used for modification of the separation processes. The incentive curve plot is presented in figure 2 using flotation data as an example.

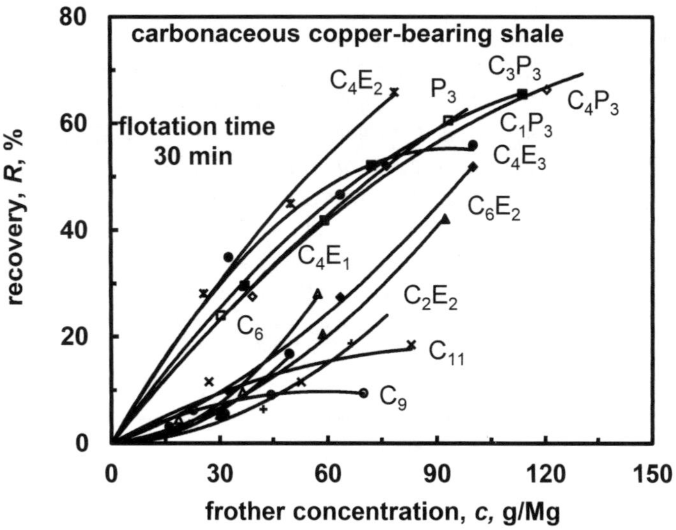

Figure 2. The incentive separation curve. After Kowalczuk et al., 2014 [8].

The incentive curves are available in the literature in many papers dealing with separation processes.

4. Limits curves

The kinetic curves provide two essential process parameters, that is k and R_{max}. Graphical relation of these parameters provides another type of curve, that is the limits curve. The available in literature limits curve indicates that in many separation processes there is a simple relation between kinetics (k) and thermodynamics (R_{max}) of the processes. Figure 3 presents a limits curve as a relation between k and R_{max} for flotation of shale.

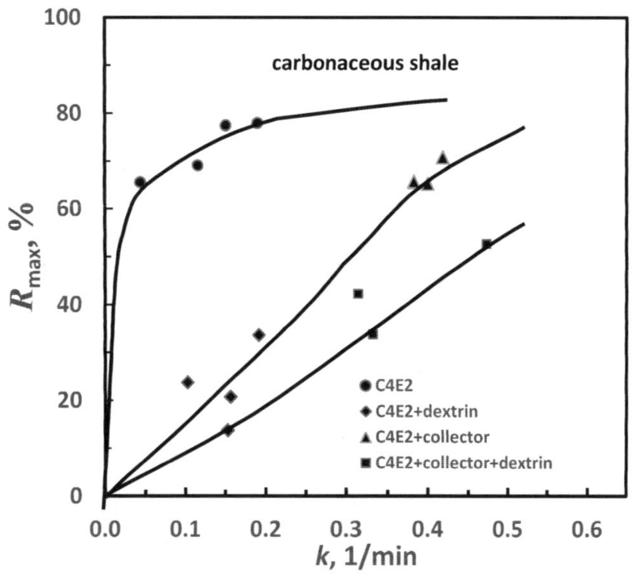

Figure 3. The limits separation curve. After Nowak and Drzymala (2017) [9].
Here k is the first order kinetic constant ($k = k_1$).

IOP Conf. Series: Materials Science and Engineering **427** (2018) 012011 doi:10.1088/1757-899X/427/1/012011

5. Arrheniusan curves

The arrheniusan curves can be created starting with the kinetic curve by replacing the process time t with a specific parameter characterizing energy E imposed to the separation system by the incentive parameter i and for x axis replacing recovery R with kinetic constant k_n, which combines process time t with recovery R. In the case of incentive parameters being concentration c of a reagent, the energy parameter is $kT \cdot \ln(c/c_o)$, where k is the Boltzmann constant, T absolute temperature, c_o is a standard concentration. The incentive parameter can be temperature, particle size, regent concentration etc., and the energy E imposed to the separation system by the incentive parameter can be found elsewhere [10]. When the experimental data points form a line, the slope of it is the activation energy E_a. The activation energy characterizes the impact of the incentive parameter on the process recovery being the most important parameter. Arrheniusan curves are shown in figure 4.

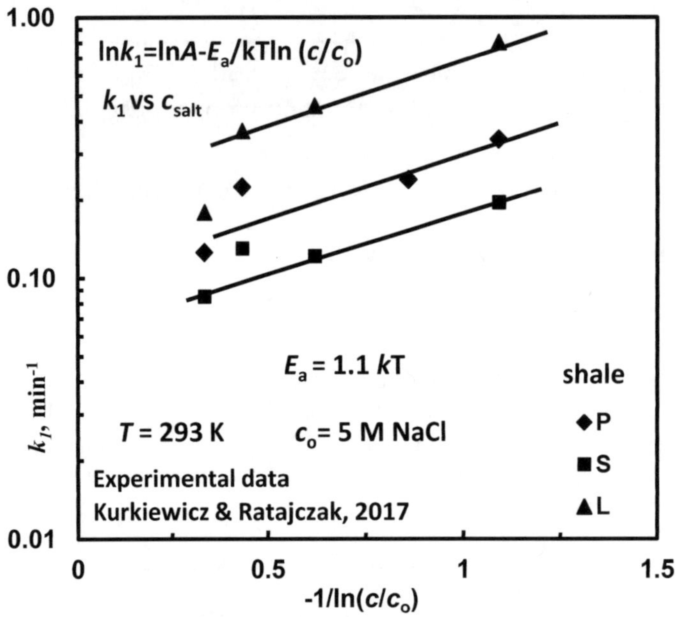

Figure 4. The arrheniusan separation curve [10], [11]. Flotation of different types, arbitrarily denoted as P,S,L copper-bearing carbonaceous shales samples. Courtesy of the Physicochemical Problems of Mineral Processing journal.

6. Conclusion

Since flotation process is very complex, appropriate procedures for its delineation and characterization are necessary. For this purpose, it is useful to use different two-parameter relations called separation curves. In the case of flotation, very beneficial are kinetic, limits, incentive and arrheniusan 2D plots. Relations between curves are presented in figure 5.

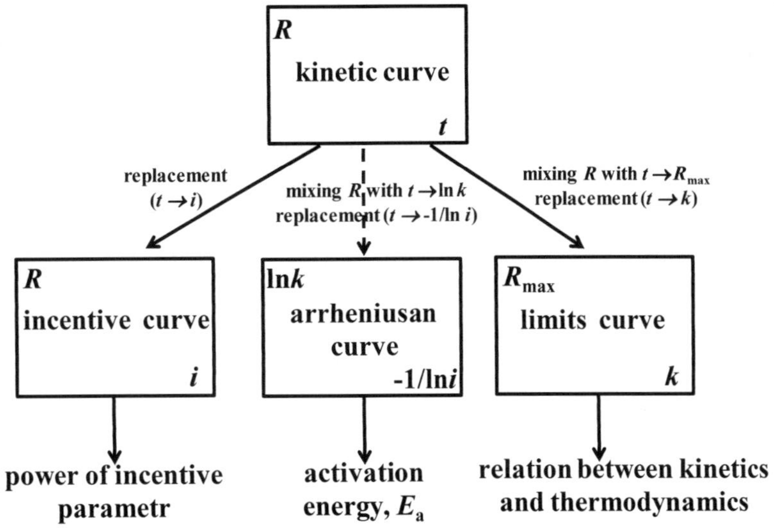

Figure 5. Relation between kinetic, limits, incentive and arrheniusan separation curves.

The most unprocessed is the kinetic curve (figure 1). By replacement of certain parameters from the plot, for instance time *t* with incentive parameter *i*, the incentive curve is created (figure 2), while mixing parameters of the kinetic curve (*t* with *R*) produces the limits curve (figure 3). Both replacement and mixing provides additional curves. One of them is the arrheniusan plot (figure 4). This plot is especially useful because it provides a new parameter called the activation energy. This parameter is able to show the dynamics of the process, that is how strongly the incentive parameter influences the system. On the other hand, the limits curves are useful for showing a relation between kinetics and process thermodynamics.

Acknowledgements
This work was financed by the Polish Statutory Research Grant 0401/0129/17.

References

[1] Wills B A and Napier-Munn T J 2006 *Wills' Mineral Processing Technology* (Elsevier: Amsterdam)
[2] Laskowski J S 1986 The relationship between floatability and hydrophobicity *Advances in Mineral Processing* (SME: Littleton) pp 189-208
[3] Laskowski J S 1989 Thermodynamic and Kinetic Flotation Criteria *Mineral Processing and Extractive Metallurgy Review* **5** pp 25-41
[4] Fuerstenau D W and Raghavan S 2007 Some aspects of flotation thermodynamics *Froth flotation. A century of innovation* (SME: Littleton)
[5] Drzymała J 2007 *Mineral Processing* (WUT: Wroclaw)
[6] Drzymala J, Ratajczak T and Kowalczuk P B 2017 Kinetic separation curves based on process rate considerations *Physicochem. Probl. Miner. Process.* **53** pp 983–95
[7] Drzymala J and Kowalczuk B P 2018 Classification of flotation frothers *Minerals* **8/53** pp 1-24
[8] Kowalczuk P B, Buluc B, Sahbaz O and Drzymala J 2014 In search of an efficient frother for preflotation of carbonaceous shale from the Kupfershiefer Stratifrom copper ore *Physicochem. Probl. Miner. Process.* **50(2)** 835-40
[9] Nowak J and Drzymala J 2017 Flotacja łupka miedzionośnego w obecności spieniacza, zbieracza oraz depresora w postaci dekstryny *Łupek miedzionośny. III* (PWr: Wroclaw) pp 118-28

IOP Conf. Series: Materials Science and Engineering **427** (2018) 012011 doi:10.1088/1757-899X/427/1/012011

[10] Drzymala J 2018 Arrheniusan activation energy of separation for different parameters regulating the process *Physicochem. Probl. Miner. Process*. **54(4)** special issue dedicated to honor Professor Janusz Laskowski
[11] Kurkiewicz S and Ratajczak T 2017 Flotometryczna hydrofobowość łupków miedzionośnych w obecności NaCl *Łupek miedzionośny III (*PWr: Wroclaw) pp 118-28

IOP Conf. Series: Materials Science and Engineering **427** (2018) 012012 doi:10.1088/1757-899X/427/1/012012

Modelling of oil agglomeration of dolomite by the use of an artificial neural network optimized using Optimal Brain Damage algorithm

M Kamiński[1] and A Bastrzyk[2]

[1]Faculty of Electrical Engineering, Wroclaw University of Science and Technology, Janiszewskiego 8, 50-377 Wroclaw, Poland

[2] Faculty of Chemistry, Wroclaw University of Science and Technology, Norwida 4/6, 50-373 Wroclaw, Poland

anna.bastrzyk@pwr.edu.pl

Abstract. This paper investigates application of Artificial Neural Network (ANN) for dolomite oil agglomeration modelling including parameters such as surfactants concentration, oil dosage, time, pH and mixing intensity. The main algorithm implemented for weights calculation was the Levenberg-Marquardt (LM) method. Common problem during process design of neural models is suitable selection of structure complexity. It is known that several connection can influence on final results after off-line training. For improvement of this stage of preparation of the net, pruning method was implemented. Analysed algorithm was based on the main theory of Optimal Brain Damage (OBD) technique. Results present high quality of process recovery prediction. Achieved outcome also shows that reduction of the applied of the neural network can lead to higher precision of calculation.

1. Introduction

Neural network is a flexible model, that can be adapted for realization of specific task in training process. It is useful attribute in a wide range of engineering applications, successful implementation of those models are noted in following fields: control (for instance of electrical drives) [1]-[3], state variables estimation [4], faults detection [5], signal prediction [6], data and images processing [7], etc. Based on papers published in journals and conference proceedings, three main – the most popular – group of structures can be mentioned: Multi-Layer Perceptron (MLP), Radial Basis Function Neural Networks (RBFNN) and Recurrent Neural Networks (RNN) [8], [9]. Characteristic feature of the first of them is one direction of input signals propagation inside of neural network. Moreover, mapping of data space is realized as global approximation using combination of several function of activation functions [10]. Neural networks with radial functions applied in hidden layer, can analyse data grouping following values in clusters (locally) and then aggregation is realized as weighted sum of signals. The division of input elements, realized with those models, is often used in the classification of data. Last significant structures of networks are RNN, those networks contain additional memory elements implemented in inner feedback connections. It is important advantage for processing of dynamic, time changeable variables.

Presently, growing number of neural network application is also being observed in chemistry. Multi-Layer Perceptron neural network can be applied for flotation process of coal fines modelling

[11]. Implementations of RBFNN are also noticed for similar task [12], moreover authors present comparison between different method of modelling used for this purpose. Other example of neural network usage is prediction of heavy metals recovery from wastewater [13]. Important problem related to neural modelling is highlighted in [14]. Final results of training and neural calculation can highly depend on initial values of weights (that are the most often selected randomly). For solution of this problem genetic algorithm was used.

In this application neural networks are used for calculation of chemical process efficiency, based on parameters and conditions of preparation. Modelling of such process can be treated as representation of data using function with one output and a few coefficients. This function and parameters are hard to simple definition, however based on experimental data representing specific behaviour, neural model of the chemical process can be prepared. For this purpose the feedforward neural networks with sigmoidal functions were selected for this task.

Adaptation of neural network for specified task is realized during training process. It concerns calculation of internal coefficients - weights. The most effective methods for this purpose are gradient-based optimization techniques. In this application the Levenberg-Marquardt algorithm was implemented. It is the second order optimization, the main point of this method is minimization of cost function using information about the Hessian matrix. It represents influence of weights changes on quality of network calculation [8]. Selected method used for coefficients of network selection give better results than algorithms based on gradients only. However it takes more time, but this disadvantage is not so much important during off-line training. One of the most important problem during design process of neural network is proper complexity of structure selection. Effective methods in real application is evolution of the network in parallel to adaptation of weights. So, at the same time values of neural network parameters and structure are modified [15]. Initial structure is defined a priori, then several connections are removed from the neural network. For correct determination of unnecessary connection additional saliency parameters (for each weights) are calculated. In application presented in this paper, decision is based on Optimal Brain Damage algorithm [16].

This paper is divided in five sections. Firstly, short introduction is presented. Then is the oil agglomeration process briefly described. In following part of article the structure of implemented neural network and method of training are shown. After presentation and analysis of results concluding remark are given.

2. Oil agglomeration process description

Oil agglomeration is a colloidal method, in which fine particles can be easily separated from mineral suspension via addition of oil to minerals slurry. A rapid mixing of such system leads to the adhesion of oil droplets into the mineral surface, where the formed liquid bridges link individual particles into larger aggregates, called agglomerates [17]. The success of the process strongly depends on the properties of the solid surface and oil/water interface. To be agglomerated the particles have to possess hydrophobic surface, in the case of hydrophilic ones after addition of oil the particles will remain un-agglomerated. This behaviour is very important especially during minerals processing, when a large amount of fine particles is produced via grinding of ore rock to reveal valuable minerals. Recovering of those valuable particles is a complex task because of the grains size (below 10 μm). Commonly used methods such as flotation, filtration or sedimentation in such condition are not very effective, and require additional steps. Oil agglomeration can be alternative for those methods, because the process makes possible to agglomerate the fine particles (below 5μm) with high selectivity using a simple equipment. Despite this, the process is not used on a large scale due to the high cost resulting from the use of large amounts of binding liquid (oil). Therefore, for several years, researchers have been trying to reduce the costs of the process. One solution is to introduce the binding liquid in a form of emulsion into the suspension [18]. It is happened because with the reduction of droplets size, the surface area increased, which leads to creation of more bridges on the mineral surface using small amount of oil. Moreover, the addition of surfactants to the emulsion lowers the interfacial tension resulting in formation of stronger agglomerates.

Modelling of such process is very difficult and time taking because it requires solving complex mathematical equations problematic in real applications. There is a lot of parameters that can affect the course of oil agglomeration, such as oil type and amount, solid content, surfactant concentration, time, pH and agitation rate. That is way to avoid this difficulties Artificial Neural Network was utilized for modelling of the process. For preparation of the ANN model we used experimental data published in [18], where oil (kerosene) was introduced to the suspension in a form of emulsion with cationic surfactant. As mineral, dolomite was used, and the surface of the grains was modified with anionic collectors. During the experiment the pH, oil dosage, rate and time of mixing, and surfactants concentration were changed and taken for modelling as inputs. The output was process recovery calculated as ratio of mass agglomerate to total mass of the feed.

3. Mathematical description of neural model, training process and Optimal Brain Damage pruning

Neural network is combination of elementary nodes that realize multiplication of input signals, then achieved values are added, in the next stage the results are arguments for activation function. Individual neurons are connected in a structure that globally performs complex function (figure 1). Exemplary (for network with one hidden layer) equation describing details of ANN calculations is presented below:

$$\mathbf{y}_{\mathbf{nn}}\left(\mathbf{x}_{\mathbf{in}}\right)=f_o\left(\mathbf{h}\right)=f_o\left(b_0+\sum_{j=1}^{J}w_j f_h\left(b_{0j}+\sum_{i=1}^{N}w_{ij}x_i\right)\right) \quad (1)$$

where: $\boldsymbol{y_{nn}}$-vector of output values, $\boldsymbol{x_{in}}$-input vector, $\boldsymbol{h}$-vector of arguments for activation function in output layer, N-number of inputs, J-number of nodes in hidden layer, w_{ij}-weights between i-th and j-th node, b_0-bias values, f_o-activation function of the output layer (the most often linear function), f_h-activation function of the hidden layer (the most often sigmoidal function).

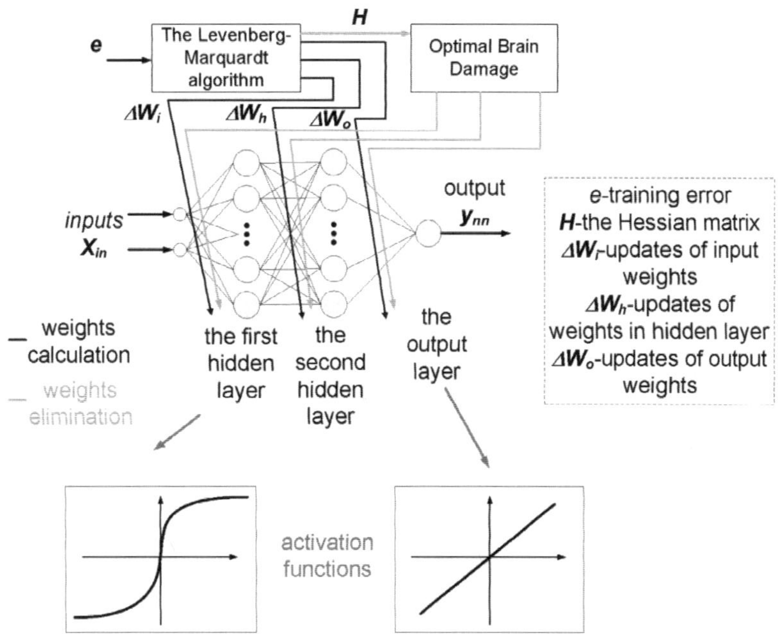

Figure 1. The topology of applied Artificial Neural Network.

One of the main elements affecting the final value of neural structure y_{nn} are values of weight coefficients. Therefore, for correct approximation of elements not included in training dataset, optimal selection of those parameters is particularly important. In this application, for this purpose the Levenberg-Marquardt algorithm was implemented [15]. In each iteration k of this method, weights are recalculated, introducing correction Δw:

$$\mathbf{w}(k+1)=\mathbf{w}(k)+\Delta\mathbf{w}(k). \tag{2}$$

Training process assumes modification of weights in order to minimization of processing error - cost function ***E***. Simplified definition of this value can be expressed as (assuming one output of neural model):

$$\mathbf{e}(\mathbf{w})=\left[y_{nni}(\mathbf{w})-d_i\right], \tag{3}$$

where: d_i – i-th reference value, y_{nni} – i-th neural network output value (obtained for d_i). The base information during calculation of weights is the Jacobian matrix, defined using following expression (for n weights and M patterns):

$$\mathbf{J}(\mathbf{w})=\begin{bmatrix} \frac{\partial e_1}{\partial w_1} & \frac{\partial e_1}{\partial w_2} & \cdots & \frac{\partial e_1}{\partial w_n} \\ \frac{\partial e_2}{\partial w_1} & \frac{\partial e_2}{\partial w_2} & \cdots & \frac{\partial e_2}{\partial w_n} \\ \vdots & \vdots & \vdots & \vdots \\ \frac{\partial e_M}{\partial w_1} & \frac{\partial e_M}{\partial w_2} & \cdots & \frac{\partial e_M}{\partial w_n} \end{bmatrix}. \tag{4}$$

According to the Gauss-Newton equation, correction of weights are calculated using formula presented below:

$$\Delta\mathbf{w}=-\left(\mathbf{J}(w)^T\mathbf{J}(\mathbf{w})\right)^{-1}\mathbf{J}^T(\mathbf{w})\mathbf{e}(\mathbf{w}). \tag{5}$$

In equation (5) the Hessian matrix ***H*** is defined using following equation:

$$\mathbf{H}(\mathbf{w})=\left(\mathbf{J}(w)^T\mathbf{J}(\mathbf{w})\right), \tag{6}$$

then, combining (5) and (6):

$$\Delta\mathbf{w}=-\left(\mathbf{H}(w)\right)^{-1}\mathbf{J}^T(\mathbf{w})\mathbf{e}(\mathbf{w}). \tag{7}$$

For simplification of inverse matrix $\boldsymbol{H}^{-1}$ calculation, approximation of ***H*** is introduced:

$$\mathbf{G}(\mathbf{w})=\mathbf{H}(\mathbf{w})+\mu\mathbf{I}\,. \tag{8}$$

where: ***I***-the identity matrix. The inverse of the matrix ***G*** is possible to calculate if:

$$(\lambda_i+\mu)>\mathbf{0}\,, \tag{9}$$

where: λ_i – i-th element of the eigenvalue vector of ***H*** matrix. Concluding above presented considerations, update of network coefficients are calculated using following expression:

$$\Delta\mathbf{w}=-\left(\mathbf{J}(w)^T\mathbf{J}(\mathbf{w})+\mu\mathbf{I}\right)^{-1}\mathbf{J}^T(\mathbf{w})\mathbf{e}(\mathbf{w})\,. \tag{10}$$

In presented work, the Levenberg-Marquardt algorithm was combined with Optimal Brain Damage method implemented for weights elimination. The OBD defines coefficients calculated for each parameter of neural network:

$$S_i = \Delta E = \frac{1}{2}\sum_i h_{ii}[\Delta w_{ii}]^2 \tag{11}$$

The main information in OBD is also the hessian matrix, it should be mentioned that the ***H*** matrix is diagonally dominant, so only diagonal elements are used in equation (11). Details are presented in [17]. Based on this information network can be reduced (coefficients with the smallest saliency values S_i). It should be noted that some information used for calculation of Si index can be taken after the Levenberg-Marquardt algorithm processing.

Following stages of OBD algorithm operation are presented in figure 2. Algorithm starts with predefinition of initial parameters: initial structure of model, number of weights removed in each sequence, number of all removed weights, conditions of training, stopping criteria, etc. Firstly, complete training is realized (according to (10)). Then, selection of weights designed for elimination should be done. The decision about replacing actual value to zero is made based on value of coefficients S_i (calculated for several coefficients of neural network). In following step, weights are eliminated and overall error is calculated. If closing conditions are not achieved, the whole process is repeated.

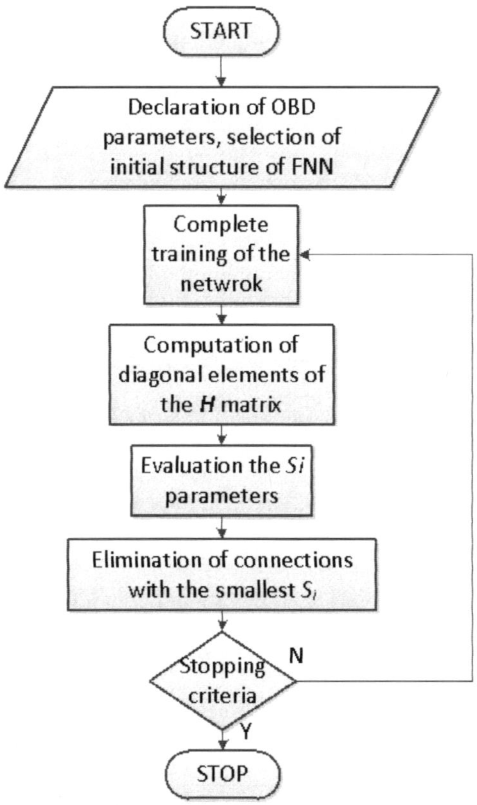

Figure 2. Flowchart presenting data processing in OBD algorithm.

IOP Conf. Series: Materials Science and Engineering **427** (2018) 012012 doi:10.1088/1757-899X/427/1/012012

4. Results

Based on theoretical description of neural model and training process, including structure optimization, presented in previous parts of this article, special code was prepared. Then tests were done. The database was prepared in chemical laboratory, so input vector of ANN contains (respectively): mixing rate, time of mixing, anionic surfactant amount (mg/g), cationic surfactant amount (mg/g), kerosene amount (ml/g), pH. As anionic and cationic surfactant used sodium oleate and dodecylammonium hydrochloride, respectively, which are commonly used in mineral processing as collectors. Output of the neural network model is recovery. The number of samples (input-output elements) included in training vector was equal 52.

The first research presents general trend of error fluctuation under changes of neural network topology (figure 3). Only exemplary tests are shown, other conditions of training were assumed. However the goal was observation of overall tendency of error changes. Each value was calculated using following formula:

$$Err = \frac{1}{M}\sum_{i=1}^{M}\left|y_{nni} - d_i\right|. \qquad (12)$$

In following iteration of OBD algorithm, the connections are removed (weights set to zero). It can be observed that initial number of hidden nodes was not selected properly. After elimination of some group of weights, quality of calculation was improved (optimal structure was obtained). This is in line with theory of neural networks.

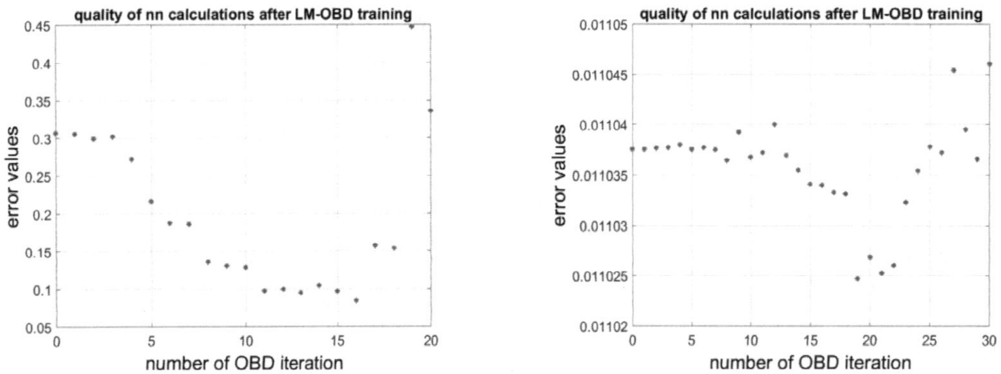

Figure 3. Exemplary values of processing errors during connection reduction.

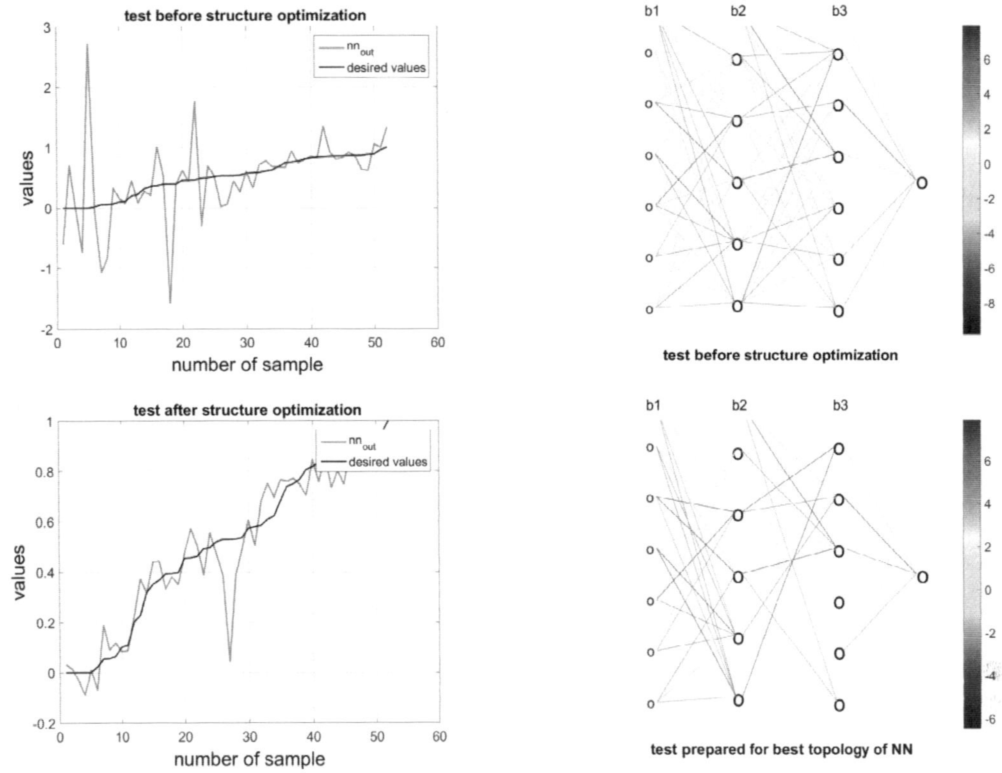

Figure 4. Tests presenting influence of structure optimization on recovery calculation.

Subsequent pictures (figure 4) show details of exemplary training process and results of structure optimization. Next to the results, presenting real and calculated values of recovery, in the right column changes of neural topology is shown. The colours of lines responds to values of weights. Initial structure was {6-5-6-1}, it means: 6 inputs, 5 neurons in the first hidden layer and 6 nodes in the second layer, 1 output. The whole OBD optimization, in this test, was repeated 30 times, in each iteration 3 connections were reduced. Best results were obtained after removing 33 links. However, it should be mentioned that, in the best topology whole nodes were eliminated. Implemented optimization method modifies the ANN, it gives significant improvements of results.

One of important expectation for neural networks is high quality of processing, under input values not included in training process. In the presented work, training data was not used in the tests. Generally, it is possible to divide samples into two groups, but in the case of the task being analyzed this is problematic due to the use of experimental data (limited number of samples). For this reason, in all tests, the measured values were appropriately disturbed by introducing noise.

Best results achieved in this research are presented in figure 5. The LM-OBD algorithm has started with following neural network: {6-15-16-1}, number of all removed connections was equal 60, number of removed connections for the best results was equal 57. It should be also noted that each training using the Levenberg-Marquardt method takes 1000 iteration (this number was selected experimentally, for dataset used in this project, based on observation of validation error and gradient), and total time of calculations using LM-OBD was 22.3157s.

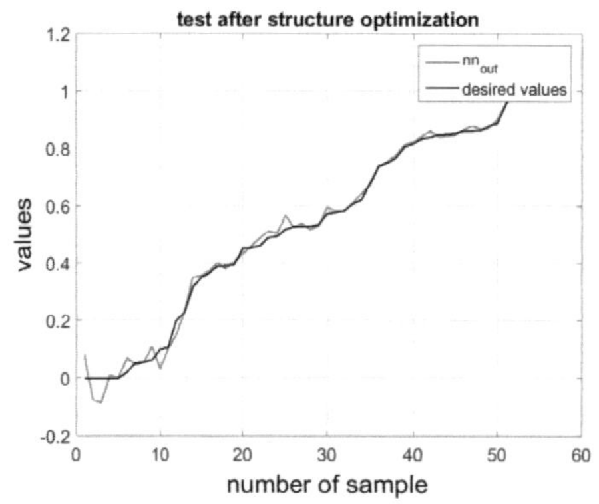

Figure 5. Example of ANN calculation – the best results.

5. Conclusions

This paper is focused on dolomite particles oil agglomeration modelling using neural networks. Based on obtained neural model, recovery of the process, for introduced input parameters, can be estimated. In next step, optimization of the process, based on prepared neural model can be done (assessment of the required input parameters in order to achieve the assumed performance). Precision of calculation of obtained neural model was very high. It is worth to mention that all calculations were done for real data collected in laboratory experiment. Important part of this work includes optimization of the ANN's structure. For this purpose Optimal Brain Damage method was implemented. Results confirm theoretical considerations related to relationship of neural network complexity and results of processing.

Acknowledgments

This work was financially supported by a statutory activity subsidy from the Polish Ministry of Science and Higher Education for the Faculty of Chemistry (A.B.) and for the Faculty of Electrical Engineering (M.K.) of Wroclaw University of Science and Technology.

6. References

[1] Nowopolski K, Wicher B, Łuczak D and Siwek P 2017 *Proc. Int. Conf. on Methods and Models in Automation and Robotics (Miedzyzdroje)*, p 576
[2] Pajchrowski K, Zawirski K and Nowopolski K 2015 *IEEE T. Ind. Inform.* **11(2)** p 560
[3] Cai JP, Xing L and Zhang M 2017 *IEEE Access* **5** p 15839
[4] Merabet A, Tanvir AA and Beddek K 2016 *IET Renew. Power Gen.* **10(10)** p 1597
[5] Khomfoi S and Tolbert LM 2007 *IEEE Trans. Power Electron* **22(3)** p 1062
[6] Chen CI and Chen YC 2015 *IEEE Trans. Power Del.* **30(3)** p 1577
[7] Yoo B, Kwak Y, Kim Y, Choi C and Kim J 2018 IEEE Signal Process. Lett. **25(6)** p 808
[8] Bishop CM 1995 *Neural Networks for Pattern Recognition* (Oxford: Oxford University Press)
[9] Mandic D and Chambers J 2001 *Recurrent Neutral Networks for prediction: learning, algorithms, architectures and stability*, 1st edition (USA: Wiley)
[10] Lawrence S, Chung Tsoi A and Back AD 1996 *Proc. Australian Conf. on Neural Networks* p 16
[11] Kalyani VK, Pallavika, Sanjay Choudhuri, Gouri Charan T, Haldar DD, Kamal KP, Badhe YP, Tambe SS and Kulkarni BD 2007 *Min. Proc. Ext. Met. Rev.* **29(2)** p 130
[12] Nakhaei F and Irannaiad M 2013 *Physicochem. Probl. Miner. Process.* **49(1)** p 255
[13] Allahkarami E, Igder A, Fazlavi A and Rezal B 2017 *Physicochem. Probl. Miner. Process.* **53(2)** p 1105
[14] Allahkarami E, Nuri OS, Abdollahzadeh A and Rezal B 2017 *Physicochem. Probl. Miner. Process.* **53(1)** p 366.
[15] Hassibi B, Stork DG and Wolff GJ 1992 Optimal brain surgeon and general network pruning *IEEE Int. conf. on neural Networks* **1** p 293

[16] Le Cun Y, Denker JS and Solla SA 1990 *Adv. Neural Inform. Proces. Syst.* **2** p 598
[17] Cebeci Y and Sönmez I 2004 *J. Colloid Interface Sci.* **273(1)** p 300
[18] Bastrzyk A, Polowczyk I., Sadowski Z. and Sikora A 2011 *Sep. Purif. Technol.* **77(3)** p 325

IOP Conf. Series: Materials Science and Engineering **427** (2018) 012013 doi:10.1088/1757-899X/427/1/012013

Application of a vision systems for assessment of particle size and shape for mineral crushing products

D Krawczykowski

AGH University of Science and Technology, Faculty of Mining and Geoengineering, Department of Environmental Engineering and Mineral Processing, Krakow

dkrawcz@agh.edu.pl

Abstract. Morphological properties of mineral particles, in particular their size and shape, play a fundamental role in the course of mineral raw materials processing. They affect the efficiency of grain classification and downstream beneficiation operations. The article presents the results of the size and shape distribution for the particles after the comminution process. The measurements were made with using a vision system with dynamic image analysis. The aim of investigations was to evaluate the applicability of this technique for control of the particle size of raw material products crushed in various laboratory crushing devices.

1. Introduction

The size and shape of particles are the basic geometric parameters used in description of mineral resources. The sieve analysis and the qualitative description of the shape of particles along with the determination of their shape made on the slotted screen, are the most commonly used methods for determination the particle size and shape of mineral raw materials. In recent years, however, there has been observed a rapid development of vision methods based on computerized image analysis of particles. The first particle analyzers using simple image processing algorithms were developed in 1985 [1], and the first advanced methods of quantitative analysis applied in basic research based on image analysis were proposed by Brawman and Sukumaran [2,3].

Currently, there are many measurement technologies that combine high quality imaging equipment with artificial intelligence software for determination the size and shape of particles with using the vision technology in laboratory and industrial conditions, i.e. QICPIC, Analysette 28 ImageSizer, CAMSIZER, Split-Online, VisioRock, Plant Vision, WipFrag Momentum. In mineral processing operational practice it is essential the knowledge concerning the properties of material, especially particle size, shape, useful mineral content, particle size composition, mineralogical composition etc., because on the basis of this information it is possible to adjust and optimize both crushing and grinding operations [4,5,6], as well as downstream beneficiation processes [7], in terms of their more effective course. The problem is significant for both for mineral aggregates production and ore beneficiation.

2. Materials and methods

The main aim of investigations presented in the paper was to verify the possibility of application a dynamic image analysis for control the particle size distribution of dolomite after crushing process and to determine the impact of the crushing device on geometric properties of the particle. The research programme included the stage of raw material preparation through crushing and screen classification as well as morphological analysis of individual particles in crushing products by means of the vision analysis system for particle size and shape. The scheme of investigations was presented in figure 1.

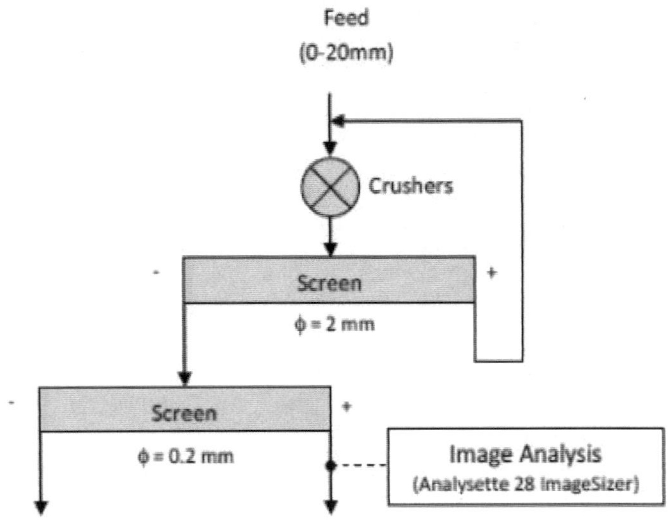

Figure 1. The scheme of investigations of particle size and shape analysis for crushing products.

The samples of the feed material were crushed below 2 mm in various laboratory crushers in a closed circuit with classification. Fine particles below 0.2 mm were removed and the particle size class from 0.2 to 2mm was obtained for further geometrical tests of particles.

The vision dynamic image analysis technique was used for evaluation of the shape of particles. The shape factor Aspect Ratio, defined as the ratio of the minimum to the maximum particle diameter according to Feret ($k = d_{Fmin}/d_{Fmax}$), was determined, as well as the particle size distribution of the crushing products according to the minimum Feret diameter. The crushing process was carried out in laboratory crushers utilizing following forces affecting the feed material:

- an impact force caused by movable working elements of the crusher mounted on the shaft,
- an impact force caused by vibration of the working elements of crusher. The reduction in size of feed takes part in several stages by gradually detaching the outer layers of the particle,
- centrifugal force generated by the vertically rotating rotor. This force gives the particle a high speed, and thus the kinetic energy, which is spent on the destruction of its structure during a collision with a stationary plate,
- compression force derived from the pressure of the working elements of the device.

The structure, principle of operation and the distribution of forces acting on particles determine the main classifications of the crushing devices. This is an issue widely described in literature [8,9,10].

The Analysette 28 ImageSizer vision analyzer was used for measurement of geometric properties of particles [11]. The measurement in this device is carried out with the dry technique in the particle size range from 0.02 mm to 20 mm. The analyzer consists of a vibratory feeder, a camera that records an image of particle and a source of strobe light (figure 2). The principle of measurement is based on the dynamic image analysis (DIA) method. The camera intercepts the image of particles moving in a measuring cell, and then the computer software analyzes the shape and size of each particle separately. Ten thousand of images per minute, approximately, is analyzed. An exemplary report is presented in figure 3.

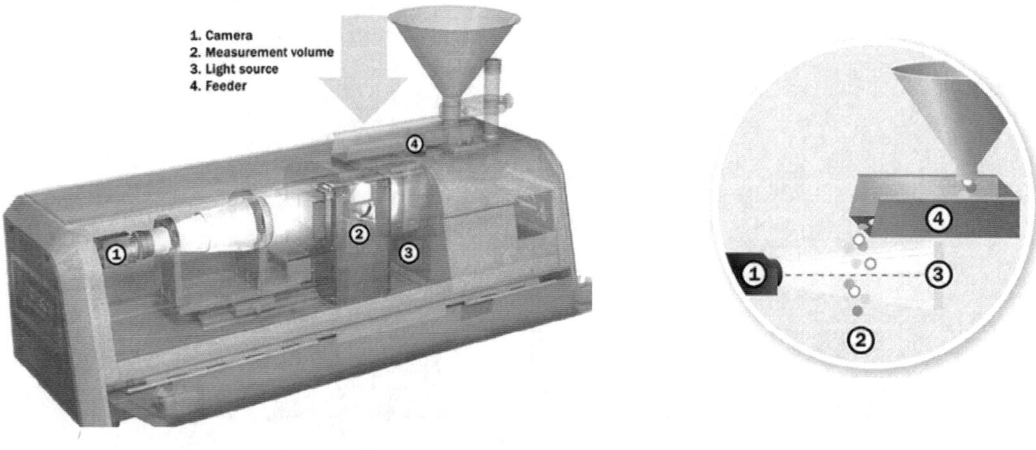

Figure 2. Construction and principle of operation of a dynamic particle size and shape analyzer: 1) camera 2) measurement volume , 3) light source, 4) feeder [11].

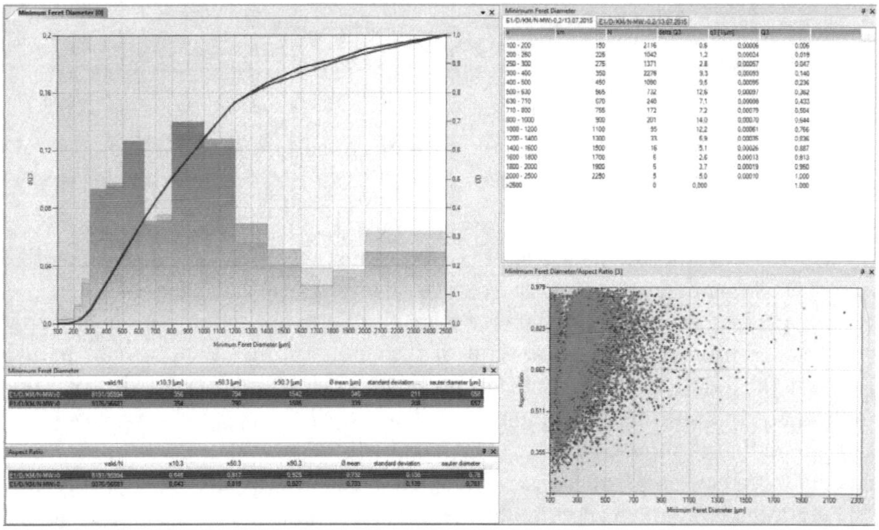

Figure 3. Exemplary report obtained from the Analysette 28 ImageSizer.

3. Analysis of results

In order to compare the operation of individual crushing devices in terms of generation of elongated (irregular) particles, the results were grouped and averaged. Figure 4 presents the average values of elongation coefficients in the tested samples obtained for different crushers, while in table 1 the distribution of the coefficient k is shown for the distribution values 10, 50 and 90%, respectively. The analysis of the following results indicates that the more elongated particles have been generated in centrifugal crusher (lower value of k index), while more the centric ones – in impact crusher (figure 4). The differences in both cases, however, are not very significant. Generally, the particles after the crushing process show a more circular shape.

Analyzing the particle size distribution of crushing products expressed by the diameters of characteristic particles d_{10}, d_{50} and d_{90}, as well as an average diameter (d_{mean}) of the particle (arithmetic average) according to Feret, it can be seen that for fine particles (d_{10}) the most efficient operation was achieved for the crusher that utilizes the pressure as the destructive force. For coarse particles (d_{90}), in turn, an impact crusher appeared to be more effective (table 2). It confirms operational practice, because crushing of fines by impact forces is of a lower efficiency due to the small inertia forces acting on these particles.

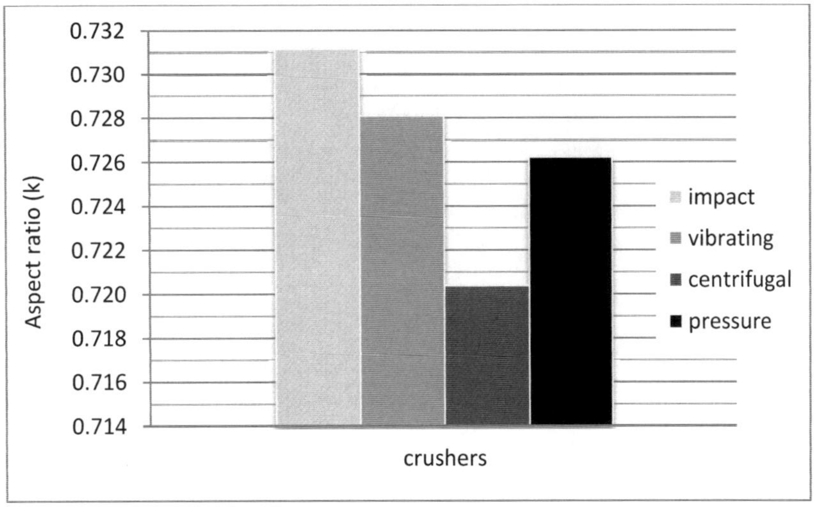

Figure 4. Average values of shape index for crushed particles determined by means of dynamic image analysis (DIA) method.

Table 1. Distribution of shape index values in crushing products for individual crushing devices.

Crusher	Aspect ratio		
	k_{10}	k_{50}	k_{99}
impact	0.648	0.818	0.925
vibrating	0.642	0.813	0.920
centrifugal	0.639	0.813	0.926
pressure	0.641	0.813	0.923

Table 2. Distribution of particle size in crushing products for individual crushing devices.

Crusher	Population	d_{10} (μm)	d_{50} (μm)	d_{90} (μm)	d_{mean} (μm)
impact	9680	352	845	1593	658
vibrating	11195	355	989	1861	707
centrifugal	6118	409	1068	2148	819
pressure	13388	317	823	1695	617

4. Conclusions

Results of investigations over geometrical properties of particles for the given type of mineral raw material with using of a video system utilizing the dynamic image analysis technique confirm the usefulness of this measurement method in evaluation of the crushing devices operation and control of the size and shape of mineral particles.

Analysis of the particle shape expressed by the elongation ratio k (Aspect Ratio) showed no significant differences in the shape of the product particles obtained in the tested crushing devices. The most regular shapes were obtained in an impact crusher, while the more elongated particles were

IOP Conf. Series: Materials Science and Engineering **427** (2018) 012013 doi:10.1088/1757-899X/427/1/012013

produced in a centrifugal crusher. Particles of the dolomite (tested material) after the crushing process are rather of a centric shape (k>0.5) than the elongated one (k <0.5).

Analysis of the particle size distribution of crushing products expressed by the average minimum diameter according to Feret indicates that the finest product was obtained in a crusher that utilizes pressure force coming from the working elements, while the coarsest one was produced in impact crusher, where centrifugal crushing forces predominate.

The research topic is important and interesting, and definitely worth continuing with application to other rocks or minerals.

References

[1] Yacher L, Gonz′alez C, Von Borries G and Nobile R 1985 Coarse particle size distribution analyze *Instrumentation in the Mining and Metallurgy Industries, Instrument Society of America* **12** pp 71–9

[2] Bowman E Soga K and Drummond W 2001 Particle shape characterization using Fourier descriptor analysis *Geotechnique* **51(6)** pp 545–54

[3] Sukumaran B and Ashmawy A 2001 Quantitative characterization of discrete particles *Geotechnique* **51(7) pp** 619–27

[4] Gawenda T Naziemiec Z and Walerak A 2013 Investigations on selected properties of aggregates from ceramic waste and concrete produced with their participation *Annual Set of Environmental Protection* **15 (3)** pp 2003-21

[5] Gawenda T and Saramak D 2018 Assessment of the process efficiency for metallurgical zinc waste screening in vibrating screen *Inżynieria Mineralna – Journal of Polish Mineral Engineering Society* **37(1)** pp 189-194

[6] Naziemiec Z and Saramak D 2015 Application of partition curves in the assessment of mineral products classification processes *Mining Science* **22** Special Issue **1** pp 119-29

[7] Saramak D, Młynarczykowska A and Krawczykowska A 2014 Influence of a high-pressure comminution technology on concentrate yields in copper ore flotation processes *Archives of Metallurgy and Materials* **59 (3)** pp 951-5

[8] Gawenda T 2013 The influence of rock raw materials comminution in various crushers and crushing stages on the quality of mineral aggregates *Mineral Resources Management* **29 (1)** pp 53–65

[9] Saramak D 2015 Simulation of comminution effects in HPGR for the feed material variable content of fine particles *Mineral Resources Management* **31(2)** pp 123–36

[10] Sidor J and Klich A 2018 *Współczesne maszyny do rozdrabniania – kruszarki i młyny* (Instytut Techniki Górniczej KOMAG: Gliwice)

[11] www.fritsch-international.com

IOP Conf. Series: Materials Science and Engineering **427** (2018) 012014 doi:10.1088/1757-899X/427/1/012014

Investigation of post-industrial pollutions' immobilization in a hydraulic self-solidifying clay-cement binder

A Stempkowska[1], Ł Wójcik[2], P Izak[2], M Staszewska[1] and J Mastalska-Popławska[2]

[1] Department of Mining and Geoengineering, AGH University of Science and Technology, al. Mickiewicza 30, 30-059 Krakow, Poland

[2] Department of Materials Science and Ceramics, AGH University of Science and Technology, al. Mickiewicza 30, 30-059 Krakow, Poland

stemp@agh.edu.pl

Abstract. Antifiltration barriers are used to secure deep excavations and flood embankments as well as to prevent the penetration of harmful substances into the ground and groundwater, also as part of the environmental protection system of waste landfills and production plants. The barriers are usually formed from a liquid suspension with binding properties which, after being pumped to the destination, passes into a solid with a fixed strength. One of the main problems of degraded post-industrial areas, preventing their introduction into the secondary economic cycle, is pollution of soil and the impact of pollution on groundwater. The interaction process of contaminated sites is complicated and the transport of pollutants from the place of danger to usable aquifers underground levels passes through a series of varied course. The article presents the possibilities of applications ground binder to reduce the emission of pollutant ions from soils in order to develop new mixtures of binders. The carried out research has shown that in the case of highly contaminated soils (eg Glasgow), almost 100% of heavy metals can be accumulated using clay-cement composites, and also significantly reduce the penetration of aromatic and aliphatic hydrocarbons. Some of the aliphatic hydrocarbons that allowed to enter to the penetrating solutions during measurements also did not exceed the standards.

1. Introduction

Soil contamination can be divided into two groups, one of them includes inorganic pollution, mainly heavy metals, and the other group constitutes of organic impurities, mostly in the form of petroleum-derived substances [1,2].

Decontamination of soils from heavy metals and petroleum substances is carried out by "*in situ*" methods (directly in contaminated areas) or "*off site*" methods (by moving contaminated soil into a place designed for this purpose), eg by using live microorganisms or by washing the soil with special solutions [3, 4, 5]. However, these processes, are long-term, often requiring additional work related to further processing of pollutants.

One of the method of immobilization of pollutants is a hydraulic self-solidifying mixture designated to locate environmentally hazardous post-industrial waste, prepared on the basis of a mineral binder with the addition of hazardous waste and stabilizing substances [4,5,6,7,8,9].

Due to their anti-filtration properties, such mixtures are also used to prevent the penetration of harmful substances into the ground and groundwater as part of the environmental protection system of waste landfills and production plants. These barriers are usually formed from a liquid suspension with binding properties which, after being pumped to the destination, passes into a solid with a fixed strength.

A characteristic feature of clay-cement binders is that in the entire bonding range they behave like viscoplastic fluids and there is not formed stiff crystalline structure, eg characteristic for concrete. Thanks to this combination of the sealing suspension with rock or soil (during excavation shaft works or other environmental works) does not harm their structure, and the fine-disperse clay fractions contained in the suspension ensure high tightness [10]. Changes of composition also allow modification of rheological and structural properties of clay-cement suspensions in a wide range.

In general the formation of the internal structure of the suspensions in the light of physicochemistry of colloids, can be explained as a specific aggregation of particles, as a result of intermolecular interactions. According to the DLVO theory [11], these interactions are the resultant of two phenomena:

- attraction - resulting from van der Waals forces (F_A),
- repulsion - resulting from electrostatic and spatial effects (F_R) (equation 1.):

$$F_{DLVO} = F_R + F_A \tag{1}$$

where: F_{DLVO} –force DLVO, F_R – repulsive electrostatic force, F_A – attractive force van der Waalsa.

As a result of these forces, spatial structures of the suspension spontaneously arise and their stiffness depends on the mutual ratio of these forces, which results from particles shape. The reversible internal structure of the suspension determines its rheological properties. The clay-cement slurries are a special case of a system in which the reversible processes associated with the thixotropic behavior of clay minerals overlap with the irreversible physico-chemical processes that accompanying the hydration process of cement. These processes, which include changes in the composition of the solution and the formation of solid reaction products, have a significant impact on the rheology of clay-cement slurries.

The cement hydration process is a complicated reaction, during which comes to the overlapping influence, reacting with water and forming the Portland clinker phases. This process is multi-stage, and individual clinker phases retain their specific reaction. Due to the above properties, clay-cement suspensions seem to be a good material for immobilization of inorganic as well as organic impurities.

The research involved a series of clay-cement mixtures with various composition, to determine the leaching of hazardous components bound in the structure of the clay -cement barrier. Immobilization was carried out with the participation of heavily contaminated soil containing hazardous ingredients eg heavy metals. The idea of the research was the possibility of both binding them in waterproofing barriers, as well as limitations of getting beyond the barrier.

2. Materials and methods

A landfill in Glasgow is post-industrial homogeneous contaminated soil. The provided samples were characterized by a very alkaline reaction and contained metal ions mainly Fe, but also Pb, Cr, Ni, Cu, Zn. However, the biggest ecological risk to the environmental to the Glasgow landfill is the content of hydrocarbons, especially aromatic ones, which have exceeded the applicable standards at least twice.

For the needs for the study, a series of mixtures with contaminated soil was prepared, the composition of which are presented in table 1. The content of soil in the samples is from 30-70%, the rest is a clay-cement binder with a cement content of 10 to 20%. For example, sample G 4.10 means Glasgow soil, binder 40% content of cement in a binder 10%. The mass of samples were similar and amounted to approx. 110-120 g

The samples were placed in pipes with a diameter of φ=50 and after concretion a distilled water flow was forced through them under pressure of 1 atm at a volume of 100 ml. The water permeability of concreted samples was on the level 10^{-10} m/s.

The following research methods were used:

- determination of pH and sulphates concentration- multifunctional WTW inoLAB multi 9310 device equipped with suitable electrodes including ion-selective,
- determination concentation of metals in binders - XRF method (Olympus X-5000 analyzer)
- determination concentration of leached out metals concentration in the filtrates- ASA method (using Thermo Scientific iCE 3500)
- determination of hydrocarbons' concentration - IR method (FTIR spectrometer from BIO-Rad model FTS-165)

Table 1. Characteristics of pollutants in a soil sample from Glasgow.

Component	Concentration	Unit
Humidity	18	(%)
Arsenic, As	3.5	(mg/kg)
Boron, B	0.38	(mg/kg)
Cadmium, Cd	<0.50	(mg/kg)
Chromium, Cr	26	(mg/kg)
Copper, Cu	27	(mg/kg)
Iron, Fe	23000	(mg/kg)
Lead, Pb	72	(mg/kg)
Mercury, Hg	<0.25	(mg/kg)
Nickel, Ni	40	(mg/kg)
Zinc, Zn	65	(mg/kg)
Soluble sulphates	0.090	(%)
pH	8.6	-
Aliphatic hydrocarbons (C5-C44)	45	(mg/kg)
Aromatic hydrocarbons (C5-C44)	840	(mg/kg)

The measuring system set-up forcing filtration is shown in figure 1. The hydraulic gradient was generated by compressed air with stabilized pressure. The filtrate was subjected to analytical tests. The total content of soluble (salinity), pH and potential Eh as well as concentrations of Na, Fe, Pb, Cr, Ni, Cu metals (selection determined by the composition of primary samples), and the concentration of aliphatic and aromatic hydrocarbons were also determined. Additionally, the filtration coefficients were determined and the scale of pollutants spread migration the year was calculated.

IOP Conf. Series: Materials Science and Engineering **427** (2018) 012014 doi:10.1088/1757-899X/427/1/012014

Figure 1. Filtration system.

3. Results and discussion

Compositions of mixtures' series containing contaminated soil from Glasgow with clay-cement binder are shown in table 2 and pH values of the filtrate are presented in table 3.

Table 2. The masses of individual components in mixtures.

Sample	Weight of sample, (g)	Mass of contaminated soil, (g)	Contribution of binder, (% wt)
G-4.10	110.2	66.1	40
G-4.15	122.4	73.4	40
G-4.20	117.6	70.6	40
G-5.10	111.2	55.6	50
G-5.15	112.3	56.2	50
G-5.20	126.7	63.4	50
G-6.10	119.8	47.9	60
G-6.15	112.6	45.0	60
G-6.20	105.3	42.1	60
G-7.10	112.3	33.7	70
G-7.15	119.1	35.7	70
G-7.20	107.0	32.1	70
G-8.10	109.5	21.9	80
G-8.15	115.4	23.1	80
G-8.20	119.2	23.8	80

Background parameters: Salinity 6.06 mg/l, pH 6.99

Table 3. The value of pH , potential Eh, sodium concentration in filtrates of series with contaminated soils.

Sample	Participation of contaminated soil, (% wt)	pH	Salinity (mg/l)	Eh (mV)	c_{Na} (mg/l)
G 4.10	60	10.03	106.5	22	5.47
G 4.15	60	10.29	138.0	45	5.21
G 4.20	60	10.44	170.8	35	5,.36
G 5.10	50	10.27	115.4	-12	4.07
G 5.15	50	10.48	163.6	13	4.46
G 5.20	50	10.48	166.1	-24	5.36
G 6.10	40	10.37	138.1	-22	4.39
G 6.15	40	10.54	185.1	7	4.98
G 6.20	40	10.25	113.2	34	4.42
G 7.10	30	10.43	155.7	15	3.60
G 7.15	30	10.25	114.2	-23	3.10
G 7.20	30	10.34	139.4	-16	3.44
G 8.10	20	10.15	100.4	10	2.37
G 8.15	20	10.23	114.3	8	2.29
G 8.20	20	10.58	217.0	13	3.11

The pH values of the filtrates in all cases was above 10 which exceeds the standard [maximum allowable/permissible value 6.5-9] [12] and also indicates the presence of a large amount of alkalies. The average salinity was about 140 mg/l and no changes in leaching depending on the composition of the samples was observed. The value of salinity is in consistence with the Regulation from 2014 [12] which determines the acceptable quantity at the level of 1000 mg/l. The migration of ions depends on the type of binder's structure and filtration coefficient.

Large positive or negative Eh values or oxidation-reduction potential, indicate an unstable chemical system, where large positive values mean above 590 mV, and in the case of negative values below - 150 mV. Pure water in equilibrium with oxygen has oxidizing properties. The Eh potential in all analyzed cases is weakly reducing. Under natural conditions, the Eh potential changes in the oxidizing direction, which in the case of barriers may affect the surrounding soils [13].

In primary samples alkaline metals were not determined, but since pH value was high, the sodium concentration was taken into consideration (table 3). It is the fastest migrating element, actually a not subject to sorption. Its concentration in the filtrate depends in proportion to the content of contaminated soil the filtrate is determined in proportion to the value of the soil contaminated, however, in all cases it is in the permissible concentrations (800 mg/l) [12]. But if the sodium cations are not compensated by the anions of acidic residues (low salinity, the presence of insoluble compounds), this results in the formation of active OH^- groups. Such a system causes a potential corrosive effect and reducing the character of Eh.

The concentration of nickel, copper and chromium in the soil are at the limit of the requirements of the standard (Ni, Cu) and lead concentration is exceed them (Pb). The permissible iron content is not specified. However, these concentrations have been referred to the most restrictive conditions, or for permeable soils, in the case of impermeable soils, the concentrations in the entire range meet the standards.

The concentration of selected metals was tested before and after the filtration process (limit of detection was $0.2 \cdot 10^{-4}$ mg/l). In the case of Ni, Zn, Pb, Fe, the cation content in the filtrate is below the sensitivity of the diagnostic devices. For content of Cu and Cr, a minor migration to the solution was observed [14].

After the filtration process, it was observed that the content of selected metals cations is almost in 100% accumulated in the binders, it means that they are stable and insoluble in water. Therefore, there is no risk of these harmful cations passing into soils and groundwater. The leaching of transition and heavy metals from the tested binders is also much lower than the limit values (the sum should not be higher than 14.0 mg/l). Therefore, there are no threats to the migration of harmful cations to soil and groundwater.

Table 4. Copper concentration before and after filtration.

Sample	c_{Cu} before filtration, (mg/kg)	c_{Cu} in filtrate, (mg/l)	c_{Cu} after filtration, (mg/kg)	Accumulation of contaminations in the mixture, (%)
G 4.10	18	0.0026	17.9974	99.985
G 4.15	16	0.0022	15.9978	99.986
G 4.20	19	0.0056	18.9944	99.970
G 5.10	15	0.0037	14.9963	99.975
G 5.15	13	0.0054	12.9946	99.958
G 5.20	13	0.0026	12.9974	99.980
G 6.10	11	0.0045	10.9955	99.959
G 6.15	11	0.0023	10.9977	99.979
G 6.20	10	0.0031	9.9969	99.969
G 7.10	9	0.0045	8.9955	99.950
G 7.15	9	0.0012	8.9988	99.986
G 7.20	8	0.0070	7.9930	99.912
G 8.10	5	0.0027	4.9973	99.946
G 8.15	5	bs	-	100
G 8.20	5	0.0016	4.9984	99.968

bs - below sensitinity

Table 5. Chromium concentration before and after filtration.

Sample	c_{cr} before filtration, (mg/kg)	c_{Cr} in filtrate, (mg/l)	c_{Cr} after filtration, (mg/kg)	Accumulation of contaminations in the mixture, (%)
G 4.10	16	0.0327	15.9673	99.795
G 4.15	16	0.0276	15.9724	99.827
G 4.20	15	0.0270	14.9730	99.820
G 5.10	13	0.0324	12.9676	99.750
G 5.15	12	0.0341	11.9659	99.715
G 5.20	12	0.0160	11.9840	99.866
G 6.10	10	0.0290	9.9710	99.710
G 6.15	10	0.0120	9.9880	99.880
G 6.20	11	0.0287	10.9713	99.740
G 7.10	8	0.0233	7.9767	99.708
G 7.15	7	0.0247	6.9753	99.647
G 7.20	8	0.0233	7.9767	99.708
G 8.10	5	0.0333	4.9667	99.334
G 8.15	4	0.0322	3.9678	99.195
G 8.20	4	0.0286	3.9714	99.285

4. Hydrocarbons

Concentrations of aliphatic and aromatic hydrocarbons were determined in binders before and after the process of forced filtration. The results are shown in tables 6 and 7.

Table 6. The concentration of aliphatic hydrocarbons before and after the filtration process.

Sample	Concentration of aliphatic hydrocarbons before filtration, (mg/kg)	Concentration of aliphatic hydrocarbons in filtrate, (mg/l)	Concentration of aliphatic hydrocarbons after filtration, (mg/kg)	Accumulation of contaminations in the mixture, (%)
G 4.10	27.78	5.415	22.365	80.507
G 4.15	28.12	5.180	22.940	81.579
G 4.20	25.87	4.670	21.200	81.948
G 5.10	22.76	3.169	19.591	86.076
G 5.15	23.29	4.470	18.820	80.911
G 5.20	21.17	4.903	16.267	76.840
G 6.10	18.09	5.010	13.080	72.305
G 6.15	18.20	3.698	14.502	79.681
G 6.20	17.61	3.475	14.195	80.607
G 7.10	14.59	3.326	11.264	77.203
G 7.15	14.45	3.321	11.129	77.017
G 7.20	13.02	2.159	10.861	83.417
G 8.10	9.91	2.514	7.396	74.631
G 8.15	9.34	2.403	6.937	74.272
G 8.20	8.87	3.930	4.940	55.693

Table 7. The concentration of aromatic hydrocarbons before and after the filtration process.

Sample	Concentration of aromatic hydrocarbons before filtration, (mg/kg)	Concentration of aromatic hydrocarbons in filtrate, (mg/l)	Content of aromatic hydrocarbons after filtration, (mg/kg)	Accumulation of contaminations in the mixture, (%)
G 4.10	504	bs	unchanged	100
G 4.15	567	bs	unchanged	100
G 4.20	487	bs	unchanged	100
G 5.10	420	bs	unchanged	100
G 5.15	421	bs	unchanged	100
G 5.20	399	bs	unchanged	100
G 6.10	336	bs	unchanged	100
G 6.15	376	bs	unchanged	100
G 6.20	319	bs	unchanged	100
G 7.10	252	bs	unchanged	100
G 7.15	290	bs	unchanged	100
G 7.20	234	bs	unchanged	100
G 8.10	168	bs	unchanged	100
G 8.15	155	bs	unchanged	100
G 8.20	187	bs	unchanged	100

* LOD/limit of detection is at approx. 0.017 mg/l
bs - below sensitinity

The total concentration of aromatic hydrocarbons in the binders greatly exceeds the permissible standards (15 mg/l), even for deeply located soils. In typical waterproofing barriers, there should be no more than 50% of contaminated soil for a depth of injection below 2 m. On the other hand, aliphatic hydrocarbons in filtrates meet the requirements of the standard for impermeable soils (15 mg/l).

The filtrate samples were extracted with tetrachlorethylene and after separation of the phases in the solution, the contents of particular compounds were determined by FTIR spectroscopy. No aromatic hydrocarbons were found in the filtrates (below the limit of detection). Aliphatic hydrocarbons only slightly leached out but still meet the requirements of the standard (permissible concentration of 15 mg/kg). Accumulation of these substances was stated in the binder.

Table 8. The sum of pollutants passing to migrating solutions.

Sample	Concentration of aliphatic hydrocarbons in filtrate, (mg/l)	Concentration of heavy metals in filtrate, (mg/l)	The sum of pollutants, (mg/l)
G-4.10	5.415	0.038	5.453
G-4.15	5.180	0.033	5.213
G-4.20	4.670	0.036	4.706
G-5.10	3.169	0.039	3.208
G-5.15	4.470	0.043	4.513
G-5.20	4.903	0.022	4.925
G-6.10	5.010	0.037	5.047
G-6.15	3.698	0.017	3.715
G-6.20	3.475	0.035	3.510
G-7.10	3.326	0.031	3.357
G-7.15	3.321	0.029	3.350
G-7.20	2.159	0.033	2.192
G-8.10	2.514	0.039	2.553
G-8.15	2.403	0.036	2.439
G-8.20	3.930	0.033	3.963

The filtration coefficient was determined according to the Darcy law from the formula below (equation 2) [15]:

$$k = Q / F \cdot I \tag{2}$$

where k- the filtration coefficient, (m/s), Q – quantity of water (m^3/s), F – section area (m^2), I- hydraulic gradient.

The filtration coefficients of the tested binders and phenomena possible migration of pollutants' during the year are presented in table 9. Filtration time was measured by using system shown on figure 1.

Table 9. Determination of the filtration coefficient and possible migration of pollutant during the year.

Sample	Filtration time (s)	Filtration coefficient (m/s)	Migration (m)	Comments
G-4.10	485710	$7.13 \cdot 10^{-10}$	0.023	-
G-4.15	512800	$5.90 \cdot 10^{-10}$	0.019	-
G-4.20	598750	$5.65 \cdot 10^{-10}$	0.018	-
G-5.10	98760	$4.00 \cdot 10^{-10}$	0.013	-
G-5.15	90140	$3.20 \cdot 10^{-09}$	0.101	-
G-5.20	89980	$4.07 \cdot 10^{-09}$	0.128	-
G-6.10	9780	$3.67 \cdot 10^{-08}$	1.160	-
G-6.15	142500	$2.72 \cdot 10^{-09}$	0.086	-
G-6.20	139840	$2.83 \cdot 10^{-09}$	0.090	-
G-7.10	4260	$9.08 \cdot 10^{-08}$	2.860	-
G-7.15	5487	$6.65 \cdot 10^{-08}$	2.100	-
G-7.20	6240	$7.05 \cdot 10^{-08}$	2.220	-
G-8.10	3480	$1.97 \cdot 10^{-07}$	**6.210**	The possibility of unsealing the barrier
G-8.15	3870	$1.58 \cdot 10^{-07}$	**4.980**	The possibility of unsealing the barrier
G-8.20	4100	$1.46 \cdot 10^{-07}$	**4,600**	The possibility of unsealing the barrier

Investigations have shown that the values of the filtration coefficient are strongly dependent on the content of contaminated soil and the amount of cement.

5. Conclusions

- The pH values of the filtrates do not meet the standards, because in all cases it was above the maximum allowable limit which also indicates the presence of alkalies.
- The systems initially are chemically unstable, the Eh potential in all cases is weakly reducing, but it will change with time in the oxidizing direction. In the case of barriers, this is an adverse (corrosive) impact on surrounding soils.
- The Na content is in the permissible concentrations (800 mg/l), but the cations are not compensated by anions (low salinity, the presence of insoluble compounds) which results in the formation of active OH^- groups, and hence the potential corrosive effect and reducing the character of Eh. The amount of leached out sodium depends on the primary composition of the sample.
- Concentrations of heavy metal cations do not exceed standards, and even remain on the limit of determination.
- In bulk of the binders about 80% of all aliphatic hydrocarbons stay accumulated, approx. 20% are leached out. However, they are still below the permissible standard requirement (15 mg/kg).
- The content of aromatic hydrocarbons in filtrates is below the limit of detection. Accumulation capacity of the binder in all cases very high. There are no threats caused by the possible migration of aromatic hydrocarbons to groundwater and soils.
- Migration of pollutants strongly depends on the filtration coefficient and the initial concentrations of the tested substances. In extremely unfavorable conditions, the spread of analyzed pollutants it could be as high as 6m per year.

Summary

- Generally the tested binders show a very good level of immobilization of dangerous substances in the soil, mixtures with soil contaminated from Glasgow can be used in barriers, under several conditions:
- The percentage of soil should not be higher than 50% due to environmental standards the injection should be at a depth below 2m depth according to Regulation from 2014 [12].
- The problem may be strong the sodium's migration through barrier, however, its filtration coefficient of 10^{-10} m/s significantly limits it. The permeability of the binder usually increases if it is dried (additional monitoring).

Research conducted on heavily contaminated ground from the Glasgow landfill site does not limit the use of clay-cement binders to protect other landfills. However, due to their individual character, testing should always be carried out.

Acknowledgment

Badania wykonano w ramach projektu "Ocena przydatności technologii spoiw mineralnych dla obiektów hydrotechnicznych oraz inżynierii środowiska na podstawie dotychczasowych i proponowanych rozwiązań" UDA.POIG.01.03.01-00-083/12

References

[1] Kaczyńska G, Borowik A and Wyszkowska J 2015 *Water Air Soil Pollut.* **226** pp 372
[2] Rakowska J, Radwan K, Ślosorz Z, Pietraszak E, Łudzik M and Suchorab P 2012 *Usuwanie substancji ropopochodnych z dróg i gruntów* (Państwowy Instytut Badawczy: Józefów)
[3] Cebula J and Rajca M 2014 Oczyszczanie gleb i gruntów (Wydawnictwo Politechniki Śląskiej: Gliwice)
[4] Liu L, Li W, Song W and Guo M 2018 *Sci. Total Environ.* **633** pp 206-19
[5] Wuana RA and Okieimen FE, 2011 *ISRN Ecology* **402647**
[6] Radziemska M 2018 *Catena* **163** pp 123-9
[7] Ou J, Li H, Yan Z, Zhou Y, Bai L, Zhang Ch, Wang X and Chen G 2018 *Sci. Rep.* **8 4618**
[8] Maiviya R and Chaudhary R 2006 *J. Mater. Cycles Waste Manag.* **8** pp 78 - 87
[9] Kuś R, Janik G, Izak P, Wójcik Ł and Słowikowski D *Patent* PL 216864 B1
[10] Izak P 2015 *Reologia w ceramice* (Kraków: Wydawnictwa AGH)
[11] Sonntag H 1982 *Koloidy* (Warszawa: Państwowe Wydawnictwo Naukowe)
[12] Rozporządzenie Ministra Środowiska z dnia 18 listopada 2014 r. w sprawie warunków, jakie należy spełnić przy wprowadzaniu ścieków do wód lub do ziemi, oraz w sprawie substancji szczególnie szkodliwych dla środowiska wodnego
[13] Merkel B and Sperling B 1998 *Hydrogeochemische Stoffsysteme* **II** Komissionsvertrieb Wirtschafts- und Verlagsgesellschaft Gas und Wasser GmbH.: Bonn)
[14] Król A 2007 *Przemysł Chemiczny* **10** pp 971- 3
[15] Kulma R 1995 *Podstawy obliczeń filtracji wód podziemnych* (Kraków Wydawnictwo AGH)

IOP Conf. Series: Materials Science and Engineering **427** (2018) 012015 doi:10.1088/1757-899X/427/1/012015

Impact of HPGR pressing force on the level of product recovery

D Saramak

AGH University of Science and Technology, Faculty of Mining and Geoengineering, Cracow, Poland,

dsaramak@agh.edu.pl

Abstract. Results of investigations show that HPGR operating pressure is proportional to the metal recovery, however too extensive pressure values have relatively lower impact on the increase of metal recovery. Particle size distribution of HPGR feed has a lower impact on the recovery level than HPGR pressure which was also investigated in the paper. Flotation tests of each HPGR product were also performed in order to investigate potential impact of various conditions of HPGR process runs on the metal recovery.

1. Introduction

The high-pressure grinding technology, invented and developed by Professor Schoenert in 70's [1], is currently regarded as one of the most efficient methods of hard ore comminution. Since then the technology was introduced into raw materials treatment circuits in many mining and mining-related industrial sectors, including iron and non-ferrous metal ores processing (copper, platinum, gold), as well as cement and limestone flour production [2]. Considerable number of works concerning modeling and optimization of High-Pressure Grinding Rolls devices (HPGR) operation were also presented [3,4,5,6,7]. Nowadays high-pressure grinding rolls devices are considered as well know technology with over 500 sold machines, according to various sources.

Results of various investigations show that the major benefit from industrial application of high-pressure comminution devices is the lower energy consumption . There are also observed some positive effects in terms of greater comminution degree values, as well as higher throughput, when the feed material is treated by HPGR instead of conventional crushing devices [8,9].

Results of various investigations show that higher recoveries of useful mineral in flotation operations can be expected, for products crushed upstream in HPGR devices. Positive effects of HPGR can be observed in leaching [10,11], flotation [12,13] and gravitational separation [14].

The aim of this study is to investigate the impact of operational pressing force on comminution effects in HPGR. Effect of operational pressure was investigated through comminution degrees, as well as through flotation recoveries obtained for HPGR comminution products.

2. Operational principles of HPGR

The HPGR operation is similar to the roller crusher work. The device consists of two counter-rotating rolls; the one (fixed roll) is settled in the frame in the fixed position, while the second one (floating roll) is allowed to move horizontally. The rolls are driven by two separate motors connected to the roll shafts

IOP Conf. Series: Materials Science and Engineering **427** (2018) 012015 doi:10.1088/1757-899X/427/1/012015

with the use of gear reducers. The comminution process in HPGR is evoked by the pressing the floating roll towards the fixed one. During the movement of rolls, the feed material is choke-fed and dragged into the nip zone of working chamber (figure 1). Together with decreasing of the distance between rolls, the pressing force significantly increases, causing the breakage of the individual particles structure and formation of micro-cracks. This results in a large reduction of the product particle size, which comes out during the next processes, usually the tertiary grinding in mills.

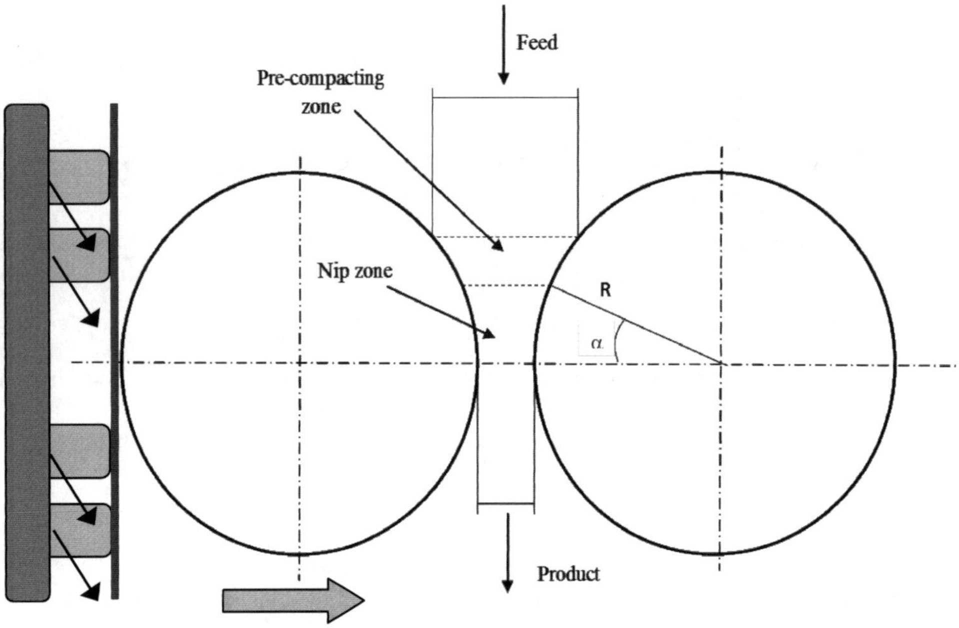

Figure 1. Scheme of comminution in HPGR.

The main parameter determining the process results is the operating pressure of HPGR. For high values of pressing force, the grater comminution intensity is observed, but too high pressure causes a formation of harder flakes which do not get de-agglomerated on the screen after HPGR. In operational practice the Fsp index (specific pressing force) is calculated on the basis of pressure value. It depends of the pressure and dimensions of HPGR device, and can be calculated from following formula:

$$F_{sp} = \frac{F}{D*l*1000} = \frac{\frac{d^2*\pi}{4}*n*P_h}{D*l*1000} [N/mm^2] \quad (1)$$

where:
F – pressing force [kN]
D – roll's diameter [m]
d – pistons' diameter [m]
n – number of pistons (usually four)
Ph – hydraulic pressure [kPa]

The speed of rolls is the second main operating parameter influencing the process. It can be measured both in [m/s] and in the number of revolutions per minute. This parameter has a crucial influence on the process capacity, determining the throughput of HPGR [9].

$$Q = l * s * v * \rho * 3600 [Mg/hour] \quad (2)$$

where:
Q – throughput [Mg/hour]

l – working length of the roller [m]
s – gap [m]
v – roller speed [m/s]
ρ - product density [Mg/m^3]

There are also a number of secondary operational parameters of HPGR device, but in this paper only the pressing force is taken into account. Results of various investigations and operational practice of HPGR show that their influence on crushing process is of lower significance.

3. Material and methods

The purpose of the paper was to investigate the effectiveness of HPGR operation under various values of Fsp. The HPGR effectiveness can be evaluated through the obtained comminution ratio of crushing products. In HPGR products that achieve a greater fineness the more intense liberation of useful mineral from the gangue can be observed. The more intense liberation, in turn, affects the efficiency of separation process. Therefore the fineness of HPGR product is to some extent positively correlated with the process efficiency. It is possible to determine the relationship between operational pressing force and comminution degree (Sx) of HPGR product, however, individual results greatly depend on physical and mechanical properties of feed material.

Increasing of pressing force affects the press productivity, because higher force acting on the floating roll move it towards the fixed one and, as a result, the gap is decreasing, too. It can be balanced, however, with higher speed of rolls.

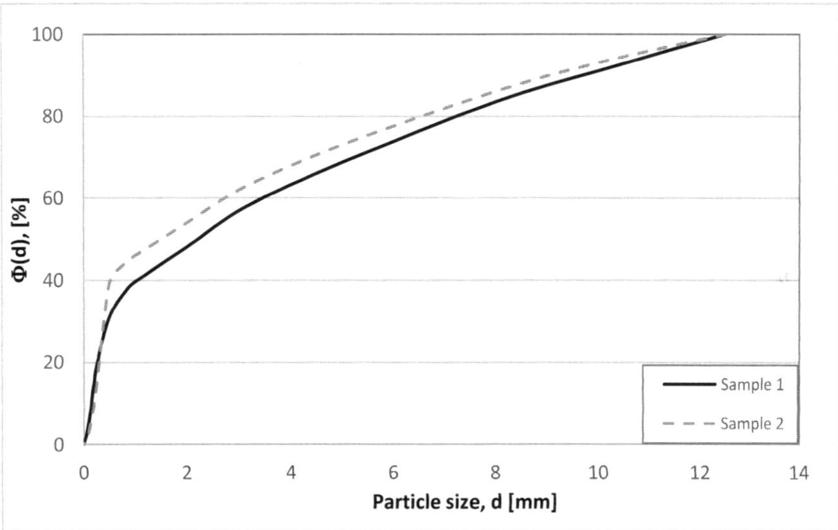

Figure 2. Particle size distribution curves of sample 1 and 2.

Sulfide copper ore was used in entire investigative programme, and two samples were comminuted in HPGR device (figure 2). Experimental programme aimed at testing the operational pressure (Fsp) of HPGR, and four values of Fsp were applied: 2, 3, 4 and 5 N/mm^2. Flotation process indicators for each HPGR crushing products were also determined.

4. Results of experiments

In the first stage of analysis the fineness of HPGR products comminuted in the press under various values of operating pressure, was evaluated. Results are presented in figures 3 – 4.

IOP Conf. Series: Materials Science and Engineering **427** (2018) 012015 doi:10.1088/1757-899X/427/1/012015

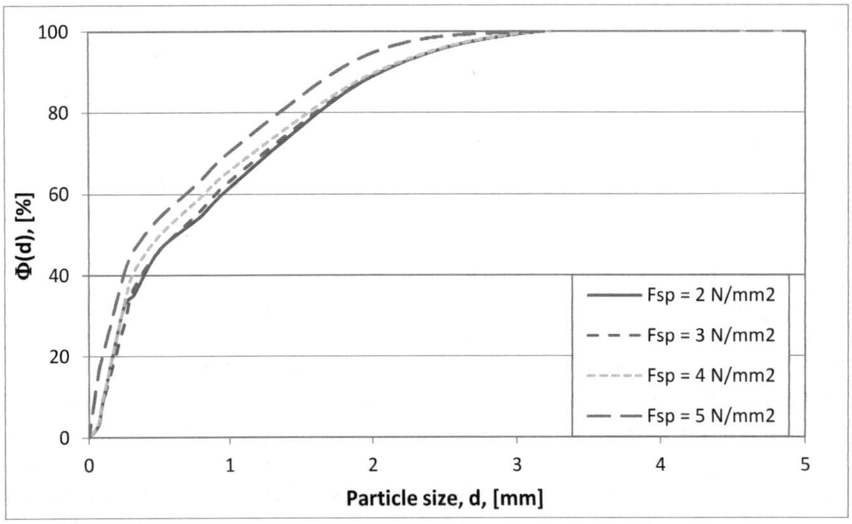

Figure 3. Particle size distribution of HPGR products for sample 1.

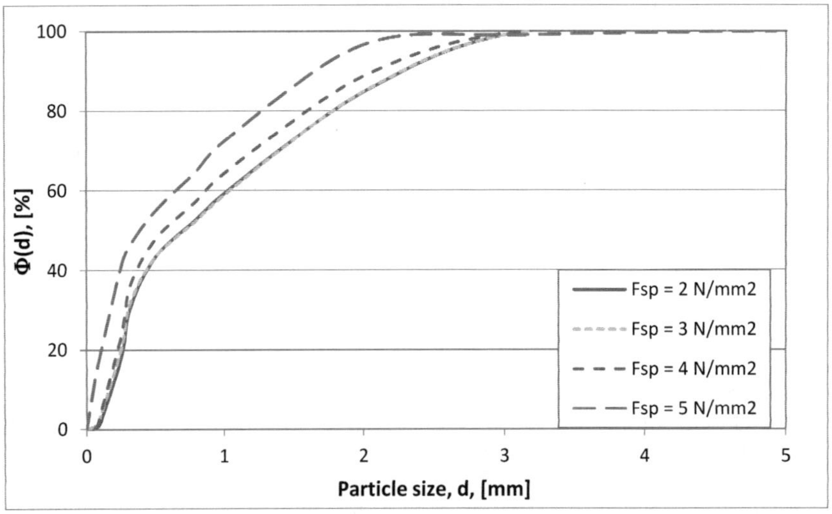

Figure 4. Particle size distribution of HPGR products for sample 2.

It can be noticed that in case of sample 1 increasing in operating pressure value results in obtaining finer particle size composition of products, but differences were not as high as in the case of sample 2. Increased yield of finest size fractions in feed helped to obtain the finer HPGR product. The more detailed information, however, can be received after determination of comminution degrees for each crushing product. Such values were determined for each HPGR product, and results are presented in figures 5 and 6.

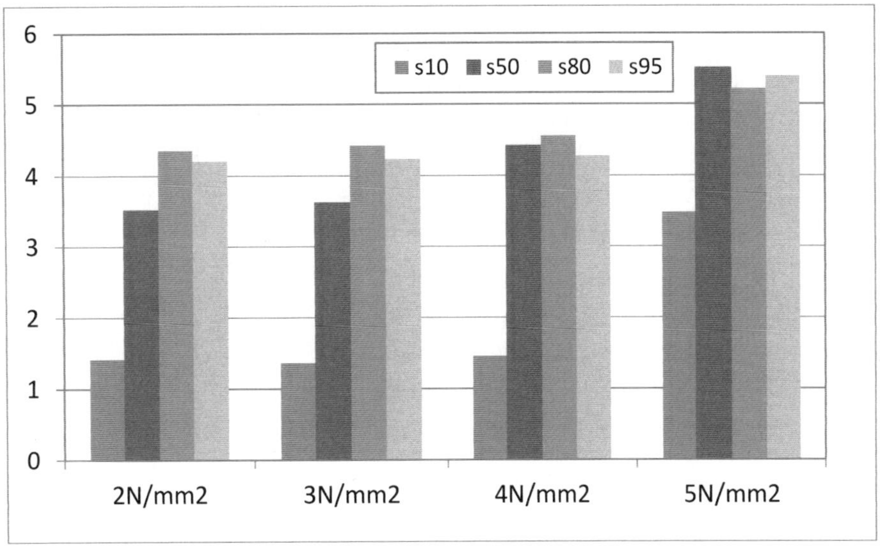

Figure 5. Comminution degree values obtained in HPGR for sample 1.

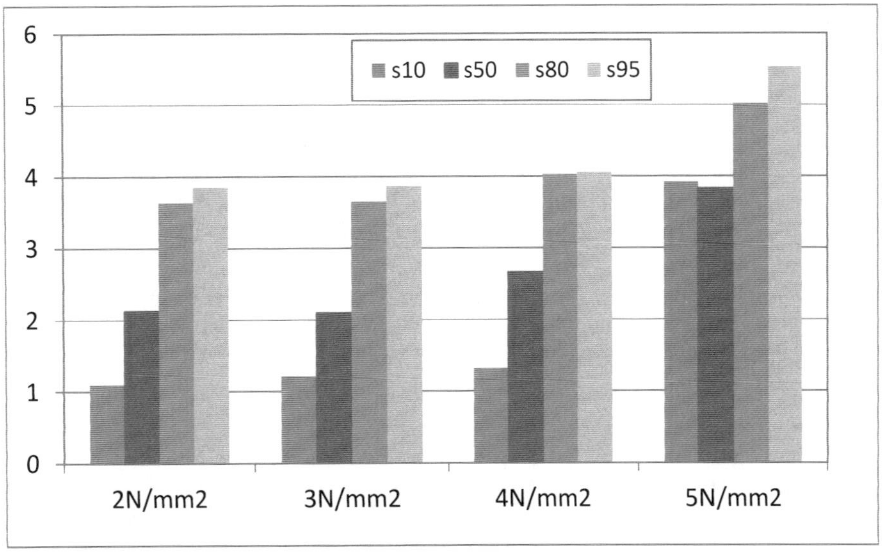

Figure 6. Comminution degree values obtained in HPGR for sample 2.

Results of investigations show that obtained values of comminution degrees are generally higher for the sample 1. Differences are especially noticeable for the average comminution degree S50. For sample 1 theses values exceed 4 and 5, for operating pressure values 4 and 5 N/mm^2, respectively, while for sample 2 analogous values were within the range 2 and 3. Only for operating pressure 5 N/mm^2, average comminution degree exceeds 4.

Two issues are worth noticing when the comminution effect of a given comminuting device is analyzed: 10 percent and maximum comminution degree values. The first one shows the potentials of breakage of the feed material into finest particles, while the second one (Smax) informs about intensity of disintegration of the coarsest particle size fractions. If the S10 value is high, that indicates a high content of fines in HPGR product. These particle size fractions may be divided from the product and directed either directly for separation operations, provided they have achieved a suitable fineness, or for re-grinding circuit. By-passing the downstream stages of crushing/grinding causes lower load of the circuit and results in decreasing the energy consumption. For sample 2, that contains higher amount of fines, S10 value obtained for 5 N/mm^2 is more favorable than for sample 1. That may indicate that for this type of ore a slight increase in fines in HPGR feed may be beneficial in terms of production of fines.

The recovery of useful mineral in flotation process was also assessed. Laboratory flotation tests were performed for selected two HPGR products from each sample: comminuted at 3 and 4 N/mm^2. The results are presented in table 1.

Table 1. Results of flotation tests

Flotation indices	Sample 1	Sample 2
	No HPGR	
ε [%]	80.4	86.6
ϑ [%]	0.23	0.15
β [%]	21.2	15.4
	HPGR product. Fsp = 3 N/mm2	
ε [%]	82.9	86.4
ϑ [%]	0.23	0.16
β [%]	19.4	15.3
	HPGR product. 4 N/mm2	
ε [%]	84.6	87.6
ϑ [%]	0.24	0.16
β [%]	15.4	15.2

Flotation tests for both samples comminuted in conventional crushers were also performed as reference tests. Analysis of date presented in Table 1 shows that more favourable values of recovery were obtained for sample 1. An increase of recovery from 80.4% to 82.9% (2.5% higher) for Fsp=3 N/mm^2 and to 84.6% (4.2% higher comparing to conventionally crushed product) for Fsp = 4 N/mm^2, were observed. Analogous values for sample 2 equaled -0.2% and 1%, respectively. Higher recoveries obtained for HPGR products are mainly due to more intense liberation of useful component. However this effect is more significant for the sample 1.

5. Summary and conclusions

The aim of the paper was to assess the impact of operational pressure in HPGR and particle size composition of feed on effectiveness of downstream separation process. The obtained results show that value of HPGR pressing force Fsp influences positively the comminution degree of HPGR products. Fineness of comminution products was higher with the increase of operational pressure, yet further increasing of Fsp resulted with relatively lower increases of comminution ratio. That may indicate, that it is possible to determine a kind of optimal value of pressure that maximizes effects of the press in technological (fineness) and economical (energy consumption) aspects. This issue will be the subject of further investigations.

Results of investigations also confirmed that exists a relationship between pressure value and recovery of useful mineral. For the type of material under investigation an application of HPGR into processing circuit may be beneficial, because the achieved recoveries of useful mineral for all HPGR products were higher. It can be also seen that the sample with average content of finest particle size fractions responds better in terms of recovery together with increasing of Fsp. However, the above issues need to be investigated more thoroughly in order to determine more general relationships and will be continued in future.

IOP Conf. Series: Materials Science and Engineering **427** (2018) 012015 doi:10.1088/1757-899X/427/1/012015

Acknowledgments
The paper was the result of AGH UST statutory work no. 11.11.100.276.

References

[1] Schönert, K 1979 Aspects of the physics of breakage relevant to comminution *Fourth Tewksbury Symposium* University of Melbourne

[2] Gawenda T Naziemiec Z and Walerak A 2013 Investigations on selected properties of aggregates from ceramic waste and concrete produced with their participation *Annual Set of Environmental Protection* **15 (3)** pp 2003-21

[3] Daniel M J and Morrell S 2004 HPGR model verification and scale-up *Minerals Engineering* **17 (11/12)** pp 1149-61

[4] Naziemiec Z and Saramak D 2015 Application of partition curves in the assessment of mineral products classification processes *MINING SCIENCE* **22** SI: 1 pp 119-29

[5] Saramak D 2012 Optimizing the performance of high-pressure grinding roll based ore enrichment circuits, *Gospodarka Surowcami Mineralnymi-Mineral Resources Management* **28 (4)** pp 87-99

[6] Saramak D and Kleiv R A 2013 The effect of feed moisture on the comminution efficiency of HPGR circuits *Minerals Engineering* **43-4** pp 105-111

[7] Tavares L M 2005 Particle weakening in high-pressure roll grinding *Minerals Engineering* **18** pp 651-7

[8] Hilden M and Suthers S 2010 Comparing energy efficiency of multipass high pressure grinding roll (HPGR) circuits *XXV International Mineral Processing Congress (IMPC), Proceedings* The Australasian Institute of Mining and Metallurgy Brisbane pp 801-11

[9] Naziemiec Z and Saramak D 2009 The analysis of changes in pressing force impact on material in HPGR compression zone (in Polish) *Górnictwo i Geoinżynieria AGH Journal* **33** (4), 221-34

[10] Klingmann H L 2005 HPGR benefits at Golden Queen Soledad Mountain gold heap leaching project *Randol Innovative Metallurgy Forum* Perth

[11] Patzelt N Knecht H and Baum W 1995 Advances in POLYCOM High-Pressure Roll Grinding of refractory gold ores *Randol Gold Forum* Perth pp 107-23

[12] Saramak D Młynarczykowska A and Krawczykowska A 2014 Influence of a high-pressure comminution technology on concentrate yields in copper ore flotation processes *Archives of Metallurgy and Materials* **59 (3)** pp 951-5

[13] Smit I 2005 Bench scale ore characterization using the High Pressure Grinding Roll *Randol HPGR Workshop 2005* Perth

[14] Gray S 2005 Gekko and HPGR – What are they doing? *Mini-presentation at Randol HPGR Workshop Perth* Perth

IOP Conf. Series: Materials Science and Engineering **427** (2018) 012016 doi:10.1088/1757-899X/427/1/012016

Influence of HPGR operation on the reduction of Bond's working index of crushing product

D Saramak and Z A Iwanów

AGH University of Science and Technology, Faculty of Mining and Geoengineering, Cracow, Poland

dsaramak@agh.edu.pl

Abstract. Article concerns the issues of ore comminution in HPGR. A series of laboratory tests were carried out for three lithologic types of sulphide ore. Four values of operating pressing force were accepted. Bond's index value, determined with using a standard Bond method, was also determined for all HPGR products. There were also determined functional relationships between Bond index and operational pressure. Results of investigation showed, that application oh HPGR for that type of feed resulted in lowering the value of Bond index. The relationship between the pressing force and Bond index can be approximated with hyperbolic function.

1. Introduction

Comminution constitute the basic operation in technological circuit of mineral processing. It's primary role is to reduce the particle size composition of the crushing product, but also the change in shape of individual particles, an increase of the specific surface area of the material, and the liberation of the useful components from the gangue. It can also affect the mechanical properties of the raw material, depending on the type of comminution device applied. The processes of raw materials comminution is the most energy-intensive operation in ore processing circuits. The amount of energy spent in crushing and grinding operations strictly depends on the properties of the raw material, a type of crushing devices applied and the adopted technology [7].

From the scope of assessment the crushing technology effectiveness, a key issue seems to be to determine the energetic demands in comminution operations. The most common energetic theories of comminution are the Rittinger, Kick and Bond laws. There are also some other methods of determination of theoretical energy demands for comminution [1], but standard Bond's method is still treated in industrial practice as an universal way of determination the energy consumption in crushing and grinding operations. The theoretical energy demand index W, is determined based on the Wi Bond index, which, in turn, determines the material's resistance to grinding. For the ball mill W_i can be determined from the following formula:

$$W_i = 1.1 \cdot \frac{44,5}{P^{0,23} G^{0,82} \left(\frac{10}{\sqrt{P_{80}}} - \frac{10}{\sqrt{F_{80}}}\right)} \quad (1)$$

where:

W_i - Bond's working index, [kWh/Mg]

P - control mesh size (3350), [μm]

IOP Conf. Series: Materials Science and Engineering **427** (2018) 012016 doi:10.1088/1757-899X/427/1/012016

G - the mass of the undersize product generated per one revolution of the mill, [g/rev]
F_{80} - eighty percent particle of the feed, [μm]
P_{80} - eighty percent particle of the product, [μm]

Following simplified formula can be also applied in industrial practice:

$$W_i = \left(\frac{16}{G^{0,82}}\right) \cdot \sqrt{\frac{P}{100}} \cdot 1.1 \tag{2}$$

Decreasing the value of Bond work index for a given type of material is possible, through application of novel comminution techniques. A high-pressure grinding technology, which is currently one of the most effective methods of ore comminution in terms of low energy consumption, can be a good example of such innovation. The HPGR crushing product has micro-cracks within the structure of individual grains, which reduces its grinding resistance in downstream grinding stages, and that contributes to lower energy consumption [2,11]. Results of various investigations show that application of the roller press devices in hard raw material processing circuits may result in a reduction of the grinding energy consumption up to 50% [2]. In general, the Bond working index value for HPGR products can be reduced from 10 to 30%, depending on the type of material. At the same time, an increase of the useful component recovery in flotation concentrates by as much as even several percent, was also noticed [3]. At the same time, these devices are characterized by lower investment and operating costs, high efficiency combined with a small footprint and low vibration and noise emission [4, 8].

The aim of the paper is to determine a relationship between operational parameters of HPGR device and the Bond work index value for the HPGR products. Determining the relationship between the value of operating pressure in the press and the Bond work index (using a standard Bond method) for the crushing product can be helpful in effective control of the grinding process in a high-pressure grinding roll device.

2. Material and methods

High-pressure comminution tests were carried out for copper ore, while three samples with different lithological compositions were under examination. Each sample contained variable contents of sandstone and carbonate fractions, while the shale contents were similar (table 1). The investigative programme included laboratory crushing tests in HPGR device, carried out in NTNU University in Trondheim (Norway).

Table 1. Lithological composition of the samples.

	Sample 1	Sample 2	Sample 3
Sandstone, [%]	48	65	27
Carbonate, [%]	41	24	62
Shale, [%]	11	11	11

The HPGR feed, with maximum particle 10 mm, was upstream crushed in a jaw crusher. Particle size composition of three samples are presented in figure 1.

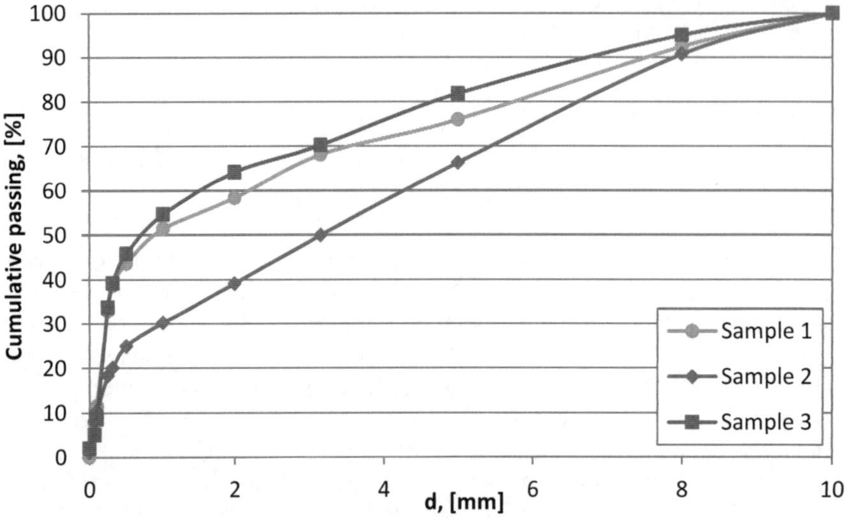

Figure 1. Particle size distribution curves for each sample.

Each sample contained variable yields of individual lithological fractions. The sandstone fraction in material 1 constituted 48%, while in samples 2 and 3: 27% and 65%, respectively. Each material sample was then split into six smaller samples, 6 kg, each. The base sample was used for determination the Bond index value for the feed, the four next samples were crushed in HPGR device under various values of operational pressing forces. The remained sixth sample was stored as a reserve. The obtained results of Bond's values for three types of feed are presented in table 2.

Table 2. Values of Bond's work indices for three types of feed used in investigations.

	Sample 1	Sample 2	Sample 3
Bond's Index Wi, [kWh/Mg]	14.1	11.2	14.9

As it can be seen from the table 2, the value of Bond's index shows strong relationships with the sandstone content in the feed material. However even though the correlation coefficient value is high (r = 0.967) and it is statistically significant on the probability level $1 - \alpha = 0.90$,the statistical sample should contain larger number of measurements, in order to formulate stronger conclusion. Nevertheless, the content of sandstone fractions in the material might influence on its comminution characteristics.

3. Results of investigations

Comminution tests in HPGR were carried out for four different values of the pressing force F: 4, 6, 8 and 10 kN. The rolls diameter was 200 mm, while the width: 100 mm. Four comminution products were obtained for each material sample. Particle size distribution curves of HPGR product for each sample were presented in figures 2 -4.

IOP Conf. Series: Materials Science and Engineering **427** (2018) 012016 doi:10.1088/1757-899X/427/1/012016

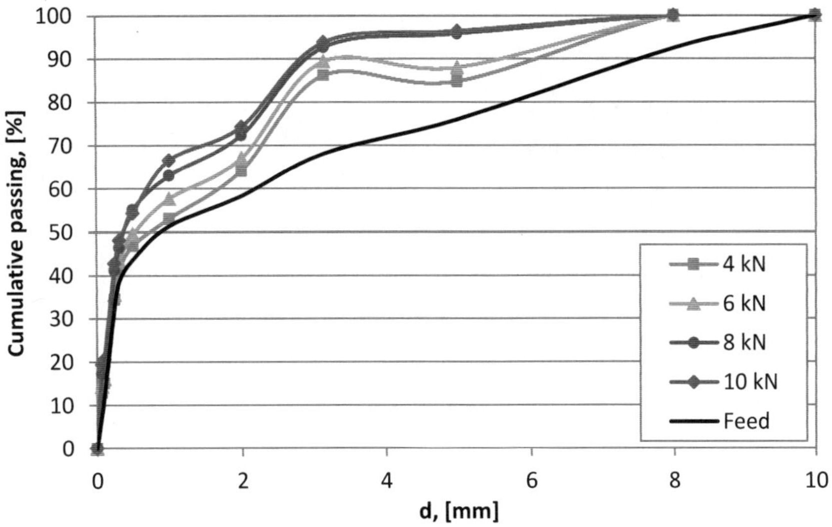

Figure 2. Results of comminution in HPGR for sample 1.

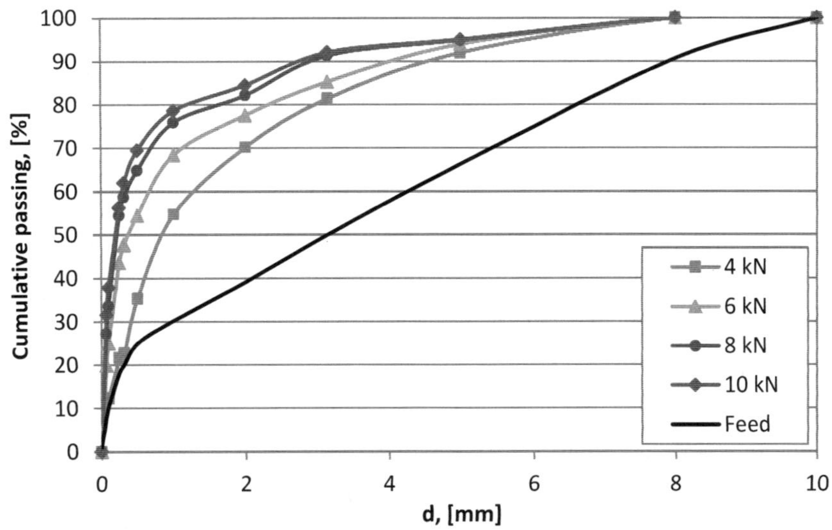

Figure 3. Results of comminution in HPGR for sample 2.

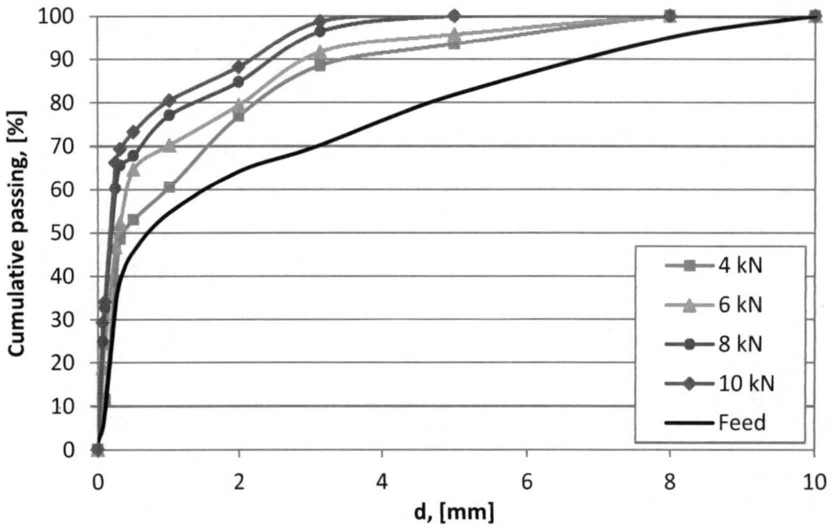

Figure 4. Results of comminution in HPGR for sample 3.

In the next stage of investigations the values of Bond's work indices were determined for each HPGR samples. Results of the twelve values of Bond's work indices are presented in table 3.

Table 3. Values of Bond's work indices for all HPGR products.

F [kN]	Sample 1	Sample 2	Sample 3
4	12.6	11.2	12.9
6	11.4	10.3	11.8
8	10.9	9.8	11.3
10	10.2	9.0	10.9

Comparing values presented in tables 2 and 3 it can be seen, that the most favorable savings on the grinding energy were obtained for the material sample 1 (27.7%), but for all samples the energy savings were higher than 20%. For the material sample 2, the energy savings equaled 24.1%, and for the material 3: 26.8%. In general, the more favorable energy savings were obtained for the materials with increased content of sandstone fractions.

In the last stage of investigations there were determined the regression models describing the Bond's values of each material as a function of operational pressing force F. Approximation was performed with using the hyperbolic function:

$$y = \frac{a}{x^b} \tag{3}$$

where:
y – Bond index value,
x – HPGR operational pressing force,
a, b – coefficients.

The type of approximation function used for determination of the models have its justification in literature [9,10]. The results are presented in table 4 and figure 5.

IOP Conf. Series: Materials Science and Engineering **427** (2018) 012016 doi:10.1088/1757-899X/427/1/012016

Table 4. Functional models of Bond values as a function of operational pressing force in HPGR.

Type of material	Model	R2 value
Sample 1	$y = \frac{14.17}{x^{0.19}}$	0.992
Sample 2	$y = \frac{11.31}{x^{0.15}}$	0.957
Sample 3	$y = \frac{14.87}{x^{0.20}}$	0.997

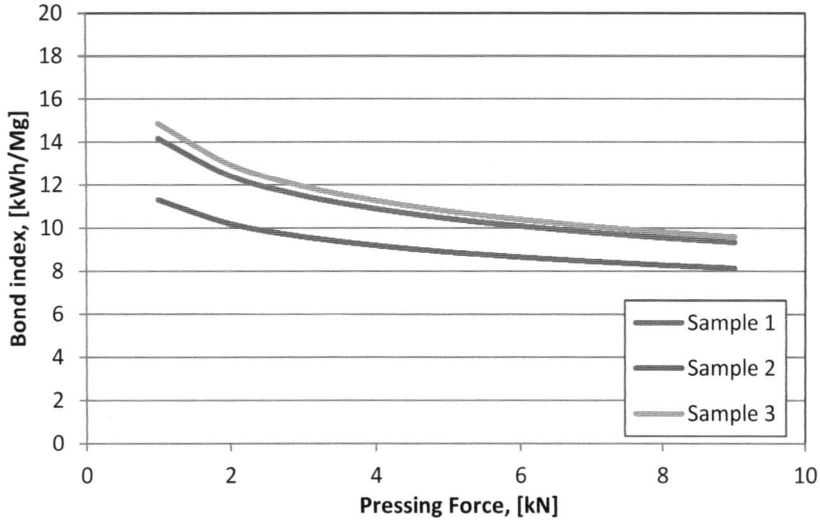

Figure 5. Functional models of Bond values.

All approximation functions describe the phenomenon well enough, R-square values are higher than 95% and they are even higher for the samples with higher content of sandstone fractions. There are also visible some relationships between values of a and b coefficients and the type of material. Together with an increase of sandstone fraction content, the coefficient increases too, while the b coefficient decreases. The above relationships are true for all materials under examinations.

4. Results and conclusions

The results of investigations show that operating pressure in HPGR significantly influences on reduction of Bond's work index. Similar results were obtained for both types of material, however the content of sandstone fraction in feed results in various values of Bond's working index Wi. Higher reduction in Bond's index value were observed for the material with higher sandstone content. For the material crushed in HPGR at pressing force 10 kN, Wi reduction achieved 28%, while for the sample 2, where the content of sandstone was the lowest, the reduction of Wi achieved 20%. Nevertheless, the results of investigations show that application of HPGR device for that type of ore can generate significant reductions in potential energy consumption in downstream grinding processes.

The results of second part of investigation showed that it is possible to build a suitable functional relationship between operating pressure in HPGR and Bond index Wi. The shape of relationship indicates that together with increasing of operating pressure in HPGR further reductions of Wi index are lower. Therefore, it can be determined an optimal value of operating pressure, for which the potential effect in terms of energy savings are the most favourable. This issue will be the subject of further investigations.

References

[1] Foszcz D, Gawenda T and Krawczykowski D 2006 Porównanie rzeczywistego i wyznaczonego teoretycznie zużycia energii dla młyna kulowego *Górnictwo i Geoinżynieria* **33/3/1** pp 79-90

[2] Gawenda T 2009 Główne aspekty rozdrabniania twardych surowców mineralnych w wysokociśnieniowych prasach walcowych *Górnictwo i Geoinżynieria* **33/4** pp 89-99

[3] Gawenda T 2015 *Zasady doboru kruszarek oraz układów technologicznych w produkcji kruszyw łamanych* (Kraków: Wydawnictwa AGH) pp 59-61

[4] Saramak D 2010 Analiza powiązań pomiędzy parametrami technologicznymi pras walcowych z wykorzystaniem analizy czynnikowej *Górnictwo i Geoinżynieria* **34/4/1** pp 251-63

[5] Fuerstenau D W, Shukla A and Kapur P 1991 Energy consumption and product size distributions in choke-fed high-compression roll mills *International Journal of Mineral Processing* **32** pp 59-79

[6] Aydogan N A and Ergun L 2011 Circuit high pressure grinding rolls (HPGR) in a cement grinding plant Part 2: Effect of operating parameters *ZGK International* **01/2011**

[7] Naziemiec Z and Saramak D 2015 Application of partition curves in the assessment of mineral products classification processes *Mining Science* **22** pp 119-29

[8] Saramak A, Naziemiec Z and Saramak D 2016 Analysis of noise emission for selected crushing devices *Mining Science* **23** pp 145-54

[9] Saramak D 2011 Technological issues of High-Pressure Grinding Rolls operation in ore comminution processes, *Archives of Mining Sciences* **56(3)** pp 517-26

[10] Daniel M J and Morrell S 2004 HPGR model verification and scale-up *Minerals Engineering* **17** pp 1149–60

[11] Barani K and Baloch H 2016 Comparative study on the effect of using conventional and high pressure grinding rolls crushing on the ball mill grinding kinetics of an iron ore *Physicochemical Problems of Mineral Processing* **52** pp 920-31

IOP Conf. Series: Materials Science and Engineering **427** (2018) 012017 doi:10.1088/1757-899X/427/1/012017

Investigations of zinc recovery from metallurgical waste

D Saramak, D Krawczykowski and T Gawenda

AGH University of Science and Technology, Faculty of Mining and Geoengineering, Department of Environmental Engineering and Mineral Processing, Cracow, Poland

dsaramak@agh.edu.pl

Abstract. Approximately 7 million tonnes of zinc is annually used for corrosion protection in processes of continuous galvanization and the hot dip processes. Two types of waste are obtained from the process: zinc ash and hard zinc and each of them constitute around 10% of the feed weight. Paper concerns the method of possible separation of metallic zinc from the non-metallic zinc compounds from the zinc ashes – the waste product of hot dip galvanization process. The investigative programme covered the comminution tests of the feed (zinc ash) in selected crushers: jaw crusher, roller crusher and impactor. Results of investigations show that application of individual crushing device results in various breakage of the feed sample. Impact force caused the highest increase in finest particles size fractions of the product, while metallic and non-metallic fractions were comminuted differently. The compression force, in turn, caused the selective comminution, i.e. the breakage of metallic zinc was observed in a higher degree.

1. Introduction and methodology

World production of metallic zinc during last decade shows a slight increase and in recent years its consumptions reach 14 million of tonnes. Main directions of zinc utilization include processes of galvanization, production of zinc alloys, brass and bronze, chemicals, as well as other semi-products based on the zinc. The major share in global zinc consumption, however, constitute galvanization process[1]. Two main types of zinc galvanization are: hot-dip galvanizing (HDG) and electro-galvanizing.

In the HDG process two kinds of residues are occurring. The zinc ash, that is formed on the surface of the HDG bath, and hard zinc, formed on the bottom of galvanizing kettle [2,3,4]. The zinc ash consists mainly of metal oxides, sulphides and chlorides and it has to be removed continuously before the galvanized goods are extracted from the bath [5]. These residues contain up to 90% of metallic zinc, depending on particle size fraction (table 1).

The ash zinc waste contain the zinc in metallic and non-metallic forms. In order to recover the metal from ash zinc, the waste is directed to smelting process [6]. However, smelting the whole charge results in higher metal loss in the process. Therefore, it is advisable to separate the metallic form of zinc from the waste, in order to reach higher effectiveness of zinc recovery in further metallurgical processes. It is also worth to remember that finer material, with too much gangue and complex compounds cannot be recovered with high effectiveness in pyrometallurgical processes. In such cases hydrometallurgical treatment might be the most suitable method of processing [7]. Both forms of zinc have different physical and mechanical properties. It was observed under microscope that light grey irregular particles of metallic zinc are bound with finer dark grey brownish areas that represent the zinc ash. On that basis it is assumed that grindability properties of both form of zinc could be different. The zinc ash particles can break more easily than particles of metallic zinc, and the crushing product may contain coarser particles of metallic zinc along with finer particles of zinc ash and non-metallic fractions. Further screen separation [8] can be applied for comminution products.

Taking into account the above, the main research aim of the paper is investigation of selective comminution of zinc ash waste, therefore comminution process should be properly adjusted in order to minimize the content of size fractions below 0.5 mm. Three different crushers were tested: jaw crusher, hammer crusher and roll crusher.

2. Experimental

The feed materials for experiments constitute the zinc ash waste from hot-dip process, with maximum particle size below 12 mm. Initial characteristics of the feed material was performed and results were presented in table 1.

Table 1. Characteristics of feed material.

Particle size fraction [mm]	Yield [%]	Metallic zinc content [%]
>3	35	>90
0.1 - 3	50	60
Dust (<0.1 mm)	15	30-35

Results of particle size analysis show that particles coarser than 3 mm constitute approximately 35-40%. Additionally, this material is of a very high purity in terms of metallic zinc content. This size fraction of material can be separated from the entire feed and proceed for the metal zinc recovery by means of pyrometallurgical methods [9]. The finest size fraction, in turn, contain 30% of zinc roughly, therefore processing of this part of feed material might be uneconomic. An average zinc content in the feed material equals 67%, approximately (figure 1).

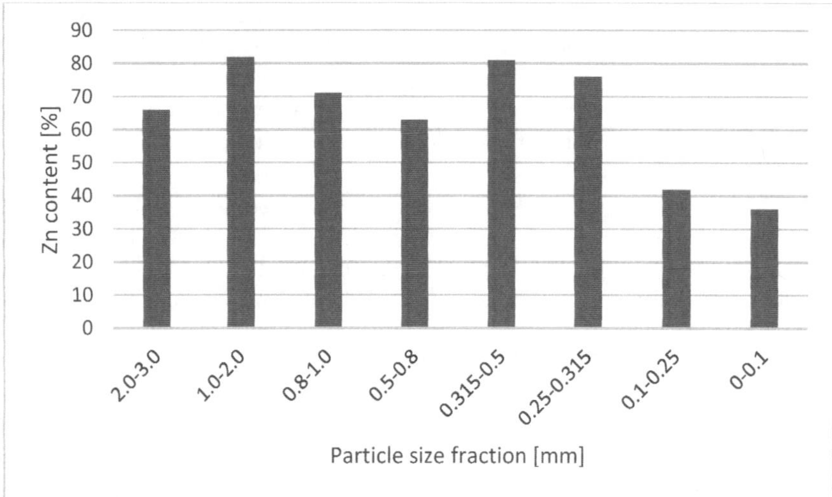

Figure 1. Metallic zinc content in individual particle size fractions in the feed material 0-3 mm.

Summing up, the particle size fraction above 3 mm can be divided from the feed. Optionally also the dusts below 0.1 mm could be removed and only the size fraction 0.1 – 3 mm processed by means of crushing.

The Zn content in individual particle size fraction is different. The lowest content of metal can be observed in two finest particle size fractions, while in coarser ones the amount on zinc is higher. It is hard to notice a clear relationship between the size of material and zinc content, however coarser fractions in general contain more metallic zinc than finer ones. Similar data can be found in literature [9].

As it was mentioned above, the research aim of the paper is to test potential usefulness of comminution operations in zinc ash waste processing. Three types of crushers were used in tests, and the scheme of investigations was presented in figure 2.

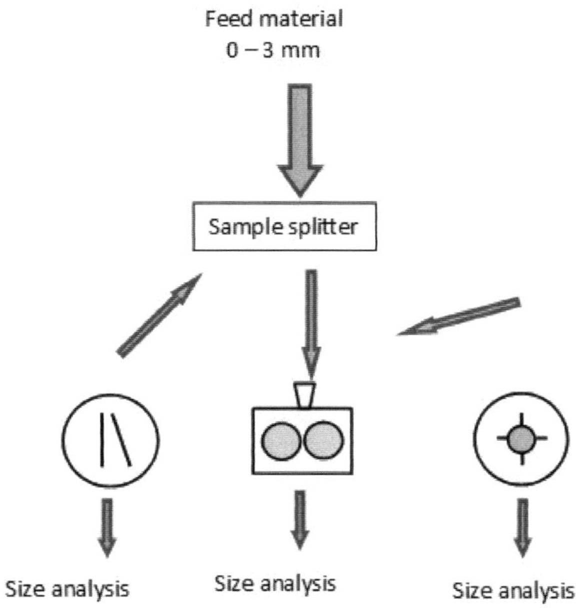

Figure 2. Scheme of investigations.

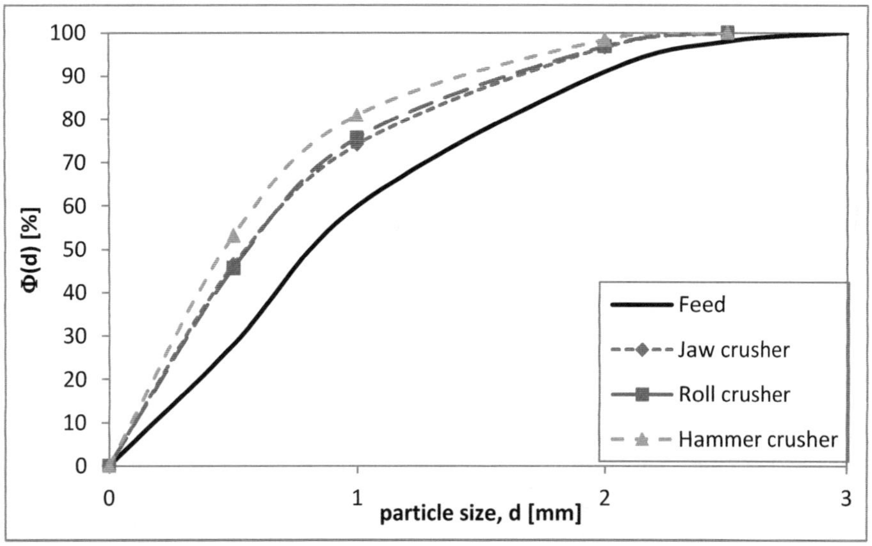

Figure 3. Crushing products from individual crushing devices.

The feed material 0-3 mm was undergone the sieve analysis and the result was presented in figure 3. Then the sample was divided into four parts, three of them were crushed in different crushers, while the fourth one was secured as spare one. Crushing tests were then performed, and size analyses of crushing products were made.

Crushing product below 0.5 mm is treated as a final waste in this case. Two issues were taken into account in the work assessment of individual crushers: the yield of product below 0.5 mm and the zinc metal content in size fraction below 0.5 mm.

If the metal zinc content in crushing product below 0.5 is lower than in feed material, that might indicate, that in crushing process mainly non-metallic particles and ash pass into fine fraction. When, in turn, the zinc content in finest crushing product increases, that may indicate that more significant amount of metallic zinc is crushed, too.

3. Results of investigations

Each crushing product was undergone the sieve analysis and the results were presented in figure 3. Inspecting the figure, it might be seen that crushing products from jaw and roller crushers have very similar particle size compositions. The product of hammer crusher, in turn (figure 4), has a slightly higher content of finer particles. Comparing all crushing products below 0.5 mm it can be stated that hammer crusher produced nearly 8% more material. However, at this point it is not possible to strictly determine which crushing device is more favorable.

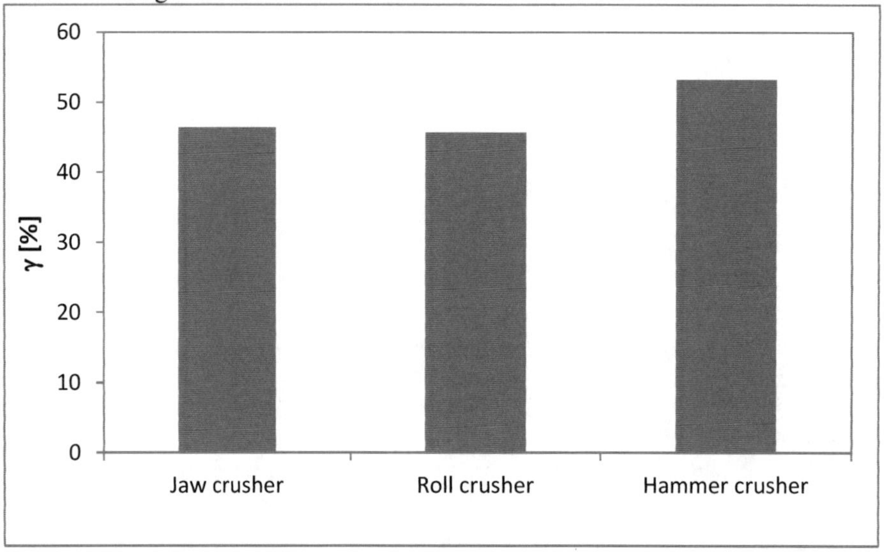

Figure 4. Yields of products below 0.5 mm obtained for individual crushers.

In the next stage of research program chemical analyses on each products were performed, and the content of metallic zinc in each sample was determined. Results are presented in table 2.

Table 2. Metallic zinc contents in individual crushing products.

Type of crusher	Jaw crusher	Roller crusher	Hammer crusher	Feed
Metallic zinc content, [%]	63.2	70.0	60.8	63.2

Calculations of average content of metallic zinc in feed are presented in table 2. Analysis of data presented in the table shows that in the case of hammer crusher the content of metallic zinc in products has decreased, while in crushing product of roller crusher the higher zinc content was recorded. For the jaw crusher the zinc content was on the same level. It may indicate that impact forces cause comminution of non-metallic fractions in a higher degree than metallic ones. In turn, the pressing force that predominates in the roller crusher, results in slight more intensive comminution of metallic zinc. Results of tests for jaw crusher show that both metallic and non-metallic fractions comminute in the same degree.

Results of investigations show that application of certain method of comminution may results in selective crushing of both zinc fractions. It is then possible to design the crushing circuit in which products with different contents of metallic and non-metallic zinc, can be obtained. The entire circuit should include sieving the material with particle size above 3 mm, crushing and size classification on a screen.

4. Summary and conclusions

Secondary resources of zinc constitute significant amount and are becoming more significant in terms of its recovery. It is especially important when the world zinc production from primary resources is on similar level since last decade and zinc demand increases. The method presented in the paper can increase effectiveness of zinc recovery from galvanizing process waste. Application of comminution operation helps in initial separation of non-metallic zinc fractions from the metallic ones. Results of investigation show that hammer crusher is the most favorable device among the three tested crushers. Impact forces increase selective comminution and, as a result, non-metallic fraction breaks more intensively, passing at the same time to the waste product.

The designed circuit can be modified, depending what cut size dT2 of the hammer crusher screen is accepted. It is possible to separate a coarser material (i.e. below 1 mm) and include further treatment of this fine size fraction in grinding operations. That would result in even more higher effectiveness of separation, but investigations within this scope will be the subject of further publications.

Acknowledgements

The article is a results of research project no. POIR.01.01.01-00-0032/16-00 "Opracowanie technologii kompleksowego zagospodarowania odpadów z ocynkowni ogniowych".

5. References

[1] Bright M A Deem N J and Fryatt J 2007 The advantages of recycling metallic zinc from the processing wastes of industrial molten zinc applications *Light Metals 2007* TMS (The Minerals, Metals & Materials Society) pp 1-9

[2] Barakat M A 2003 The pyrometallurgical processing of galvanizing zinc ash and flue dust *Journal of Metals* **August 2003** TMS pp 26 – 9

[3] Maass P and Peissker P 2011 *Handbook of Hot-dip Galvanization* (WILLEY-VCH Verlag GmbH & Co: Weinheim)

[4] Thomas HC 2000 Hot-dip galvanizing technology *Met. Finish* **98 (8)** pp 24-28

[5] Lupa L Negrea P Iovi A Ciopec M and Mosoarca G 2006 Zinc Recover from Zinc Ash by Extraction with Clorhidric Acid Solutions *Chem. Bull "POLITEHNICA" Univ. (Timişoara)* **51(65)** pp 1-2

[6] Najiba S 2009 *Recovery of Zinc from Ash of Galvanizing Plant by Hydrometallurgical Route (PhD thesis)* Department of Materials and Metallurgical Engineering (Bangladesh University of Engineering and Technology: Bangladesh)

[7] Abdel-latif M A 2002 Fundamentals of zinc recovery from metallurgical wastes in the Enviroplas process *Minerals Engineering* **15** pp 942-52

[8] Gawenda T and Saramak D 2018 Assessment of the process efficiency for metallurgical zinc waste screening in vibrating screen *Inżynieria Mineralna – Journal of Polish Mineral Engineering Society* **37(1)** pp 189-194

[9] Trpčevská J Hoľková B Briančin J Korálová K and Pirošková J 2015 The pyrometallurgical recovery of zinc from the coarse-grained fraction of zinc ash by centrifugal force *International Journal of Mineral Processing* **143** pp 25–33

IOP Conf. Series: Materials Science and Engineering **427** (2018) 012018 doi:10.1088/1757-899X/427/1/012018

Investigations over the selection of reagent dose and speed of rotor in laboratory process of ore flotation for HPGR crushing products

K Oleksik, D Saramak and A Krawczykowska

Faculty of Mining and Geoengineering, Department of Mineral Processing and Environmental Engineering, AGH University of Science and Technology, A. Mickiewicza 30, 30-059, Kraków, Poland

koleksik@agh.edu.pl

Abstract. The aim of this this paper was to investigate the effectiveness of flotation process depending on the changeable dose of reagents and variable speed of rotor in flotation device. HPGR crushing products were used as a feed in all flotation tests. Efficiency indices, along with Fuerstenau curves, were determined for each test. Both rotational speed and the dose of reagent have affected the flotation results in a lower degree that expected. However it could be possible to determine the optimal dose and rotational speed of rotor that increase the effectiveness of flotation process.

1. Introduction

Process of flotation is a base beneficiation method utilized in sulphide copper ore processing. It is a complex process, effectiveness of which depends on a number of factors, that can be divided into four main groups [1]:

a) factors related to the feed characteristics (mineralogical and particle size composition, fee grade, surface properties of individual minerals, size of mineralization, liberation rate of useful minerals) [2, 3],
b) factors characterizing the flotational suspension (pulp density, pH, particle size composition, particle shape, temperature) [4, 5, 6, 7],
c) factors resulting from the accepted reagent dosage regime and type of flotation (type and dose level of reagent, method and order of their dosage into the pulp, contact time of the reagent with the pulp, type of flotation process), [8, 9, 10, 11],
d) factors related to the construction and type of flotational device (rotational speed of rotor, aeration degree, type of the flotation chamber or the manner of collecting the foam product). [9, 12, 13, 14, 15].

Application of HPGR devices into ore comminution circuit may result in more favourable beneficiation results for a certain types of material. Results of various investigation confirm that HPGR comminution manner – a generation of micro-cracks through application of high-pressure force acting on the feed material – in general, positively affects the downstream flotational beneficiation of selected types of raw materials [16, 17].

In order to obtain a good quality concentrate, it is also necessary to determine the appropriate conditions of the flotation process course. Enrichment curves, based on the relationship between selected enrichment parameters [18], are commonly used for this purpose. Over the years, many curves have been developed, the most frequently used ones are: Henry, Mayer, Halbich and Fuerstenau curves.

The Fuerstenau curve describes the relationship between the yield and the useful component grade in beneficiation products. The curve shows symmetry in the area of its maximum bend, thanks to which it is possible to determine the selectivity index F, which characterizes approximately the total separation level. Table 1 presents the conventional scale of separation selectivity based on the F index [18].

Table 1. The level of separation selectivity based on Fuerstenau curve.

Class of separation	Selectivity index	Separation level
I	50/50 – 60/60	Very low
II	50/50 – 60/60	Low
III	70/70 – 80/80	Average
IV	80/80 – 90/90	Good
V	90/90 – 99/99	Very good
VI	100/100	Perfect

2. Scope of investigation

The scope of research was to test the possibilities of improving the efficiency of flotation depending on the change in the value of selected process parameters. Based on preliminary flotation tests [8], the following parameters were selected for investigation:

- the dose of a collecting reagent,
- rotational speed of the rotor.

Sulphide copper ore, with particle size below 20 mm, constituted the testing material (figure 1). The average feed grade α was 1.66%.

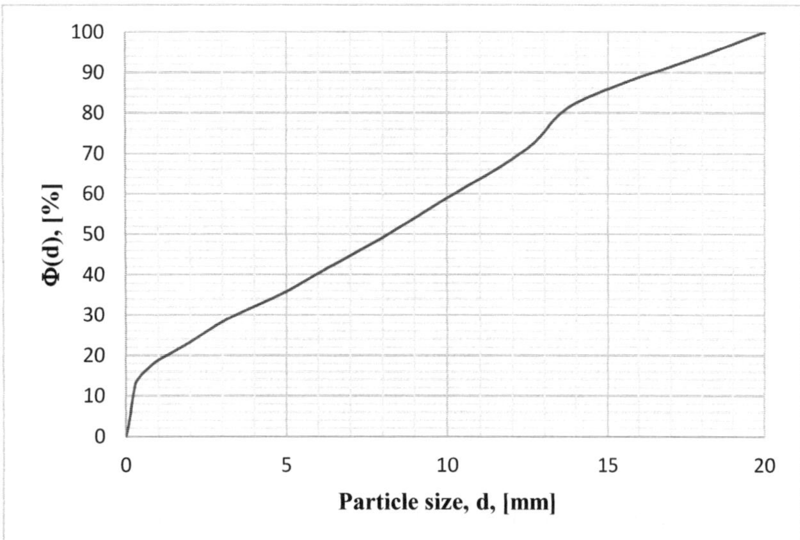

Figure 1. Particle size distribution of the feed for the first stage of comminution.

The preparation of feed material for flotation tests was carried out in a two-stage laboratory crushing circuit (figure 2), with the jaw crusher operating on the first stage and the ball mill working on the second one.

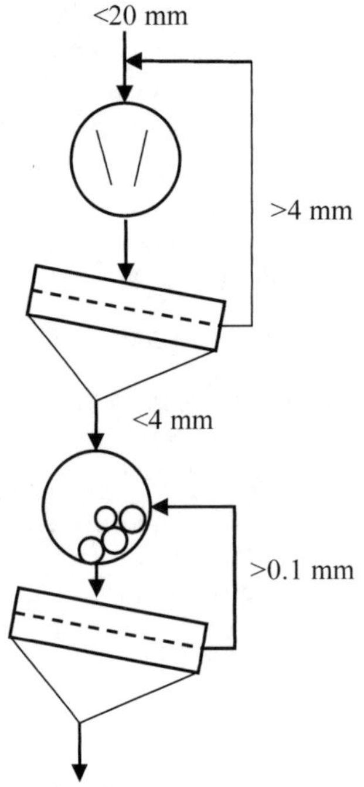

Figure 2. Scheme of comminution circuit.

The particle size composition of the final product from the circuit was below 0.1 mm. The material was then divided into representative samples, 0.45 kg each. The flotation process was carried out in a Metso's Denver D12 laboratory flotation machine (figure 3.), with the working chamber volume 1.5 dm^3.

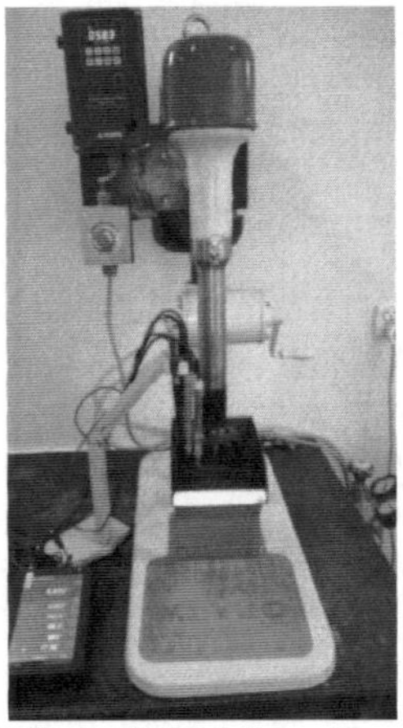

Figure 3. Metso's Denver D12 laboratory flotation machine.

Figure 4 shows the sequence of fractional flotation. The froth products were collecting after 1st, 3rd, 6th, 12th and 30th minute of the process.

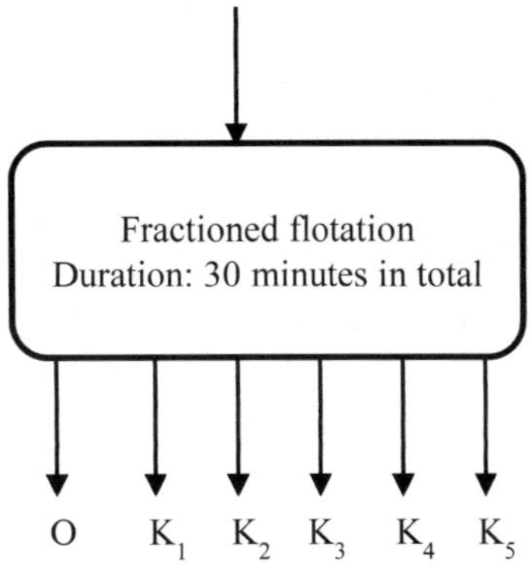

Figure 4. Scheme of fractional flotation tests.

Four flotation tests were carried out. The flotational pulp density and aeration were constant and equaled 1190 g/dm^3 and 125 dm^3/h, respectively. Two levels of rotational speed of rotor were accepted in investigations: 1350 and 1650 rpm. The ethyl xanthate sodium was used as a collector in two doses: 100 g/Mg and 130 g/Mg, while the froth reagent (Nasfroth) dose was 30 g/Mg.

3. Analysis of results

All products of flotation (concentrates and tails) were dried, weighted and prepared for further analyses in XRF device, in order to determine the useful mineral grade in individual products. The yields of concentrates (γ), concentrate grades (β), flotation recoveries (ε_k), waste recoveries (ε_r), waste recovery in concentrate (ε_{nk}), as well as tail grades (ϑ) were determined on the basis of the obtained results of analyses. Complete results are presented in tables 2 to 6.

Table 2. Values of selected technological indicators for copper ore flotation test 1: dosage of collector 100 [g/Mg], rotor speed 1350 [rpm].

Product	γ [%]	$\sum\gamma$ [%]	C_u [%]	β_{Cu} [%]	ε_k [%]	$\sum\varepsilon_k$ [%]	ε_{nk} [%]	$\sum\varepsilon_{nk}$ [%]	$\sum\varepsilon_r$ [%]
K1	6.41	6.41	11.95	11.95	45.75	45.75	5.74	5.74	94.26
K2	10.96	17.37	3.73	6.76	24.45	70.19	10.73	16.47	83.53
K3	4.67	22.04	2.11	5.78	5.89	76.08	4.65	21.12	78.88
K4	7.15	29.19	1.71	4.78	7.32	83.40	7.15	28.27	71.73
K5	10.20	39.39	0.79	3.75	4.79	88.19	10.29	38.56	61.44
O	60.61	100.00	0.33	1.67	11.81	100.00	61.44	100.00	0.00
	100		1.67				100		

IOP Conf. Series: Materials Science and Engineering **427** (2018) 012018 doi:10.1088/1757-899X/427/1/012018

Table 3. Values of selected technological indicators for copper ore flotation test 2: dosage of collector 130 [g/Mg], rotor speed 1350 [rpm].

Product	γ [%]	$\sum\gamma$ [%]	C_u [%]	β_{Cu} [%]	ε_k [%]	$\sum\varepsilon_k$ [%]	ε_{nk} [%]	$\sum\varepsilon_{nk}$ [%]	$\sum\varepsilon_r$ [%]
K1	7.93	7.93	9.42	9.42	44.82	44.82	7.31	7.31	92.69
K2	10.40	18.33	2.90	5.72	18.09	62.91	10.27	17.58	82.42
K3	5.14	23.48	1.79	4.86	5.52	68.43	5.14	22.71	77.29
K4	9.63	33.11	2.86	4.28	16.51	84.94	9.51	32.23	67.77
K5	7.64	40.74	0.60	3.59	2.76	87.71	7.72	39.95	60.05
O	59.26	100.00	0.35	1.67	12.29	100.00	60.05	100.00	0.00
	100		1.67				100		

Table 4. Values of selected technological indicators for copper ore flotation test 3: dosage of collector 100 [g/Mg], rotor speed 1650 [rpm].

Product	γ [%]	$\sum\gamma$ [%]	C_u [%]	β_{Cu} [%]	ε_k [%]	$\sum\varepsilon_k$ [%]	ε_{nk} [%]	$\sum\varepsilon_{nk}$ [%]	$\sum\varepsilon_r$ [%]
K1	8.90	8.90	8.77	8.77	47.23	47.23	8.26	8.26	91.74
K2	6.62	15.52	3.57	6.55	14.30	61.52	6.49	14.75	85.25
K3	6.14	21.66	1.82	5.21	6.76	68.28	6.13	20.88	79.12
K4	8.90	30.56	3.14	4.61	16.90	85.18	8.77	29.64	70.36
K5	6.37	36.93	0.78	3.95	2.99	88.17	6.43	36.07	63.93
O	63.07	100.00	0.31	1.65	11.83	100.00	63.93	100.00	0.00
	100		1.67				100		

Table 5. Values of selected technological indicators for copper ore flotation test 4: dosage of collector 130 [g/Mg], rotor speed 1650 [rpm].

Product	γ [%]	$\sum\gamma$ [%]	C_u [%]	β_{Cu} [%]	ε_k [%]	$\sum\varepsilon_k$ [%]	ε_{nk} [%]	$\sum\varepsilon_{nk}$ [%]	$\sum\varepsilon_r$ [%]
K1	8.83	8.83	7.09	7.09	38.00	38.00	8.34	8.34	91.66
K2	6.64	15.47	3.58	5.58	14.44	52.44	6.51	14.85	85.15
K3	4.74	20.22	2.11	4.77	6.08	58.52	4.72	19.58	80.42
K4	9.85	30.07	4.36	4.63	26.06	84.58	9.58	29.15	70.85
K5	8.06	38.13	0.65	3.79	3.18	87.75	8.15	37.30	62.70
O	61.87	100.00	0.33	1.65	12.25	100.00	62.70	100.00	0.00
	100		1.67				100		

Table 6. Summary of results for four flotation tests.

Number of flotation test	Dosage of collector [g/Mg]	Rotor speed [rmp]	α [%]	γ_c [%]	β_{mean} [%]	ϑ [%]	ε_k [%]	ε_0 [%]	ε_r [%]	γ_{c1} [%]	β_{c1} [%]	ε_{c1} [%]
1	100	1350	1.67	39.39	3.75	0.33	88.19	11.81	61.44	6.41	11.95	45.75
2	130	1350	1.67	40.74	3.59	0.35	87.71	12.19	60.05	7.93	9.42	44.82
3	100	1650	1.65	36.93	3.95	0.31	88.17	11.83	63.93	8.9	8.77	47.23
4	130	1650	1.65	38.13	3.79	0.33	87.75	12.25	62.70	8.83	7.09	38.00

In order to determine optimal values of selected flotation parameters. there were determined the Fuerstenau curves (figure 5). and for each test there were determined the separation ratio F values:

$$F_1 = (77.5/77.5) \tag{1}$$

$$F_2 = (74/74) \tag{2}$$

$$F_3 = (75/75) \tag{3}$$

$$F_4 = (75.5/75.5) \tag{4}$$

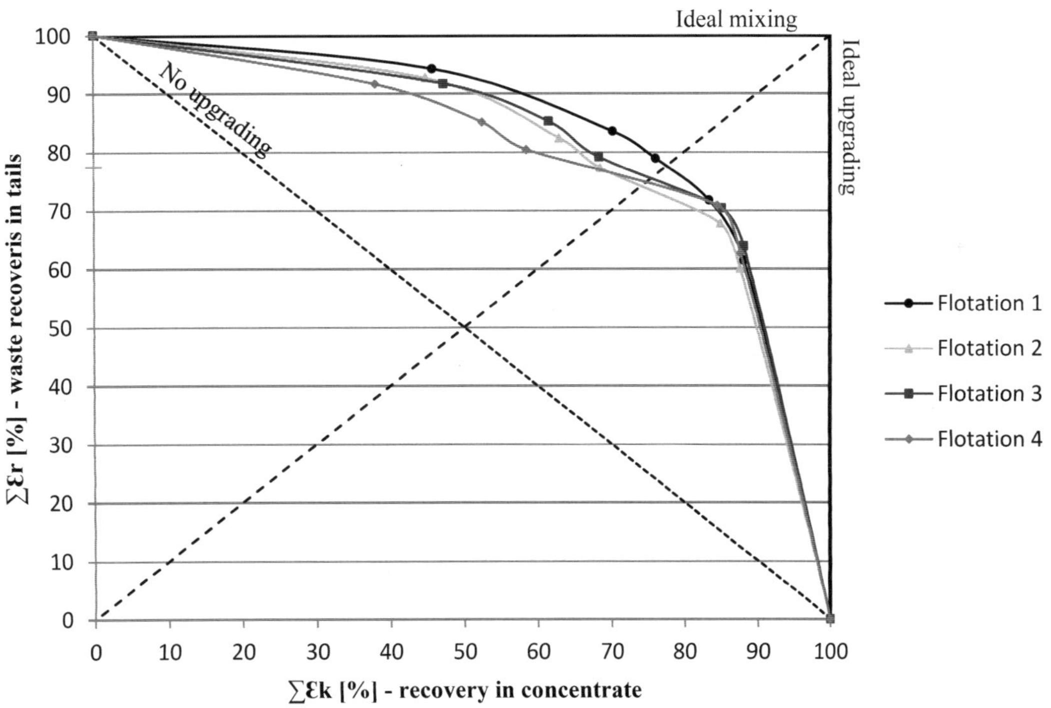

Figure 5. Fuerstenau upgrading curves.

The separation ratio F for each flotation test is located within 3rd class of separation what indicates an average level of separation effectiveness. The most favorable results were obtained for the lower value

IOP Conf. Series: Materials Science and Engineering **427** (2018) 012018 doi:10.1088/1757-899X/427/1/012018

of reagent's dose. as well as lower speed of rotor. It may result from the fact that the type of material used in investigations was of a lower flotation potential due to ore oxidation.

4. Summary

The results of investigations show that variable speed of rotor and dose of flotational reagent affect the flotational result in a rather lower degree. However, the expected results should show a more intense variation together with an increase/decrease of reagents dose and rotational speed of rotor.

However it is worth to mention that variable speed of the rotor and dosage of the collector reagent showed the most significant impact on the obtained results for the first concentrate fractions. The most favorable results were obtained for the flotation test with the rotor speed of 1350 rpm and a 100 g/Mg of collector dosage. Increasing the collector dosage to the value 130 g/Mg resulted in an increase of a concentrate yield and decrease in the values of qualitative indicators of the process. Similar effect was observed for the variable speed of the rotor. At the same dosage of collector. the increased speed of rotor caused the higher value of concentrate yield. This trend could be observed for the first three fractions of concentrates. However. after 6 minutes the flotation test duration. the kinetics stabilizes.

References

[1] Potulska A 2008 Doctoral thesis: Wpływ drobnego mielenia na flotację krakowych rud miedzi *Politechnika Wroclawska Wydzial Geoinzynierii Gornictwa i Geologii Wroclaw*

[2] Morizot G, Conil P, Durance M V and Badri FG 1997 Liberation and its role in flotation-based flowsheet development *Int. J. Miner. Process.* **51** pp 39-49

[3] Marković Z S, Janković A and Tomanec R 2008 Effect of particle size and liberation on flotation of a low grade porphyry copper ore *Journal of Mining and Metallurgy A: Mining* **44 (1)** pp 24–30

[4] Pease J D, Curry D and Young M F 2006 Designing flotation circuits for high fines recovery *Minerals Engeneering* **19** pp 831–40

[5] Göktepe F 2002 Effect of pH on Pulp Potential and Sulphide Mineral Flotation *Turkish J. Eng. Env. Sci.* **26** pp 309 – 18

[6] Dehghani F, Rahimi M and Rezai B 2013 Influence of particle shape on the flotation of magnetite, alone and in the presence of quartz particles *J. South. Afr. Inst. Min. Metall.* **113** pp 905–11

[7] Lazarov D, Alexandrova L and Nishkov I 1994 Effect of temperature on the kinetics of froth flotation *I. Miner Eng* **7** pp 503–9

[8] Konieczny J, Pawlos W, Ksiezniak K, Krzeminska M, Kasinska-Pilut E and Piwowar P 2015 Badania wpływu gęstości zawiesiny flotacyjnej oraz prędkości obrotowej wirnika na wzbogacalność urobku w KGHM Polska Miedź S.A. *CUPRUM.* **2(75)** pp 87-95

[9] Oleksik K, Saramak D and Młynarczykowska A 2017 Evaluation of flotation process course on the example of sulphide ores *E3S Web of Conf.* **18** pp 1-4

[10] Ali S, Khan M M, Tufail M and Yahya K M 2009 Optimization of Operating Process Parameters of Copper Flotation by Using Statistical Techniques *Journal of the Chemical Society of Pakistan* **31** pp 193-200

[11] Mpongo M K and Siame E 2006 Effect of collector, frother and depressant addition on the copper recovery and concentrate grade of the nchanga underground scavenger circuit of konkola copper mine - Zambia *AJST.* **7(1)** pp 8-11

[12] Pogorely A D 1962 Limits of the use of the kinetic equation proposed by K.F. Beloglazov *Invest Uysskikh Ucheb. Zavendenii. Tsvetn. Metall.* **5** pp 33-50

[13] Yang X-S and Aldrich C 2006. Effects of impeller speed and aeration rate on flotation performance of sulfide ore *Trans. Nonferrous Metal Soc. China* **16** pp 185–90

[14] King R P 1972 Flotation research work of the N.I.M research group and the Department of Chemical Engineering, University of Natal *J. South Afr. Inst. Min. and Metall.* **72** pp 135-45

[15] Laplante, A R. Toguri J M and Smith H W 1983 The effect of air flow rate of the kinetics of flotation. Part 1: The transfer of material from the slurry to the froth. *Int. J. Min. Process.* **11**

IOP Conf. Series: Materials Science and Engineering **427** (2018) 012018 doi:10.1088/1757-899X/427/1/012018

pp 203- 19
[16] Smit I 2005 Bench scale ore characterization using the High Pressure Grinding Roll *Randol HPGR Workshop 2005* Perth
[17] Saramak D, Młynarczykowska A and Krawczykowska A 2014 Influence of a high-pressure comminution technology on concentrate yields in copper ore flotation processes *Archives of Metallurgy and Materials* **59** pp 951-55
[18] Drzymała J 2009 *Podstawy mineralurgii* Wyd. 2 (Wrocław: Ofic. Wyd. PWr.)

IOP Conf. Series: Materials Science and Engineering **427** (2018) 012019 doi:10.1088/1757-899X/427/1/012019

Management of waste from energy production – waste combustion in Poland

A Uliasz-Bocheńczyk and P Bąk

AGH University of Science and Technology, Faculty of Mining and Geoengineering, al. Mickiewicza 30, 30-059 Krakow, Poland

aub@agh.edu.pl

Abstract. Energy production processes generate mineral waste – products of the combustion of solid fuels and waste from after-treatment systems. This is classified as group 10 waste – waste from thermal processes. For many years such waste has either been recovered or disposed of in landfills (D10 process), according to waste management hierarchy. The most important areas in which combustion waste is utilized are: the production of construction materials, macro-levelling, the rehabilitation of degraded land, and backfilling. Certain types of combustion waste are currently classified as a by-product. Energy mineral waste management is implemented in accordance with the provisions of the Waste Act of 14 December 2012. In spite of the examples of proper combustion waste management given in this paper, the amount of disposed waste had increased by 2016 while at the same time recovery on a national scale had declined. In 2016 both fly ash and compounds of fly ash and calcium-based reaction waste from flue-gas desulfurization in solid form were no longer disposal and the volume that was recovered had increased. The key problem are dust-slag compounds resulting from the wet treatment of furnace waste, almost 90% of which are landfilled.

1. Introduction

Because of the raw material conditions prevailing in Poland, the basic forms of fuel powering the country's energy industry are hard coal and lignite. In the last few years, due to the obligation to use renewable energy sources, the commercial power industry makes use of solid biomass [1].

Coal is one of the most important energy sources in the world and provides almost 30% of its non-renewable primary energy [2]. In the case of Poland, at current consumption levels there are sufficient industrial coal sources to last more than 40 years [3]. Coal will thus continue to be the country's basic energy raw material. Hence, constant restructuring of the mining industry in many areas is justified, as is investment in new technologies for exploiting this raw material. Mining is a highly capital-intensive and specific industry, which involves some degree of mining-geological risk associated with extraction activity [4].

Besides emitting gaseous pollutants, the combustion of solid fuels also generates waste with certain properties depending on the energy raw material and type of boiler used. According to the Regulation of the Minister of Environment of 9 December 2014, such waste is classified in the waste catalogue (Journal of Laws 2014, item 1923) [5] as Group 10 waste – waste from thermal processes.

The main types of waste generated in energy production processes are: dust-slag compounds from the wet treatment of furnace waste (10 01 80), fly ash from coal (10 01 02), compounds of fly ash and

IOP Conf. Series: Materials Science and Engineering **427** (2018) 012019 doi:10.1088/1757-899X/427/1/012019

calcium-based reaction waste from flue-gas desulfurization (10 01 82), and slag, waste incineration and boiler dust (10 01 01) [6].

For many years mineral waste from the power industry has been treated using recovery processes, although some of this waste has been disposed in landfills.

However, a significant percentage of waste incineration in other countries is recovered [7,8].

In Germany and Denmark, 100% of ash and slag is recovered while in the USA, Great Britain and China it is up to 70%. In turn, Russian cogeneration plants produce approximately 1.3 billion tons of ash, but out of these only 3 million tons (10%) are recovered [8].

Eight hundred Mt of fly ash alone is produced by global coal combustion, primarily in the following countries: China, India, the USA and EU [9]. Ash recovery ranged in different countries from 3% up to 90%; in the USA – 50%, the EU – over 90%, India – 60%; China - 67% [10]. For many years fly ash from coal combustion has been used in different branches of the economy, among other things as sorbents for cleaning waste gases and as agents for removing toxic components from waste, in zeolite synthesis, as well as in the production of construction materials, road construction, and mining [11].

One form of waste that is difficult to utilize is biomass waste incineration [12]. For example, in Lithuania a significant percentage of ash from biomass combustion (25.000 – 30.000 tons) ends up in landfills [13].

The aim of the article is to present the current situation in the management of waste from energy production in Poland in light of the current market trends.

2. Waste from energy production processes– furnace waste and waste from gas treatment systems

The latest data on the amount of waste generated by the power industry together with the methods used to handle such waste are given by the Central Statistical Office [14], including fly ash (table 1), for dust-slag compounds from the wet treatment of furnace waste (table 2), and for fly ash and calcium-based reaction waste from flue-gas desulfurization in solid form (table 3).

The amount of fly ash produced over year has changed over the last 9 years, ranging from a maximum of 4.6 million tons in 2012 down to 3.3 million tons in 2016 and 2017. The amount of recovered waste and landfilled waste has also fluctuated. According to data from the Central Statistical Office [14] for 2016, the ash from the combustion of coal was not disposed in landfills (table 1), while the amount of waste deposited in landfills has declined.

Table 1. Different methods of fly ash treatment [14,15].

Years	Fly ash generated during the year					Waste landfilled (accumulated) to date (as of the end of the year) in million tons
	grand total (million tons)	recovered	disposed			
			total	of which landfilled	temporarily stored	
2007	4.5	98.7	0.4	0.3	0.9	18.8
2008	4.2	90.0	8.0	0.0	2.0	18.7
2009	3.7	89.5	10.0	0.0	0.5	18.7
2010	4.0	89.3	6.1	0.2	4.6	18.5
2011	4.5	92.5	4.6	0.1	0.3	19.0
2012	4.6	84.4	14.7	0.1	0.9	27.2
2013	4.5	87.0	10.6	0.1	2.2	27.2
2014	3.8	0.1	0.1	0.0	0.5	26.9
2015	3.3	3.3	6.6	0.1	0.8	26.3
2016	3.3	86.7	11.3	0.0	2.0	26.0

In the case of dust-slag compounds from the wet treatment of furnace waste (table 2), in 2016 the amount of recovered waste declined while the amount of waste stored in landfills has increased to 303.600 tons.

Table 2. The different methods for dealing with dust-slag compounds from the wet treatment of furnace waste [14].

Years	Dust-slag compounds from the wet treatment of furnace waste generated during the course of a year (% of waste generated)					waste landfilled (accumulated) to date, (as of the end of the year) in million tons
	grand total (million tons)	recovered	disposed			
			total	of which landfilled	temporarily stored	
2007	6.8	23.3	71.9	71.9	4.8	241.8
2008	7.4	28.2	65.1	65.0	6.7	243.1
2009	8.9	17.9	80.1	80.1	2.0	253.5
2010	9.0	13.6	78.4	78.4	8.0	253.7
2011	10.4	12.8	82.5	82.5	4.7	258.0
2012	10.6	11.6	85.1	85.1	3.3	266.9
2013	11.5	10.3	86.7	86.7	3.0	276.6
2014	12.0	0.0	10.4	10.4	0.2	285.9
2015	12.0	0.6	87.4	87.4	1.2	294.1
2016	11.4	8.6	87.5	86.5	0.9	303.6

Ninety percent of fly ash and calcium-based reaction waste from flue-gas desulfurization in solid form (table 3) is recovered, not landfilled and only temporarily stored.

Table 3. The Different methods for dealing with fly ash calcium-based reaction waste from flue-gas desulfurization in solid form [14].

Year	Compounds of fly-ash and calcium-based reaction waste from flue-gas desulfurization in solid form(% of waste generated)					Waste landfilled (accumulated) to date, (as of the end of the year) in million tons
	grand total (million tons)	recovered	disposed			
			total	of which landfilled	temporarily stored	
2007	3.9	96.2	3.7	0.1	0.1	-
2008	4.0	99.9	0.1	0.1	-	-
2009	3.4	99.9	0.1	0.1	0.0	0.0
2010	3.8	100.0	-	-	0.0	0.0
2011	4.2	100.0	-	-	0.0	0.0
2012	3.9	99.9	0.0	0.0	0.1	0.0
2013	3.8	100.0	-	-	-	0.0
2014	3.2	0.2	-	-	-	0.0
2015	3.2	1.2	-	-	0.0	0.0
2016	3.0	92.8	0.1	-	0.5	0.0

Despite the considerable amount of waste from energy production that is recovered, the National Waste Management Plan 2022 [6] identified a number of waste management problems. The authors of the plan [6] underline a relatively large amount of waste from energy production compared to the total waste, the decline in the amount of waste that is recovered and an increase in the amount of waste that is disposed of, as well as the large volume of waste stored in landfills.

The combustion waste from energy generation is primarily used in the production of construction materials (including cement), the construction of roads and mining (table 4). Two problem areas are fly ash from biomass combustion 10 01 03, which is used in small amounts, as well as biomass waste incineration 10 01 82, which is only landfilled [15].

Table 4. The use of combustion waste for economic purposes in 2012 [16], in 1000 tons.

Year	Amount captured	Utilization type				
		Construct. materials	Cement	Road construction	Mining	Others
bottom ash, slag and boiler dust (excluding boiler dust mentioned in 10 01 04) (10 01 01)	1728.7	182.2	1.4	595.5	245.5	212.5
coal fly ash (10 01 02)	5 025.0	1901.1	968.9	191.8	698.6	484.6
fly ash from peat and untreated wood (10 01 03)	27.0	-	-	-	-	6.0
calcium-based reaction waste from flue-gas desulfurization in solid form (10 01 05)	729.9	175.6	124.6	10	47.1	316.5
dust-slag compound from the wet treatment of furnace waste (10 01 80)	15921.8	2985.8	202.5	1625.6	29.6	3221.0
calcium-based reaction waste from flue-gas desulfurization in sludge form (10 01 07)	357.6	-	-	-	-	218.1
sand from fluidized beds (10 01 24)	559.7	35.2	12.8	135.0	20.5	344.4
compounds of fly ash and calcium-based reaction waste from flue-gas desulfurization in solid form (dry and semi-dry methods of desulfurization and fluid bed combustion) (10 01 82)	4159.5	27.6	21.7	270.2	792.1	812.2
fly ash from co-incineration other than those mentioned in 10 01 16	652.1	428.8	223.1	-	-	0.1

3. Examples of waste management in plants operating in the commercial power industry in Poland

Power stations and cogeneration plants in Poland burning solid fuels conduct waste management in accordance with the provisions of the Waste Act.

3.1. Dolna Odra Power Plant Complex

The incineration waste generated in the Dolna Odra Power Station is comprised of the following: dust-slag compounds generated in power boilers, fly ash captured in flue gas dedusting units, synthetic gypsum from flue gas desulfurization systems, and dust from fluidized beds generated by biomass boilers. Waste incineration is partially recovered in the following cases: construction and the restoration of degraded land. Fly ash and synthetic gypsum are by-products, the entire production of which is fully utilized [17].

3.2. Opole Power Station

The following forms of waste are generated by fuel combustion flue gas cleaning technology at the Opole Power Station: slag, fly ash from coal, and gypsum.

Gypsum produced by the flue gas desulfurization unit is a raw material used by Knauff company in the production of plasterboard and is classified as a by-product. Other by-products include fly ash that meets the requirements of the PN-EN 450-1 standard.

All the waste produced in the Opole Power Station is stored selectively in special places designated for this purpose and is then transferred to specialist firms possessing legally required waste management licenses. Waste types 10 01 01 and 10 01 02 can also partially be covered by the D5 waste disposal process in the "Groszowice" waste incineration landfill [18].

3.3. Turów Power Station

In 2017 the Turów Power Station produced a total of 1 437 541.25 Mg of waste incineration, which was then treated using the R5 recovery process and supplied as an additive in construction materials. Waste incineration produced at the Turów Power Station is used to fill in adversely transformed sites [19].

3.4. PGNiG TERMIKA (Kawęczyn Heating Plant, Pruszków Cogeneration Plant, Siekierki Cogeneration Plant, Żerań Cogeneration Plant, and Wola Heating Plant)

By-products of combustion (ash and slag) at the PGNiG TERMIKA plants are regularly transferred to recipients possessing the appropriate licenses. They are used in the following areas: road construction, in the production of construction materials and as levelling material. The ash and slag generated in large volumes during peak heating period are stored and then recovered in the summer.

3.5. PGE Górnictwo i Energetyka Konwencjonalna, Kielce Cogeneration Plant branch

The following forms of waste incineration generated at PGE Górnictwo i Energetyka Konwencjonalna Kielce Cogeneration Plant Branch are: dust-slag compounds from WP-140 boiler, dust-slag compounds from boilers WR-25 and OR-50, ash and slag from biomass combustion from OS-20 boiler [21].

In PGE Górnictwo i Energetyka Konwencjonalna, Kielce Cogeneration Plant branch waste is collected selectively, gathered in specially designated and suitably prepared sites and then subjected to either recovery or disposal processes.

Dust-slag compounds from WP-140 boiler following its hydraulic transportation via pipelines is then disposed of in the „Gruchawka" landfill.

Dust-slag compounds from boilers WR-25 and OR-50, ash and slag from biomass combustion from OS-20 boiler, 22 277.0 Mg of which was generated in 2017 has been taken from temporary waste storage by physical persons.

Ash from biomass combustion (523.4 Mg produced in 2017) is disposed of or recovered [21].

3.6. ENERGA Kogeneracja Sp. z o.o.

One hundred percent of the ash and slag resulting from the combustion of hard coal at ENERGA Kogeneracja Sp z o.o. is recovered and utilized in the production of concrete, the production of construction ceramics, the restoration of degraded areas, and levelling [22].

4. Conclusion

The problem connected with the waste generated by the power industry has compelled interested entities to conduct economic, legal, technical, ecological and social analyses on the possible ways of utilizing landfilled materials [23]. From an economic perspective, costs are of key importance. According to data [16], the environmental protection costs of power stations and cogeneration plants in Poland (2014) totaled PLN 2.170.332.000 which accounted for 8.4% of the total costs of generating electricity and heat. According to data for 2013 the corresponding costs amounted to PLN 2.152.434.000 which translates into 7.6% of overall costs.

The environmental protection costs of power plants and cogeneration plants include charges and fees for emissions of pollutants, the use of water, the discharge of waste water and the landfilling of waste. Moreover, these costs include penalties for air pollution, discharging waste-water in waters and land, as well as for other uses of the environment [16].

In accordance with the hierarchy of waste management methods [24], measures should be taken to prevent waste generation, which in the case of waste from energy production primarily involves the application of suitable combustion technologies [6].

Methods for reducing waste from energy production include improving the energy efficiency of installations, applying modern combustion technologies, replacing solid fossil fuels with biomass, as well as increasing the share of energy obtained from renewable sources in the overall energy balance [6].It is important to stress, however, that biomass combustion produces waste that is of limited commercial use.

Solid-fuel power stations and cogeneration plants try to recover as much waste as possible and minimize landfilling. Plants also classify their waste as a by-product, which by definition is used for economic purposes.

The current economic situation in Poland will result in an increased demand for this type of waste from the building, construction, and mining industries in the future. This is due to the ongoing restructuring process of the mining industry (including the closure of mines and reclamation of post-mining areas).

The main problem is the management of waste from biomass combustion. Therefore, further studies on the management of waste are required.

Acknowledgements

This work was done as a result the support of the AGH University of Science and Technology Research Programs no. 11.11.100.482 and no 11.11.100.693.

References

[1] Uliasz-Bocheńczyk A and Mokrzycki E 2015 Biomass as a fuel in power industry *Rocz. Ochr. Sr* **17** pp 900–13
[2] *Medium-term coal market report 2016 market analysis and forecasts to 2021* International Energy Agency 2016
[3] Mokrzycki E and Uliasz-Bocheńczyk A 2009 Management of primary energy carriers in Poland versus environmental protection *Rocz Ochr Sr* **11** pp 103–31
[4] Bąk P 2008 Financing of the investement activity based on example of coal mining industry *Miner Resour Manag* **24** pp 265–72
[5] Rozporządzenie Ministra Środowiska z dnia 9 grudnia 2014 r. w sprawie katalogu odpadów (Dz. U. 2014 poz. 1923)
[6] National Waste Management Plan 2022 (M.P. 2016 poz. 784)
[7] Hadbaatar A, Mashkin N A and Stenina N G 2016 Study of Ash-Slag Wastes of Electric Power Plants of Mongolia Applied to their Utilization in Road Construction *Procedia Eng.* **150** pp 1558–62 doi:10.1016/j.proeng.2016.07.111
[8] Menshov P V, Khlupin Y V, Nalesnik O I and Makarovskikh A V 2014 Ash and Slag Waste as a Secondary Raw Material *Procedia Chem* **10** pp 184–91 doi:10.1016/j.proche.2014.10.032

IOP Conf. Series: Materials Science and Engineering **427** (2018) 012019 doi:10.1088/1757-899X/427/1/012019

[9] Belviso C 2018 State-of-the-art applications of fly ash from coal and biomass: A focus on zeolite synthesis processes and issues *Prog Energy Combust Sci* **65** pp 109–35 doi:10.1016/j.pecs.2017.10.004
[10] Yao Z T, Ji X S, Sarker P K, Tang J H, Ge L Q, Xia M S, et al. 2015 A comprehensive review on the applications of coal fly ash *Earth-Science Rev* **141** pp 105–21 doi:10.1016/j.earscirev.2014.11.016
[11] Ahmaruzzaman M 2010 A review on the utilization of fly ash *Prog Energy Combust Sci* **36** pp 327–63 doi:10.1016/j.pecs.2009.11.003
[12] Vassilev S V, Baxter D, Andersen L K and Vassileva C G 2013 An overview of the composition and application of biomass ash. Part 1. Phase-mineral and chemical composition and classification *Fuel* **105** pp 40–76 doi:10.1016/j.fuel.2012.09.041
[13] Nagrockienė D and Daugėla A 2018 Investigation into the properties of concrete modified with biomass combustion fly ash *Constr Build Mater* **174** pp 369–75
[14] *Ochrona środowiska 2008-2017* (Warszawa: Central Statistical Office)
[15] Uliasz-Bocheńczyk A, Mazurkiewicz M and Mokrzycki E 2015 Fly ash from energy production – A waste, byproduct and raw material *Gospod Surowcami Miner/Miner Resour Manag* **31** pp 139–50 doi:10.1515/gospo-2015-0042
[16] E*mitor 2009, 2010, 2011, 2012, 2013. Emisja Zanieczyszczeń Środowiska w Elektrowniach i Elektrociepłowniach Zawodowych* (Warszawa: Agencja Rynku Energii S.A.)
[17] https://zedolnaodra.pgegiek.pl/Ochrona-srodowiska/Gospodarka-odpadami
[18] https://elopole.pgegiek.pl/Ochrona-srodowiska/Gospodarka-odpadami
[19] https://elturow.pgegiek.pl/Ochrona-srodowiska/Gospodarka-odpadami
[20] http://www.termika.pgnig.pl/gospodarka-odpadami
[21] http://www.termika.pgnig.pl/gospodarka-odpadami.
[22] https://eckielce.pgegiek.pl/Ochrona-srodowiska/Gospodarka-odpadami
[23] http://www.energa-kogeneracja.pl/s46-srodowisko
[24] Kulczycka J, Uberman R and Cholewa M 2014 Analiza kosztów i korzyści zagospodarowania odpadów z górnictwa węgla kamiennego *Stud Ekon* **166** pp 272–82
[25] Ustawa z dnia 14 grudnia 2012 r. o odpadach (Dz. U. 2013 poz. 21)

IOP Conf. Series: Materials Science and Engineering **427** (2018) 012020 doi:10.1088/1757-899X/427/1/012020

Monitoring of coal jig operation using a radiometric meter with a variable time of measurement

S Cierpisz[1] and J Joostberens[2]

[1]Institute of Innovative Technologies EMAG, Leopolda 31, 40-189 Katowice, Poland

[2]Silesian University of Technology, Faculty of Mining and Geology, Akademicka 2, 44-100 Gliwice, Poland

jaroslaw.joostberens@polsl.pl

Abstract. Authors discuss the problem of how to monitor the coal/water pulsating bed in a jig with the use of a radiation density meter. Time of measurement applied in radiation monitors is usually constant in industrial radiometric monitors. The dynamic error in the process of measuring changes in density depends on the time of measurement; its optimal value can be found for a given shape of density changes. Authors propose an alternative method of signal filtration for variable time of measurement during a cycle of pulsations. Good results can be achieved for a linear relation between the time of measurement and time derivative of the density. The time derivative of the density can be determined for a long period of time on the basis of industrial tests which establish average changes in density in subsequent cycles of pulsations. In this case, the dynamic error of measurement MSE can be reduced by half compared to the optimal constant time of the measurement.

1. Introduction

Raw coal is often beneficiated in gravitational processes, where coal grains are stratified according to their densities in a pulsating coal/water medium in jigs, as shown in figure 1. This problem was discussed in [1-6].

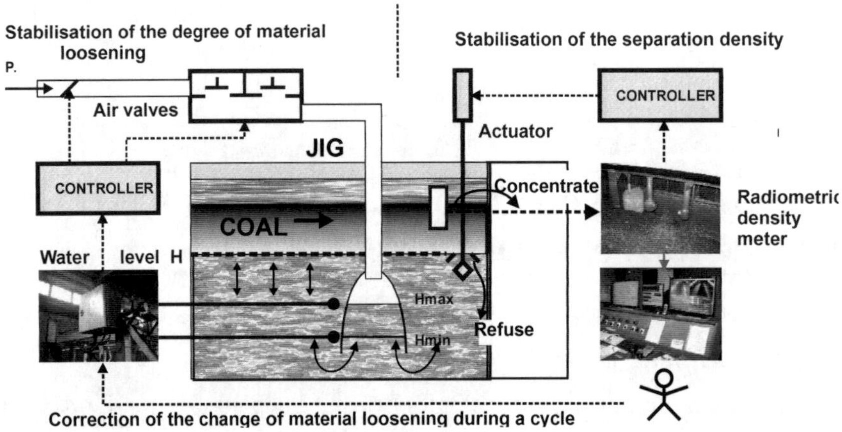

Figure 1. Control system of a coal separation process with a radiometric monitor of a separation density.

Separation of stratified material is based on a chosen separation density which is the density of the layer reporting in half to the upper product (concentrate) and in half to the discharged lower product (refuse). The refuse is removed through the discharge gate and the concentrate overflows the splitting gate. The quality of products is determined by the density of the separation layer. Its position should be monitored on-line and kept at the splitting-gate level regardless of the changes in the tonnage of the feed or changes in the washability characteristics of the raw coal. It is usually measured by a metal float of required shape and density.

The desired position of a float is stabilized through controlling the amount of the lower product discharged through the bottom gate. Float is not accurate in indicating the chosen density layer, especially with changes in the amount of the feed and varying composition of the grains. In new experimental systems, floats are being replaced by more accurate radiometric density meters which can monitor the process of material loosening/compressing during each cycle of coal/water pulsations. The output signal from a radiometric meter can be used for two purposes:

- to stabilize the shape of dynamic changes in density;
- to stabilize the separation density measured when material is compressed at the end of the cycle.

The typical dynamic change in coal/water density in a single cycle of separation process is shown in figure 2.

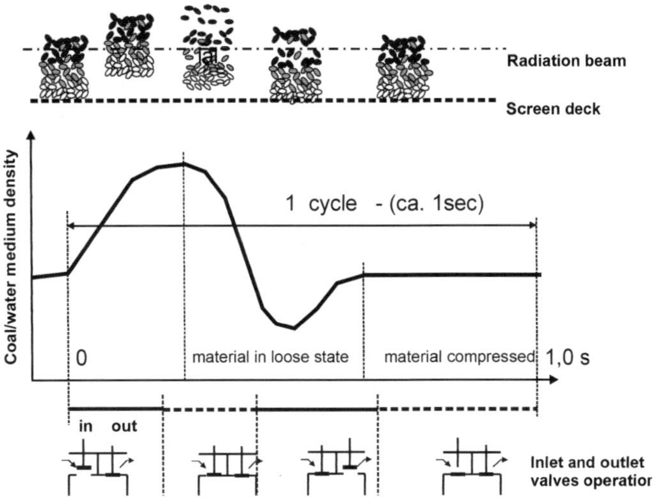

Figure.2. Changes in medium density in a single cycle of pulsations.

At first, while the inlet air valve is open, the material is lifted upwards without loosening the upper part of the coal bed. Then the material gradually separates, the inlet valve is closed and grains sink again to be consolidated at the end of the cycle. The outlet air valve is opened to speed-up the sinking. Then the outlet valve is closed to ensure the same hydraulic conditions for the next cycle.

The shape of changes in density at the level where the concentrate overflows the upper discharge gate varies due to variations in the feed tonnage, density composition of the grains and fluctuations in air pressure in the air collector. To achieve the best conditions for coal separation, i.e. the optimal stratification of grains according to their density and constant separation density in the jig, the shape of changes in density during each cycle of pulsations should be stabilized. The radiometric density monitor should reproduce changes in medium density with minimum error to achieve the best monitoring and control results.

The measuring head of a radiometric density meter consists of a radiation source (^{60}Co, ^{137}Cs) and a detector which usually is a scintillation counter. The review of detectors was presented in [7]. The output signal from the detector is always a discrete stochastic signal (Poisson distribution), regardless of the character of the input signal (i.e. density) modulating the intensity of the detected radiation beam. The longer the averaging time, the higher the statistical (static) accuracy of the monitor. At the same time, if the input signal varies, the dynamic error of the measurement is higher. This suggests that for a given shape of the input signal and a given structure of the monitor circuit, one can find an optimal averaging time of input pulses, which allows for the minimum dynamic error within accepted criteria. Furthermore, this leads to the application of a circuit with an adapting time of input pulses averaging. This problem was discussed in [8, 9]. If the input signal is, for example, a step function and the density monitor is to reproduce this change, the time of measurement should be small at the beginning of the measurement in order to speed-up the reaction of the meter and then it should increase in order for the new value of the density to be read out accurately. In the case of stochastic changes in the measured density, the signal from the detector is a doubly non-stationary Poisson process. Such processes were discussed in [10,11,12].

The problem becomes important when changes in density are frequent and rapid; if this is the case, the process requires short times of measurements. It may happen in technological systems for coal separation in a jig where stratification of coal grains based on their densities is performed in a pulsating coal/water bed to produce two or three final products: concentrate and refuse or middlings.

Radiometric density meters in which time of measurement adapts to changes in measured densities require instantaneous monitoring of these changes over short periods of time. Due to the stochastic nature of the signal coming from the radiation detector, it is difficult to determine the derivative of the density signal as a function of time without significant errors. Also, software applied to measure the changes in density in order to adjust the period of time for the next measurement must be efficient enough to operate on-line in short periods of time (i.e. 1-2 ms). The aim of this paper is to propose an alternative method to adjust the time of measurement on-line. This method is based on an a priori determination of the change in time of measurement on the basis of the average shape of changes in density in subsequent cycles of the coal bed pulsations.

2. The method of signal filtration from the digital radiometric meter

The output signal $s(t)$ from the scintillation detector can be processed in an analogue integrator or as a moving average number of pulses $u(t)$ counted during the time of measurement t_s as it is shown in figure 3. The relation between the measured medium density $\rho(t)$ and the mean intensity of registered pulses $n(t)$ is theoretically exponential but for small changes of density can be approximated by a linear equation. The value of the signal u at the output of the density meter at *j-th* period of time of measurement t_s is determined from the following equations:

$$u[i] = \frac{1}{k_n} \cdot \left(n_0 - \frac{1}{t_s} \cdot k[i]\right) \quad (1a)$$

$$k[i] = \sum_{l=1}^{t_s/\Delta t} S\left[\frac{t_s}{\Delta t} - l\right] \quad (1b)$$

k_n – coefficient,
n_0 – intensity of pulses for the reference density (e.g. water),
S – number of pulses (0 or 1) in the elementary period of time,
Δt – elementary period of time (determined by the dead time of the radiation detector) $\Delta t < t_s$.

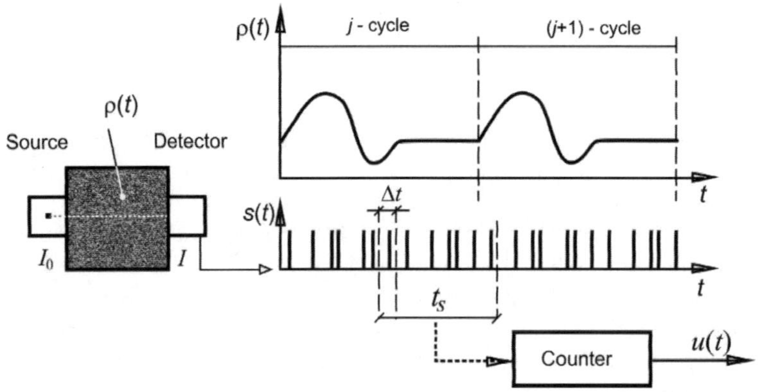

Figure 3. Digital filtration of stochastic signal from the detector.

Extending the measurement time t_s reduces the statistical error of the measurement but at the same time deteriorates the dynamic properties of the density meter. On the other hand, the short measurement times improve the dynamic properties of the meter, but increase stochastic fluctuations in the signal $u(i)$. An example of simulated changes in the signal $u(i)$ as a response to changes in the bed density in a jig for short and long measurement time t_s is shown in figure 4.

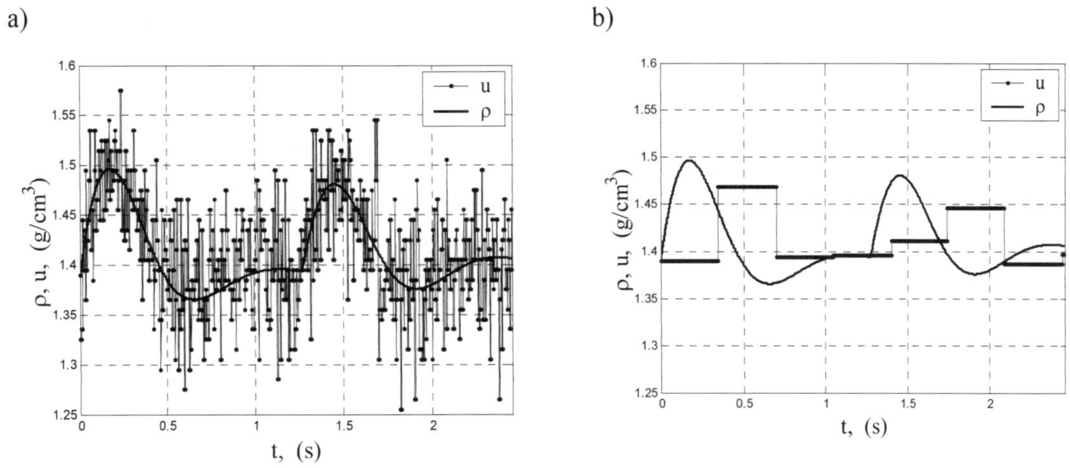

Figure 4. Changes in the measured density signal $\rho(t)$ and changes in the output signal $u(t)$ for (a) short and (b) long measurement time t_s.

The time of measurement t_s is constant in typical industrial radiometric monitoring systems. The problem of determining its specific value lies in a compromise between the expected static accuracy of the meter and the dynamic error caused by inertia resulting from the measurement time t_s. This problem is especially important in monitoring fast changes in the density of the coal/water bed in a jig. Reproducing a rapid change in density requires a short time of measurement, which makes it more difficult to measure the instantaneous density value accurately. The optimal time of measurement t_s, for a known change in the media density $\rho(t)$, can be determined from the minimum value of the mean square error:

$$\mathrm{MSE}_c = \frac{1}{N_i}\sum_{m=1}^{N_i}(u[i] - \rho[i])^2 \tag{2}$$

N_i – number of data used for calculation of t_s.

As mentioned earlier, better results of the dynamic measurement of the bed density in a jig may be expected when the time of the measurement t_s depends on the speed of changes in density.

For example, in the case of a step change in density, the time of measurement at the beginning should be short to speed-up the reaction of the meter and then should increase for a more accurate read-out of the density value in the steady state. Radiometric monitors with times of measurement adapting to changes in the measured signal were discussed in [8]. The discussed systems applied a differential filter to detect a change in a previously smoothed signal and to use it to immediately adjust the share of signals from three filters with different times of measurement (fast, medium, long) in the output signal $u(t)$. These adaptive filters give better results than filters with the constant time of measurement t_s determined by the criterion (2). The main problem in this case lies in the accuracy of determination of the change (derivative) in density as the signal from the radiation detector is the stochastic series of pulses whose mean intensity is modulated by the measured density. The advantage of the adaptive method lies in the possibility of applying it to any shape of change in the measured density. The disadvantage, in the case of radiometric meters, lies in (a) significant noise in the differential signal and (b) the short time required for the execution of the algorithm to process the signal and predict the time of measurement t_s for the subsequent step.

In this paper we propose another method to filter the signal from the radiation detector which can be used for monitoring special signals similar to periodic changes in the density of the coal/water bed in a jig. A typical shape of a change in bed density in a jig is shown in figure 3 and figure 4. This change is repeated in the subsequent cycles of bed pulsation. Depending on conditions, changes in density may differ in absolute values of ordinates but their character remains similar from cycle to cycle. The mean ("typical") change in density can be presented by a function of time and this function can be further used to adjust subsequent times of measurement t_s according to the accepted function. This concept of the density signal processing is expected to give smaller dynamic errors of measurement in comparison to the method with the constant time of the measurement; in some cases it may give better results than adaptive algorithms. In this paper we limit our analysis to the determination of optimal parameters to process a "typical" periodic signal of changes in the bed density in a jig.

Mathematical models of changes in coal/water bed density, based on experimental data from two coal mines, were presented by in [9]. A typical change in density can be described by equation (3):

$$\rho(t) = A_j \cdot e^{-\alpha_j \cdot t} \cdot \sin(\omega_j t - \psi_j) + \rho_{uj} \tag{3}$$

A_j, α_j, ω_j, ψ_j, ρ_{uj} – parameters of the equation (3) of *j-th* cycle of the pulsation,
t – time of jig operation.

The derivative of the media density as a function of time is as follows:

$$\frac{d\rho(t)}{\mathrm{dt}} = A_j \cdot e^{-\alpha_j \cdot t}\left(\omega_j \cos(\omega_j t - \psi_j) + \alpha_j \sin(\omega_j t - \psi_j)\right) \tag{4}$$

Equation (4) can be used for determination of the relation between the time of measurement t_s and the derivative $d\rho/dt$:

$$t_p = f\left(\frac{d\rho}{\mathrm{dt}}\right) \tag{5}$$

Let us normalize the parameters in equation (5) in the range (0,1):

$$t_N = \frac{t - t_0}{t_{\max} - t_0} \tag{6}$$

$$\left(\frac{d\rho}{\mathrm{dt}}\right)_N = \left(\frac{d\rho(t)}{\mathrm{dt}} - \left(\frac{d\rho}{\mathrm{dt}}\right)_{\min}\right) \cdot \left(\left(\frac{d\rho}{\mathrm{dt}}\right)_{\max} - \left(\frac{d\rho}{\mathrm{dt}}\right)_{\min}\right)^{-1} \tag{7}$$

where:
t_N – normalized time [0,1],
t_0 – the beginning of the cycle pulsation,
$t_{\max}$ – the end of the cycle pulsation,
$(d\rho/dt)_N$ – normalized derivative $d\rho/dt$,
$(d\rho/dt)_{\min(\max)}$ – expected (minimum/maximum) value of $d\rho/dt$ during the pulsation cycle.

The time of measurement t_s should depend on the value of the derivative $d\rho/dt$ and not on its sign. Hence the equation (4) should be rewritten as follows:

$$t_{sN} = f\left(\left|\left(\frac{d\rho}{\mathrm{dt}}\right)_N\right|\right) \tag{8}$$

The time derivative $d\rho/dt$ expressed by equation (4) before and after the standardization is shown in figure 5.

Values of the time of measurement ts can be calculated from normalized values (equation 8) as follows:

$$t_s = \left(t_{s(\max)} - t_{s(\min)}\right) \cdot t_{sN} + t_{s(\min)} \tag{9}$$

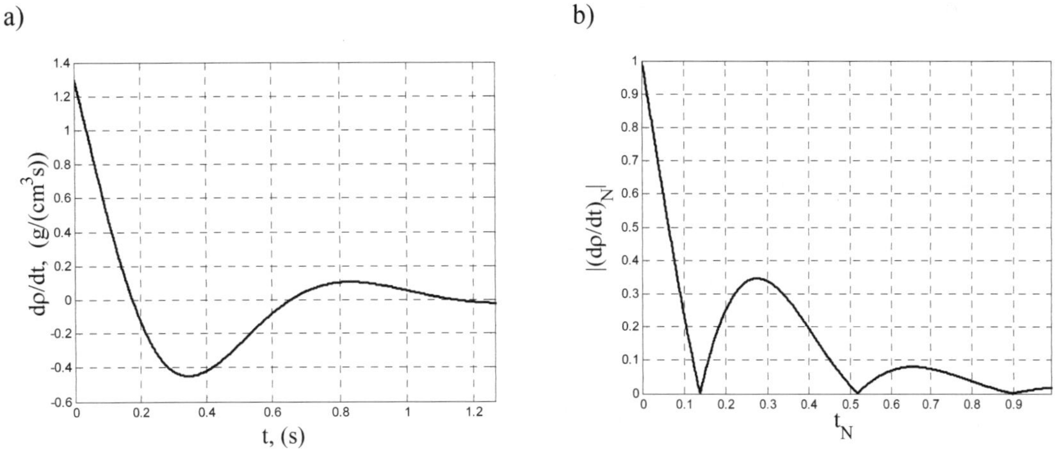

Figure 5. The derivative *dρ/dt* as a function of time (a) before and (b) after standardization in the range [0,1].

3. Simulation analysis

The simulation analysis was performed using the Matlab software package. The radiometric density meter was modeled by a generator of the discrete Poisson noise (GDPN) with the mean value of the intensity of pulses modulated by the measured density [9, 13]. The block scheme of the simulation model is shown in figure 6.

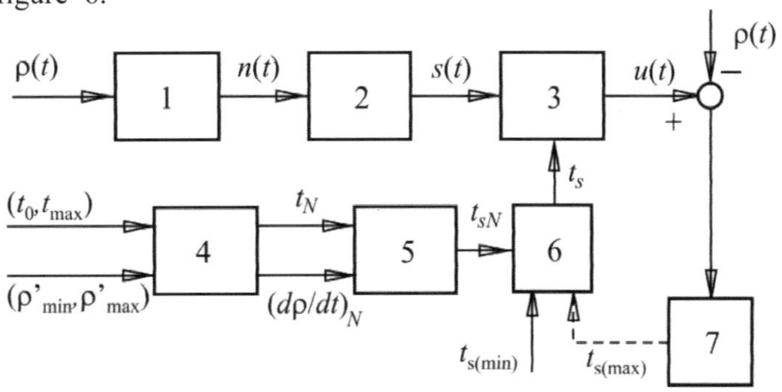

Figure 6. The block scheme of the simulation model
1 – measuring head, 2 – generator of a discrete Poisson noise, 3 – counter of pulses, 4 – normalization, 5 – computing t_{sN} using one of the equations (10)-(13), 6 – denormalization, 7 – computing of the criterion value (eq. 2).

To establish a proper relation between the time of measurement and the time derivative of the density, the following relations were tested:

$$t_{sN} = \left(1 - \left|\left(\frac{d\rho}{dt}\right)_N\right|^2\right)^{0.5} \tag{10}$$

$$t_{sN} = -\left|\left(\frac{d\rho}{dt}\right)_N\right|^2 + 1 \tag{11}$$

$$t_{sN} = -\left|\left(\frac{d\rho}{dt}\right)_N\right| + 1 \tag{12}$$

$$t_{sN} = e^{-a\cdot\left|\left(\frac{d\rho}{dt}\right)_N\right|} \tag{13}$$

a – coefficient (range 2 - 5).

Relations (10) to (13) are presented in figure 7.

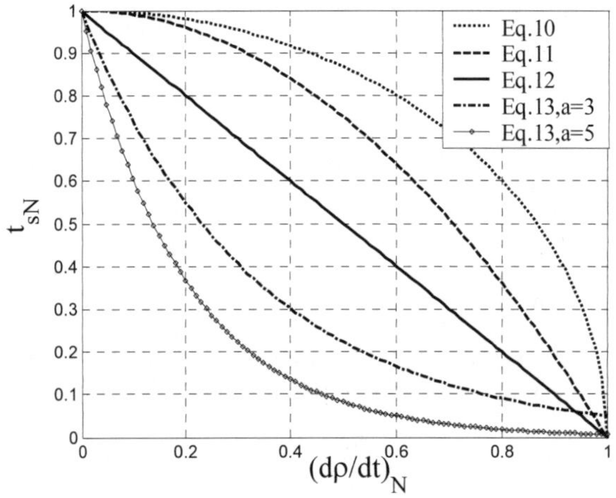

Figure 7. Functions used for calculation of the normalized time of measurement t_s.

The simulation analysis was performed for two cycles of pulsations in a jig. The generator of a non-stationary discrete noise of the Poisson distribution (GDNP) was used as a source of pulses at the output of the radiation detector. The detailed description of the simulation model was presented in [9, 13]. The series of pulses (N_i = 2456) modeled density changes during two cycles of pulsations of the coal/water bed. Changes in density during two cycles were described by equation (3) with the following parameters: $A_1 = 0.2$, $\alpha_1 = 3.04$, $\omega_1 = 6.50$, $\psi_1 = 0$, $\rho_{u1} = 1.39$ (for the first cycle in the time range 0 - 1.270 s) and $A_2 = 3.74$, $\alpha_2 = 2.60$, $\omega_2 = 6.77$, $\psi_2 = 2.32$, $\rho_{u2} = 1.40$ (for the second cycle in the time range 1.247 - 2.457 s). The minimum time of measurement was set to t_s = 11 ms and the maximum time was in the range 11 - 230 ms. The elementary time of simulation was Δt = 1 µs, but the sample time necessary for calculation of MSE_c value was assumed as 1ms. The results of simulation are summarized in table 1 and figure 8. Figure 8a shows values of MSE_c (criterion 2) as a function of constant time of measurement t_s and figures 8b-f show values of MSE_c as functions of $t_{s(\max)}$ for variable time of measurement t_s.

Table 1. Summary of simulation results

t_s (ms)	$t_{s(\max)}$ (ms)	Equation	MSE_c, ($10^{-4}\cdot g^2/cm^9$)
37	-	-	2.770
-	80	(10)	1.191
-	73	(11)	1.158
-	108	(12)	1.104
-	78	(13), a=2	1.252
-	139	(13), a=3	0.999
-	189	(13), a=4	1.067
-	256	(13), a=5	1.048

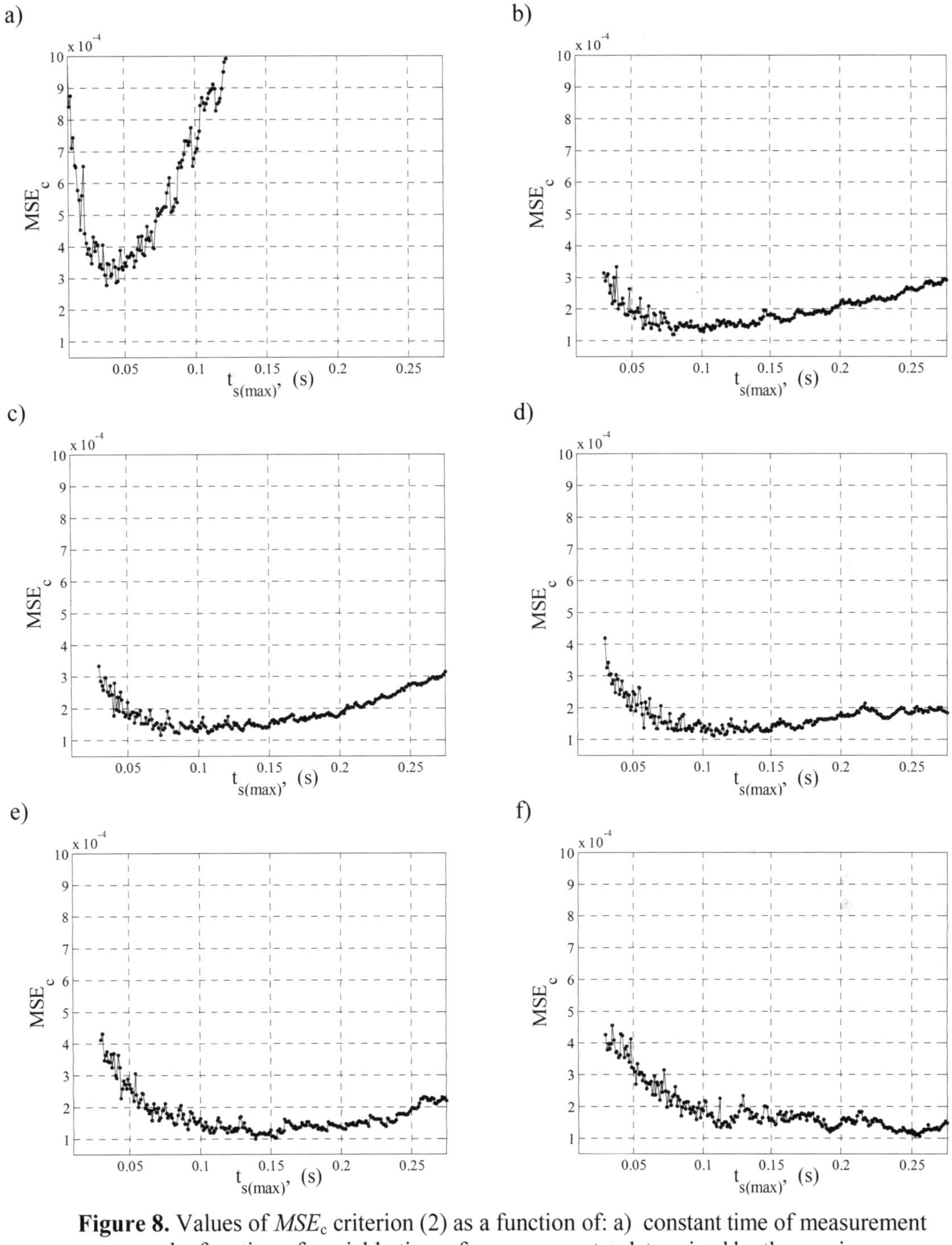

Figure 8. Values of MSE_c criterion (2) as a function of: a) constant time of measurement t_s and a function of variable time of measurement t_s determined by the maximum value $t_{s(\max)}$ according to equation b) (10), c) (11), d) (12), e) (13), a = 3, f) (13), a = 5.

The results of simulation show better accuracy in the dynamic measurement of density changes for the variable time of measurement t_s than for the constant time t_s (for all tested functions $t_{sN} = f(|(d\rho/dt)_N|)$). The best results have been obtained for exponential equations (13) and for the linear equation (12). Figure 9 shows the simulated signals $\rho(t)$ and $u(t)$ for the constant (figure 9a) and variable times of measurement t_s (figures 9b,c,d). Simulated functions $u(t)$ differ slightly in the analyzed cases. For the constant time of measurement (figure 9a) a higher shift of the phase between $u(t)$ and $\rho(t)$ can be observed compared to figure9b and figures 9c,d. Fluctuations of $u(t)$ are lower for the variable time of measurement (figure 9b and figures 9c,d).

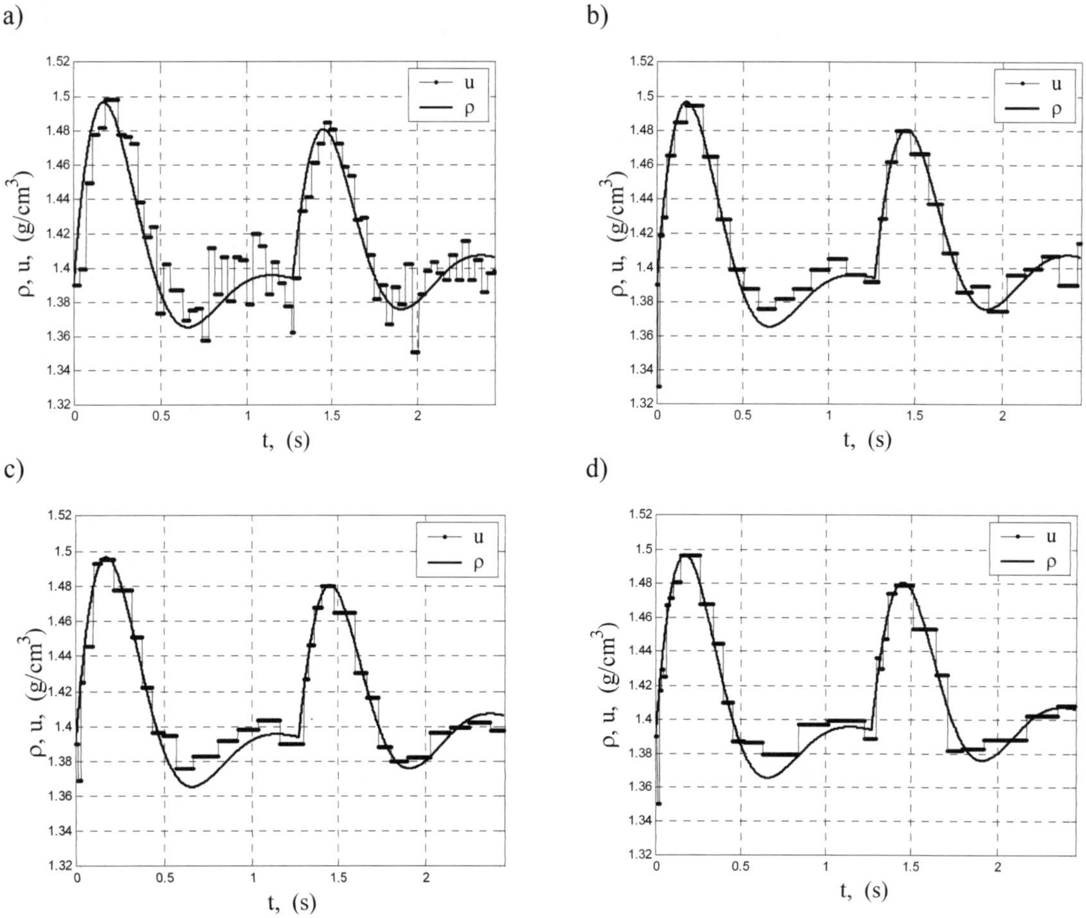

Figure 9. Output signals $u(t)$ for a) constant time of the measurement $t_s = 37$ms and time of the measurement determined form b) eq. (12) at $t_{s(\max)} = 108$ ms c) eq. (13) at $t_{s(\max)} = 139$ ms and $a = 3$ and d) eq. (13) at $t_{s(\max)} = 256$ ms and $a = 5$.

4. Conclusions

Coal separation process in a jig to a great extend depends on the dynamics of coal particles moving in the coal/water bed resulting in changes in the bed density. These changes can be monitored by a radiometric density meter placed at the edge of the upper gate discharging the coal concentrate. The shape of density changes in one cycle of pulsations (ca. 1.0 - 1.5 s) can be usually approximated by a sinusoid function with the exponentially decreasing amplitude and is repeated in the subsequent cycles of pulsations. The shape of changes in density differs from cycle to cycle, but this process is slow enough to use the information gained during a particular cycle to process data in several subsequent cycles.

The signal from the radiation detector is a non-stationary discrete process described by the Poisson distribution. The number of counted pulses during the time of measurement t_s is calibrated in density units. The time of measurement is usually constant in industrial radiometric monitors. The dynamic error of measurement in the case of changes in density (how accurately the changes are reproduced) depends on the time of measurement t_s and can be minimized for the optimal t_s. The optimal time of measurement depends on the shape of changes in density. Better results can be achieved for discrete filters with parameters adapting on-line to changes in density during a cycle. The basic problem of these methods is how to determine time derivative for a highly noisy non-stationary signal from the radiation detector.

An alternative method of signal filtration, discussed in this paper, is proposed for the variable time of measurement during a cycle of pulsations. Good results can be achieved for a linear relation between the time of measurement and time derivative of the density. The density time derivative can be determined for a long period of time on the basis of industrial tests of average changes in density in subsequent cycles of pulsations. In this case, the dynamic error of measurement MSE_c can be reduced by half compared to the optimal constant time of measurement.

The above results suggest that it is possible to further improve the presented method of density signal filtration by using the identified change in density in one cycle to adjust parameters of the algorithm generating the variable time of measurement in the next cycle or several cycles. This adaptive method will be the subject of future research.

Acknowledgments

Acknowledgement goes to the Rydultowy Mine for providing plant data.

References

[1] Lyman G J and Jonkers A 1992 *Forschungshefte* **53** pp 131-6
[2] Lyman G J 1992 *Coal Preparation* **11**, pp 41-72
[3] Jonkers A, Lyman G J and Loveday G K 1998 *The XIII ICPC (Brisbane)* Advances in modeling of stratification in jigs
[4] Jonkers A and Cameron P M 2000 *After 2000 –The Future of Mining Annual Conf (Sydney)* JigScan – A case study in product research, development and commercialization p 174-78
[5] King R P 2001 *Modeling & simulation of mineral processing systems* (Oxford: Butterworth-Heinemann)
[6] Diendonne V and Jonkers A 2006 An approach to confidently predicting jigging performance *The Journal of the South African Institute of Mining and Metallurgy* **106**
[7] Knoll G F 2000 *Radiation Detection and Measurement*, (New York: John Wiley & Sons)
[8] Cierpisz S and Joostberens J 2016 *Measurement*, **88** pp 147-52
[9] Cierpisz S and Joostberens J 2017 *Inżynieria Mineralna - Journal of the Polish Mineral Engineering Society* **12** pp 119-26
[10] Snyder D R and Miller M 1991 *Random Point Process in Time and Space* (New York: Springer - Verlag).
[11] Leveille E and Hamel P 2018 *Methodology and Computing in Applied Probability* **20** pp 353-68
[12] Picinbono B 2014 *Communications in Statistics - Simulation and Computation* **43(7)**
[13] Li Z 2011 *Proc. of the Fourth International Workshop on Advanced Computational Intelligence IWACI (Wuhan, China)* The stability of delays BAM neural networks with variable-time and Poisson arrival impulses.

IOP Conf. Series: Materials Science and Engineering **427** (2018) 012021 doi:10.1088/1757-899X/427/1/012021

Water content in froth products depending on the salinity level of pulp flotation

K Tupek-Murowany and A Młynarczykowska

AGH, University of Since and Technology, Faculty of Mining and Geoengineering Department of Environmental Engineering and mineral Processing, al. Mickiewicza 30, 30-059 Kraków, Poland

ktupek@agh.edu.pl

Abstract. This paper presents the results of the evaluation of the coal flotation efficiency in the presence of an inorganic salt in the pulp. It was shown that change of salt contains (NaCl) in pulp affects to the stability and mineralization degree of floatation foam. These results have confirmed to the effectiveness of enrichment. The quantitative measure of foam quality was the water recovery in this flotation product.

1. Introduction

Flotation is one of the many most complex methods of enrichment. It allows the separation of components useful from gangue in the smallest grades of particles. The effects of the process depend on a number of closely related parameters. The most important aspect for the distribution of foam flotation material is the formation of a float foam layer in the upper part of the pulp.

The foam is formed as a result of floating of particle - gas bubble aggregates to the surface of the pulp surface. Its construction and stability have a significant impact on the quality of the obtained concentrate [1,2].

According to the definition, foam stability is related to the effectiveness of bubbles for foaming, their durability and limited susceptibility to coalescence [2]. The stability of the floatation foam generated in the flotation process depends on type and dose of foaming agents used, gas dispersion, suspension viscosity, size and particles number suspended in the flotation chamber, induction time and the water quality used of the process [1-6]. It turns out that fine particles, mainly grains of gangue have a greater impact on the stability of froth floatation, and thus lead to an increase of water amount in the foam [2,3,6,7]. Similarly, a larger amount of water in the foam means higher content of gangue in concentrate [7,8]. Therefore, very important during flotation it is to drain water from foam, which makes it possible to break off the gangue grains and return back of them to the pulp, thus increasing the quality of concentrate. Greater stability allows better concentrate yield, and its dewatering takes place mainly in lower part of foam, so that the amount of water in froth has been linked to at the speed of receiving the concentrate, depth of its stripping and flotation time [7].

Foam stability is also strongly associated with the salinity of technological water. An addition of electrolyte to the enrichment process results in the production of smaller and more strongly round and dense air bubbles, and also inhibits their coalescence due to the increased repulsive force between them. [3,4,5,6,9,10]. It should be borne in mind, that not all inorganic salts are able to inhibit bubble coalescence [3,6]. Thanks to the production of fine air bubbles, flotation in saline water leads to better yields of the combustible substance in concentrate. The salt present in the flotation process causes the aggregation of grains, including gangue trapped within the coal grains, thereby increasing its entrainment, which in turn increases the ash content of concentrate and foam stability [2,4,9,11].

In addition, the presence of electrolytes in pulp flotation has a significant impact on the frother. According to Castro et all, 2013 [3], relatively weak surfactants become stronger as a result of increased water salinity, which in the initial stages leads to improved foaming. Nevertheless, at high concentrations of electrolytes, the solubility of frother is blocked. Wei et all, 2016 [5] investigated the effect of common salt in pulp flotation and the frother on foam stability, and they found that salt is a more important stimulus in stabilizing foam.

In summary, it can be concluded that the presence of very stable foam is undesirable, because by mechanical gangue flotation the quality of concentrate is lower [2,12].

The aim of the tests was to determine the impact of salt content in process water on the amount of water in foam products. The relationship between yields and water content in products has been evaluated.

2. Flotation tests

The tests were carried out on a 32.2 type carbon slurry with a particle size below 1 mm, of which almost 50% of grains constituted a particle class below 0.063mm. The average ash content of the examined coal was 26.55%.

The flotation tests were performed in a Denver type pneumo-mechanical laboratory machine with a working chamber volume of 1dm^3. Both the flotation pulp density and rotor speed for each flotation test was constant and amounted to 80 g / dm^3 and 1850 rpm respectively. The RF55 (Reaflot-55) reagent from the glycol group was used as a collector and frother in the flotation process. A variable parameter of experiments was the salt content in process water considered as the content of sodium chloride. One of the test was carried out with mine water taken from one of the Polish mines whose salinity was 0.23M. Water with 0.3 and 0.4M salinity was obtained by adding pure NaCl to mine water. While, the salinity at the level of 0.05M NaCl was obtained by mixing a coal slurry with distilled water. The final salt content (NaCl) expressed by the molar concentrations in process water was: 0.05M, 0.23M, 0.3M and 0.4M. Mixture of distilled water with a coal slurry rinsed from salt was used for flotation test number 5.

3. Kinetics of flotation

Each time, the carbon samples were flooded of saline water (the various salt concentration) and the material was conditioned to moisten it, for 5 minutes. After a set time, the floatation machine was started and mixed feed without air for 1 minute, after which the flotation reagent was added at 1016.21g per ton of dry feed and stirring continued for further 3 minutes. In the next step, the air in flotation chamber was started and the fractional flotation was performed. The total flotation time was 6 minutes except for one case of flotation test (shortening flotation resulted from the absence of mineralization froth flotation and its spontaneous quenched). The concentrates obtained during the flotation tests were weighed with water and after drying. The measurements were carried out in various time steps.

4. Results

The paper evaluates the impact of salinity pulp flotation on water content in the foam. In table 1 presented results obtained from the flotation tests for concentrates and water content in grams for individual flotation concentrates from experiments carried out with variable NaCl content in the pulp.

IOP Conf. Series: Materials Science and Engineering **427** (2018) 012021 doi:10.1088/1757-899X/427/1/012021

Table 1. Yields of concentrates and water content in concentrates at a variable salinity level of the flotation pulp, depending on the duration of the process.

Concentrate number	Flotation time t(s)	**F1**, salinity of the water 0,4M NaCl		**F2**, salinity of the water 0,3M NaCl		**F3**, salinity of the water 0,23M NaCl		**F4**, salinity of the water 0,05M NaCl		**F5**, flotation without salt	
		Mass of water in the froth (g)	Yield (%)	Mass of water in the froth (g)	Yield (%)	Mass of water in the froth (g)	Yield (%)	Mass of water in the froth (g)	Yield (%)	Mass of water in the froth (g)	Yield (%)
K1	30	302.6	72.54	379.8	71.76	274.3	66.54	239.6	62.57	399.1	81.25
K2	60	158.0	8.03	169.8	6.92	191.0	12.23	188.7	14.11	211.6	8.44
K3	120	187.6	4.40	233.8	4.90	176.1	6.62	140.2	8.38	167.9	3.48
K4	180	212.6	3.11	224.4	2.59	149.7	3.87	87.8	3.35	54.6	0.07
K5	240	105.4	1.30	151.8	1.15	61.4	0.75	66.6	1.40	-	-
K6	300	96.4	1.04	102.6	0.86	60.5	0.62	45.1	0.70	-	-
K7	360	115.2	1.04	111.0	1.15	78.5	0.37	30.8	0.28	-	-

In order to visualize the results, the graph showing the dependence of water content in individual foam products for flotation time was presented in figure 1, while figure 2 illustrates the changes of water content in each flotation concentrates depending on the salinity of the process water and reaction time.

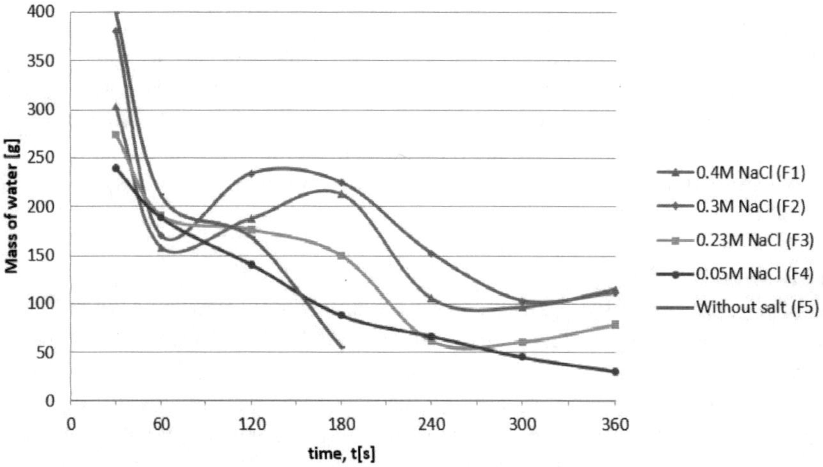

Figure 1. The water content in the float foam as a function of the process progress, for the variable salinity level of floatation pulp. F1- F5 – number of the flotation test.

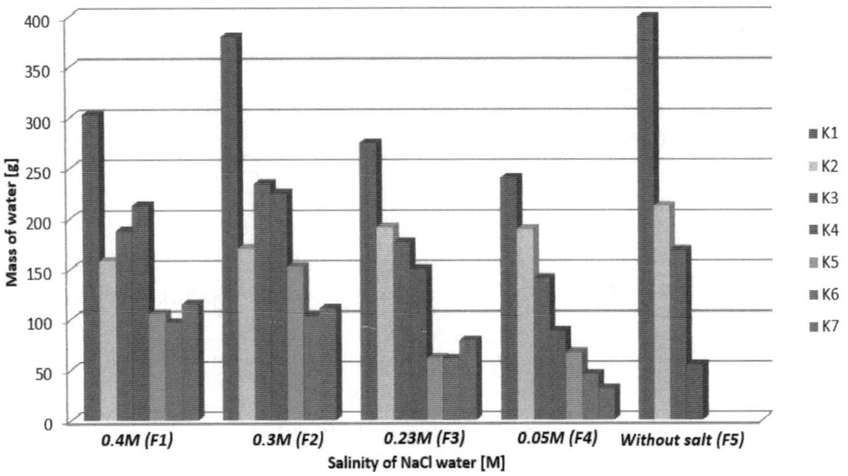

Figure 2. The water content in froth products for individual concentrate in relationship on the concentration of NaCl in flotation pulp. K1-K7 – concentrate number; F1-F5 – number of the flotation test.

The analysis of each of the floatation tests (table 1 and table 2) we may notice that the usual amount of water in the received floatation foam decreases overtime during the process. The highest amounts of water are present in the first floatation concentrates (K1) which reflect the acquired yield values. As the floatation process progresses yield values significantly drop over time along with the amounts of water in the foam. One exception is the example which shows an increase in the amount of water in the foam product of the third (K3) and fourth (K4) concentrate bearing the highest salinity level of the floatation pulp (F1 and F2) - see fig. 2. We can also noticed a minute increase in water amount for the last concentrates in flotations F1, F2, and F3 which does not thoroughly correlate with the increase in yields of these products. The probable cause might be a mechanical elevation of gangue towards the concentrate which raises the density of solid particles in the foam (decreases concentrate quality) and thus the amount of water in the foam.

Taking into account the amount of water in the foam products depending on floatation time for various concentrations of NaCl in floatation pulps (table 1 and figure 1), we can observe a gradual decrease of water in the foam of the first concentrate (K1) for flotations F2 through F4. The correlation does not apply to concentrate K1 for floatation F1 where the water amount in the foam is 77,2 grams less than in K1 floatation F2. However, the yield acquired in F1 is larger than in floatation F2 even though the quality of the concentrate is lower. In concentrate K1 for floatation F1 the ash content is 17,56% whereas in concentrate K1 for floatation F2 equals 16,53%. This tendency might be caused by insufficient solubility of the frother (foaming factor) accompanied by higher levels of inorganic salt and the mechanic elevation of gangue towards the concentrate due to particle aggregation [3], thus we observe a decrease in the amount of water in the foam. Relating to the other floatation concentrates we may observe a descent of water amount in the foam starting with floatation F2 and successively flotations F1, F3, and F4. The trend is inverted for K2 and ranks the flotations respectively F3, F4, F2, and F1. Single derogations occur for K1 F5 where the lack of salt in the water caused a production of a wet foam which correlates with the highest yield for K1.We may assume that a significant salinity of the water adversely impacts the flotation reagent. We must highlight, however, that not only the presence of NaCl but also the amount of salt in the flotation pulp directly affects the enrichment outcomes, because yields for K1 increase along with the salinity of the flotation pulp. For experiments F3 and F4 the quality of the concentrate is also increasing (the amount of ash is F3: λ_A = 14.82% and F4: λA = 15.19%) compared to the results obtained for F1, F2, and F5 (F1: λ_A = 17,56%; F2: λ_A = 16,53% and F5: λ_A = 16,46%). Furthermore, we need to highlight the fact that a large yield does not always correspond with producing a better quality concentrate. Literary sources [2,4,9,11,13] support the theory that a large amount of water in the foam might be the stimulus of raising the hydrophilic particles towards the foam layer. Similar to the aforementioned case, despite higher figures of yield the quality for concentrate K1 in flotation F5 (λ_A = 16.46%) has decreased in comparison with the results of flotation F3 (λ_A = 14.82%) and F4 (λ_A = 15.19%).

5. Summary

The article is an analysis of the effects of salt in the form of sodium chloride in water on the amounts of water in flotation foam. The analysis of water in the foam has been correlated with yield amount and quality of concentrates. On the basis of the presented results of experiments carried out we may assume that:

- increased wetness of the flotation foam allows to obtain higher yield concentrates compared to the first enrichment products, although they are of lower quality,
- usually the amount of water in the flotation foam increases with the growth of water salinity
- on average water amount in flotation foams drop as the process progresses over time,
- in the event of a lack of salt in the flotation pulp, but also with higher levels of salinity, there is correlation between the increase of yield in first concentrates, which is caused by the presence of hydrophilic particles, and the increased amount of water in the flotation foam,
- raising the level of inorganic electrolytes in the flotation pulp leads to mechanical elevation of gangue towards the concentrate.

Acknowledgments

This work was done as a part of the Faculty Dean Research Program No. 15.11.100.086 and as a part of the University of Science and Technology Research Program No.11.11.100.276

References

[1] Farrokhpay S and Zanin M 2011 Chemeca 2011: Engineering a Better World Sydney, New South Wales, Australia, 18-21 Sptember 2011 pp 1-8
[2] Farrokhpay S 2011 *Adv. Coll. Interf. Sci.* **166** pp 1-7
[3] Castro S, Miranda C, Toledo P and Laskowski J S 2013 *Int. J. Miner. Process.* **124** pp 8-14
[4] Wang B and Peng Y 2014 *Miner. Eng.* **66-8** pp 13-24
[5] Wei T, Peng Y and Vink S 2016 *Int. J. Miner. Process.* **148** pp 15-22
[6] Li G, Deng L, Cao Y, Wang B, Ran J and Zhang H 2017 *Int. J. Miner. Process.* **169** pp 47-52
[7] Liang L, Li Z, Peng Y, Tan J and Xie G 2015 *Miner. Eng.* **81** pp 96-102
[8] Warren L.J 1985 *Int. J. Miner. Process.* **14(1)** pp 33-44
[9] Wang B and Peng Y 2014 *Fuel* **134** pp 326-32
[10] Ozdemir O, Taran E, Hampton M.A, Karakashev S.I and Nguyen A.V 2009 *Int. J. Miner. Process.* **92** pp 177-83
[11] Wang B, Peng Y and Vink S 2014 *Miner. Eng.* **66-68** pp 145-51
[12] Qu X, Wang L and Nguyen A.V 2013 *Miner. Eng.* **41** pp 25-30
[13] Drzymała J 2009 *Mineral Processing* (Wrocław: Oficyna Wydawnicza Politechniki Wrocławskiej) pp 315-16

IOP Conf. Series: Materials Science and Engineering **427** (2018) 012022 doi:10.1088/1757-899X/427/1/012022

Biomineralization of calcium carbonate controlled by biomolecules produced by *Bacillus* sp.

A Bastrzyk[1], I Polowczyk[1], KJ Legawiec[1], J Chojniak[2], K Paraszkiewicz[3] and G A Płaza[2]

[1]Faculty of Chemistry, Wroclaw University of Science and Technology, Norwida 4/6, 50-373 Wroclaw, Poland

[2]Department of Environmental Microbiology, Institute for Ecology of Industrial Areas, Kossutha 6, 40-844 Katowice, Poland

[3]Faculty of Biology and Environmental Protection, University of Lodz, Banacha 12/16, 90-237, Lodz, Poland

izabela.polowczyk@pwr.edu.pl

Abstract. In this study, culture supernatant of Bacillus subtilis KP7 growing on liquid LB medium consisted of 2 % of molasses was used for calcium carbonate synthesis. Synthesis of the calcium carbonate structures was performed in a reactor by mixing $CaCl_2$ and Na_2CO_3 solutions at 25°C. The calcium carbonate particles were characterized by optical microscopy, FTIR and XRD analysis. In addition, the size distribution and the value of zeta potential were determined. It was observed that properties of obtained calcium carbonate depend on the supernatant amount. The smallest crystals of pure calcite with disordered structure were obtained when the concentration of the cell-free supernatant was 5 % (v/v).

1. Introduction

Calcium carbonate is an extremely important material both in the fundamental research and industry due to their properties, such as non-toxicity, high porosity, and surface area [1]. In nature, calcium carbonate is a biomaterial consisted of both mineral (c.a. 95 %) and organic (c.a. 5 %) phases [2]. The organic content involved in biomineralization was found to be mostly biomolecules containing the functional groups such as carboxylic, phosphate, sulfate, hydroxyl or amino [3]. Those biomolecules can act as nucleation sites, and control the nucleation and further growth of the crystals to desired shape and morphology. During the last decades a wide range of organic additives or templates have been used in the precipitation of calcium carbonate to obtain product with desired properties [1]. A very interesting issue seems to be the application of molecules produced extracellularly by microorganisms as agents controlling the growth of calcium carbonate. It is well known that in nature microorganisms can induce the mineralization of calcium carbonate via three mechanisms: as a result of microbial metabolic pathways, ions exchange through the cell membrane, and extracellular polymeric substances (EPS) mediated mineralization [4]. These EPS may influence mineral formation by trapping calcium ions and inhibit the transformation of metastable to thermodynamically stable phases [5-7].

The purpose of this study was to investigate the influence of cell-free supernatant on the course of calcium carbonate precipitation. The application of cell-free supernatant to control the morphology of calcium carbonate allows to develop a new eco-friendly technology of inorganic materials production. The culture supernatant contains not only the EPS (protein, polysaccharides, and enzymes) but also the other metabolic products, such as biosurfactants and other ions, which can affect the shape, size and polymorphs of calcium carbonate. It is happened because in living organisms not only the proteins but also the content of reaction fluid (ions, polysaccharides, amino acids, and other metabolic products) takes part in biomineralization of calcium carbonate.

2. Materials and Methods

Bacillus subtilis KP7 was obtained from a collection owned by the Department of Industrial Microbiology and Biotechnology, University of Lodz. Production of bio-products by Bacillus sp. was carried out as described previously in Paraszkiewicz et al. [8]. The culture was grown on a media containing 2 % of beet molasses. The strain was cultivated aerobically for 96 h in a shaker (120 rpm) and at a constant temperature, 30 °C. Next, to remove the bacterial cells from the media the culture solution was centrifuged at 10.000 x g for 20 min and then filtered using a cellulose nitrate membrane filter (0.22 μm diameter of pore; Sartorius). The surface tension of the supernatant was measured using a ring method (type K100, Krüss). The pH of supernatant was found to be 8.0±0.1.

Calcium carbonate was synthesized using a semi-batch method at 25 °C. Firstly, 125 ml of calcium chloride (purity > 99 %, Across Organics) and di-sodium carbonate (p.a., Avantor) solutions at 0.01 M were separately prepared. Then, the calcium chloride solution was placed in a reactor vessel equipped with a mechanical stirrer (type OS20 S, ChemLand, Poland). The reaction was started when the Na_2CO_3 solution was dozing (Masterflex L/S, Cole Parmer) to the reactor vessel with constant rate, 150 ml/min. The mixture was stirred at a constant speed for 1 h. After that the precipitated calcium carbonate was removed by centrifugation (MPW, Poland) at 4000 rpm for 10 min. The precipitate was washed several times with deionized water and once with ethanol (96 %, POCh). Next, the solid was dried at 40 °C and kept in a desiccator for further analysis. In the experiment with biomolecules, the calcium ions were contacted with cell-free supernatant at concentration 2.5 and 5 % for 24 h.

The shape of calcium carbonate was characterized by optical microscopy (AxioImager M 1, Zeiss). The polymorphs of the precipitate was analysed by X-ray diffraction (D8 ADVANCE, Bruker, Massachusetts, USA). The size of crystals was analysed using a laser diffractometer (Mastersizer2000, Malvern, UK). The zeta potential of the crystals was measured using a Zetasizer2000 apparatus (Malvern, UK).

3. Results and discussion

The morphologies of calcium carbonate precipitated in the presence of cell-free supernatant are presented in figure 1. From figure 1a-b it can be seen that in the control sample synthesized without biomolecules a mixture of rhombohedral calcite and spherical vaterite crystals were formed. Addition of various amount of cell-free supernatant produced by Bacillus subtilis KP7 resulted in formation of only one phase, calcite crystals with some deformation on the surface. From literature it is known that biomolecules can affect the morphology of crystals by presence of some irregularities (e.g. terraces and depletion) on the surface, and distribution of the crystals' edges [1,7]. The crystals' structure was more distorted when more supernatant were added to the reaction system.

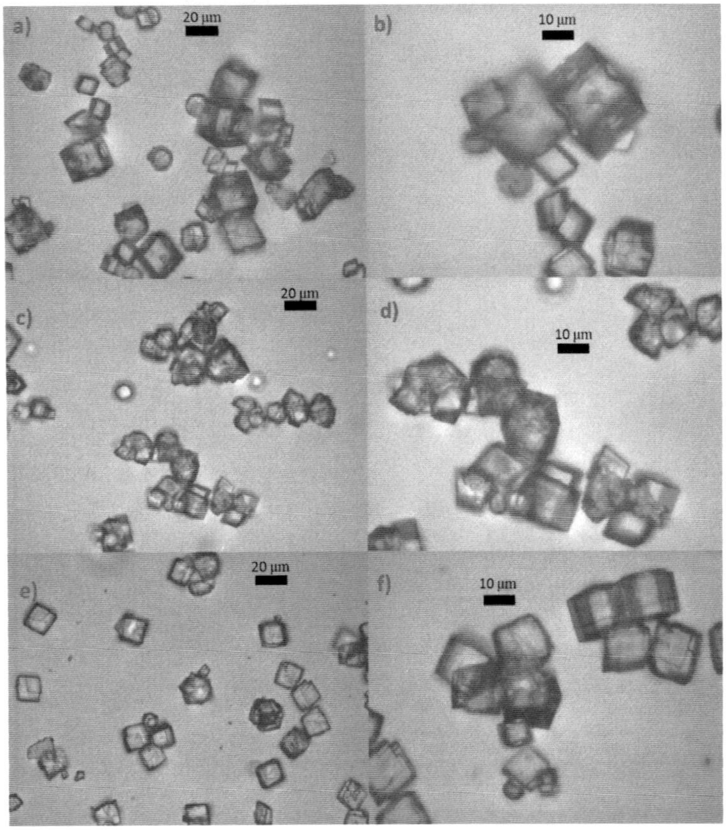

Figure 1. Optical images of calcium carbonate crystals obtained in the presence of biomolecules produced by *Bacillus subtilis* KP7. (a-b) control sample (without biomolecules), (c-d) 2.5 %, (e-f) 5 %.

The polymorphs of obtained crystals were analysed by X-ray Powder Diffraction technique, and the XRD patterns are shown in Figure 2. The presence of calcite indicates the peaks located at 2θ values of 23.0; 29.4; 36.1; 39.4; 43.1; 47.5; 48.5°. Whereas, the formation of vaterite in the system is evidenced by peaks at 24.9, 27.2; 32.9; 43.1; 50.2° (JCPDS 00-033-0268). Based on the data presented in Figure 2, the phase composition of precipitates was calculated using the Rietvield method. It was found that the control sample contains 15.1 % of vaterite and 84.9 % of calcite. The sample synthesized with 2.5 % of the supernatant is consisted of 1.2 % of vaterite and 98.8 % of calcite. While addition of 5 % of the supernatant leads to the formation of pure calcite. These data suggest that biomolecules present in the supernatant favours the formation of calcite crystals in the reaction medium. It is worth to mention that the supernatant contained not only the EPS but also (bio)molecules (fermentable sugars, D- and L-lactic acid, short-fatty acids, phenolic compounds and products of sugar caramelization) delivered from the molasses [9]. All those biomolecules can possess functional groups which can bind calcium ions and can be involved in the mineralization of calcium carbonate in nature. This interaction can affect the nucleation rate as well as the crystals growth [1].

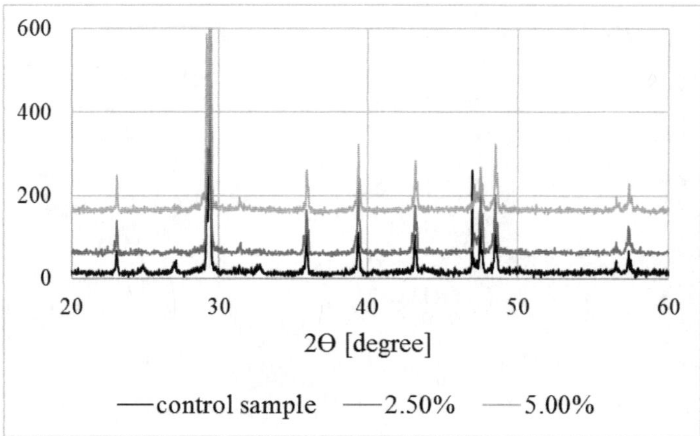

Figure 2. XRD patterns of calcium carbonate crystals obtained in the presence of bio-products produced by *Bacillus subtilis*.

The size distribution and the diameters of crystals are presented in figure 3 and table 1. It can be seen that biomolecules in the supernatant reduce the size of formed crystals. The mean diameter (d_{50}) was 27.7, 17.7 and 12.1 μm for crystals obtained at 0, 2.5 and 5 % of the supernatant concentration, respectively. The values of d_{10} and d_{90} decreased with an increase in the supernatant content in the reaction mixture from 16.8 to 6.1 and from 43.5 to 22.2 μm, respectively. We believe that the biomolecules can react not only with calcium ions but also with the crystals surface inhibiting the growth of crystals. For this reason, the small particles are formed in the presence of cell-free supernatant.

IOP Conf. Series: Materials Science and Engineering **427** (2018) 012022 doi:10.1088/1757-899X/427/1/012022

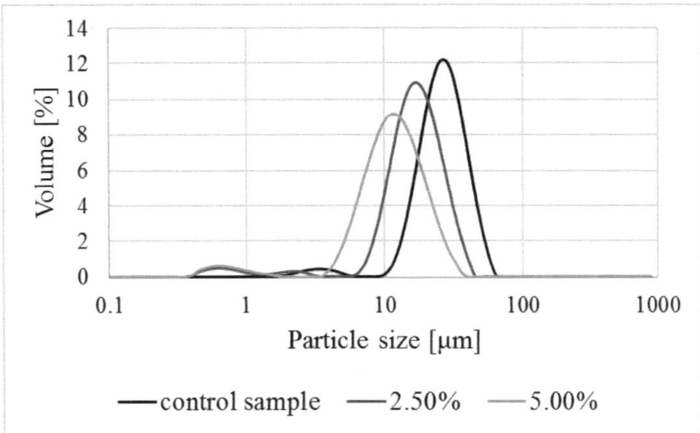

Figure 3. Size distribution of calcium carbonate crystals obtained in the presence of bio-products produced by *Bacillus subtilis* KP7.

Table 1. Effect of cell-free supernatant concentration on the crystal's size (The standard deviation is lower than 1 µm).

Concentration [%]	d_{10} [µm]	d_{50} [µm]	d_{90} [µm]
0	16.8	27.7	43.5
2.5	10.0	17.7	29.4
5.0	6.1	12.1	22.2

The presence of the biomolecules at the surface of the crystals was proven by zeta potential measurements and the data are presented in Table 2. It was observed that the particles precipitated without the additives possess negative value of zeta potential equals -25.6 mV. The value was changing when the supernatant was added to the reaction medium and it was -30.6 and -36.4 at 2.5 and 5 % of the supernatant concentration, respectively. We suspect that the changes in the zeta potential value results from the presence of the biomolecules at the crystals surface.

Table 2. Zeta potential (ζ) values of calcium carbonate crystals precipitated in the presence of biomolecules (The standard deviation is lower than 1 mV).

Concentration [%]	ζ [mV]
0	-25.6
2.5	-30.6
5.0	-36.4

Literature data showed that *Bacillus* species are microorganisms producing extracellularly various proteins (mostly enzymes), amino acids, biosurfactants (surfactin, iturin and fengicyn) and a lot of

IOP Conf. Series: Materials Science and Engineering **427** (2018) 012022 doi:10.1088/1757-899X/427/1/012022

polysaccharides [8]. The presence of biosurfactants was proven by surface tension measurements, and it was found to be 35.03 mN/m. The biomolecules possess functional groups such as carboxyl, hydroxyl, amino and aliphatic. Among them the carboxyl and hydroxyl group are suggested to play a major role in the biomineralization.

4. Conclusions

Presented data showed that the biomolecules produced by *Bacillus subtilis* KP7 growing on media containing 2 % of molasses can be used as agent to control the growth of calcium carbonate. These biopolymers can affect the morphology, size and surface properties of the crystals. It was shown that smaller crystals with some irregularities at the surface were formed when in the reaction medium the cell-free supernatant was added. This finding encourage us to further studies on this topic. We believe that using cell-free supernatant produced by *Bacillus* sp. can develop an eco-friendly technique to produce inorganic matrices for environmental application. Moreover, the application of molasses as a source of nutrition to microorganisms growth can reduce the cost of biopolymers production.

Acknowledgments

Authors wishing to acknowledge the National Centre of Science for financial support (project no. 2014/15/D/ST8/00544).

References

[1] Boyjoo Y, Pareek VK and Liu J 2014 *J. Mater. Chem. A.* **10** p 14270
[2] Marin F and Luquet G 2008 *Mater. Sci. Eng. C* **25(2)** p 105
[3] Deng H, Shen XC, Wang XM and Du C 2013 *Front. Mater. Sci.* 7 p 62
[4] Lian B, Hu Q, Chen J, Ji JH and Teng H 2006 *Geochim. Cosmochim. Acta* **70** p 5522
[5] Bains A, Dhami NK, Mukherjee A and Reddy MS 2015 *Appl. Biochem. Biotechnol.* **175** p 3531
[6] Czemierska M, Szcześ A, Hołysz L, Wiater A and Jarosz-Wilkołazka A 2017 *Eur. Polymer J.* **88** p 21
[7] Szcześ A, Czemierska M, Jarosz-Wilkołazka A, Magierek E, Chibowski E and Hołysz L 2018 *Pysichem. Probl. Miner. Process.* **54(1)** p 142
[8] Paraszkiewicz K, Bernat P, Kuśmierska A, Chojniak J and Płaza G 2018 *J. Environ Manage.* **209** p 65
[9] Šarić LC, Filipčev BV, Šimurina OD, Plavšić DV, Šarić BM, Lazarević J.M and Milovanović IL 2016 *Food Feed Res.* **43(2)** p 135

IOP Conf. Series: Materials Science and Engineering **427** (2018) 012023 doi:10.1088/1757-899X/427/1/012023

Image histogram as a tool for a coal stream homogeneity evaluation

A Heyduk[1], J Joostberens[1], A Nowrot[1], J Pielot[1] and K Fabiańczyk[2]

[1]Department of Electrical Engineering and Automation in Industry, Faculty of Mining and Geology, Silesian Technical University, Akademicka 2, 44-100, Gliwice, Poland

[2]Polcargo International Ltd, Szczecin, Poland.

aheyduk@polsl.pl

Abstract. The paper presents selected results of the preliminary laboratory research on the application of histogram analysis to the problem of coal stream homogeneity evaluation. This is an important problem of industrial coal mine output stream quality testing. Image acquisition is fast and cheap, but needs complex analysis techniques. Histogram analysis can be considered to be one of them. There has been proposed a distance measure, and there have been presented analysis results which proved an ability to detect gangue, rock, schist, and even wooden block contamination of the homogenous coal stream. The method needs further improvement and comprehensive analysis of other available options of the RGB-to-grayscale conversion and of histogram comparison to make it more selective. These further research directions have also been described in the paper.

1. Introduction

Coal quality monitoring is an important part of the whole chain of power generation and utilization. It is important from commercial and ecological point of view. Stable coal quality is very important for the proper operation of coal-fired power plants in order to ensure stable combustion conditions and low level of pollutants emission [1]. Continuous quality testing meters (like ash-monitors, heat-quantity meters, sulphur meters) are expensive. This fact (in addition to legal regulations requiring expensive and time-consuming laboratory testing of samples taken from the coal stream) hinders their wide application and inspires the search for new relatively low-cost methods of coal quality assessment, easily integrable with an existing coal quality testing framework, based on discrete-time and limited-size sample laboratory tests. Image processing can be considered as one of this methods because one of the crucial factors is the number of samples necessary to analyze in laboratory conditions. When the coal stream is homogenous, the sample size can be reduced, therefore decreasing the cost and time of laboratory analyzes. This coal stream homogeneity is currently analyzed locally by human inspectors but the video monitoring with automatic image processing and analysis can further reduce costs and time of this analysis simultaneously improving the result (due to limited human visual perception capabilities). Visual assessment of the coal quality can be performed at various detail levels both in the time domain and in the spatial domain. In the spatial domain the most detailed analysis level may include the thorough evaluation of specific coal particles. These coal particle contours must be precisely separated - that is, the entire image must be subjected to a complex and time-consuming image segmentation process [2] The most general analysis level in the space domain may include assessment of the whole image (or

IOP Conf. Series: Materials Science and Engineering **427** (2018) 012023 doi:10.1088/1757-899X/427/1/012023

selected image fragment) as a single entity in either colour/intensity or in texture domain. This is the method analyzed in this paper. Resolution in time domain - regardless of the selected image analysis method - can vary from several random images of the whole train (homogeneity of the whole train coal load) to several images for each railway wagon (homogeneity monitoring of each wagon content). These two extreme possibilities are presented in figure 1.

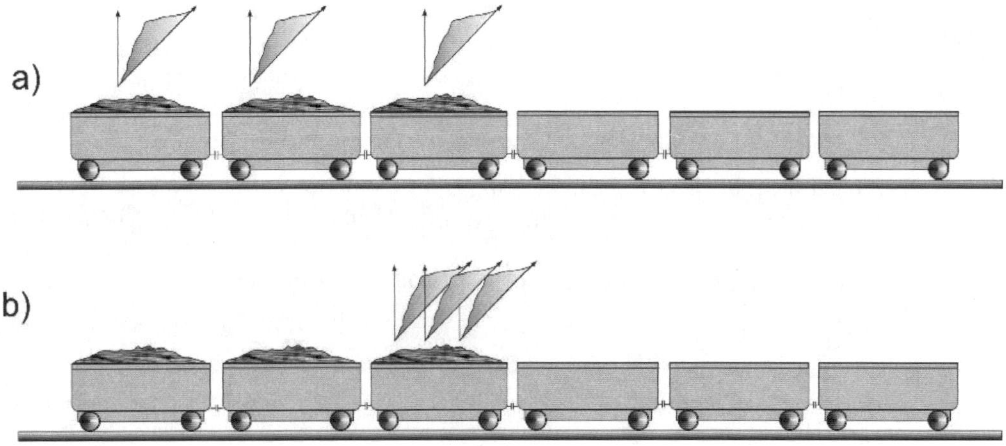

Figure 1. Two possible applications of the proposed method: a) analysis of the content homogeneity of a larger number of wagons throughout the whole train; b) detailed analysis of the coal stream homogeneity throughout one wagon.

In order to ensure stable lighting conditions, images necessary for further analysis should be recorded not directly in the train or railway wagon, but on the path of the belt conveyor transporting the coal to the loading point in a single place with a stable artificial lighting, shielded from the direct influence of the sunlight. Before implementing this concept in industrial practice, there have been conducted small-scale laboratory tests in controlled laboratory environment using selected samples of both homogenous clean coal and contaminated coal. Laboratory tests concerned two issues:

- homogeneity evaluation of the clean coal stream
- the ability to detect contaminated fragments of the coal stream

2. Image histogram acquisition

The images of coal samples have been recorded using a Logitech HD camera (2Mpix resolution). They contained several coal particles placed on a real conveyor belt. In the laboratory setup 1 pixel corresponded to approximately 1mm^2 of the real image area. Then –as a task was to analyze coal surface not a conveyor belt surface- only a small rectangular fragment of the image has been selected, containing only the particles and the minimum part of the background. Therefore the information value of the analyzed image has been significantly enhanced. The selection process is presented in figure 2. Particular attention has been paid to provide a uniform illumination of the coal surface, so a distributed system of several halogen bulbs has been used.

The image has been recorded using an RGB camera, so for each pixel three different colour component values R(x,y), G(x,y), B(x,y) have been recorded. They have been converted to single grayscale intensity value GS(x,y) using an equation corresponding to human colour perception [3][4]:

$$GS(x,y) = 0{,}299 \cdot R(x,y) + 0{,}587 \cdot G(x,y) + 0{,}114 \cdot B(x,y) \tag{1}$$

IOP Conf. Series: Materials Science and Engineering **427** (2018) 012023 doi:10.1088/1757-899X/427/1/012023

a) b) c)

Figure 2. Image fragment selection. a) whole recorded image b) selection of the meaningful part, c) selected part – zoomed.

There can be also used simplified formulas like

$$GS(x,y) = 0{,}3 \cdot R(x,y) + 0{,}6 \cdot G(x,y) + 0{,}1 \cdot B(x,y) \tag{2}$$

or even

$$GS(x,y) = \frac{R(x,y) + G(x,y) + B(x,y)}{3} \tag{3}$$

and nonlinear (max/min) formulas like

$$GS(x,y) = Max\left[R(x,y),\, G(x,y),\, B(x,y)\right] \tag{4}$$

and

$$GS(x,y) = Min\left[R(x,y),\, G(x,y),\, B(x,y)\right] \tag{5}$$

or

$$GS(x,y) = \frac{Max\left[R(x,y), G(x,y), B(x,y)\right] + Min\left[R(x,y), G(x,y), B(x,y)\right]}{2} \tag{6}$$

Advantages and disadvantages of these approaches need further research (particularly in the case of nonlinear formulas – because they use nonlinear Max/Min functions it can be said that the equation (4) enhances brighter parts of the image and equation (5) enhances darker parts of the image).

3. Image histogram analysis

3.1. Histogram similarity and dissimilarity measure

As the analyzed histograms contain the great number of bins (natural colour resolution of the RGB camera CCD matrix is usually 8-bit so it covers the range between 0 and 255) their comparison can be a complex task. Because they are unimodal (they have a single dominating value – corresponding to the most frequently occurring colour of coal particles surface) the modal value comparison can be one of the conditions of histogram similarity. But the modal value describes only a histogram location not its shape. The distance between two histograms must be calculated as a function of all bins. Sometimes it should also consider a different number of samples in each histogram, but in this case this condition can be omitted as the number of image pixels can be established as constant, or histogram bins values can be rescaled to their relative value as a part of the total number of pixels in the whole analyzed image area. As a distance measure for two histograms $h_1[n]$ and $h_2[n]$ there has been proposed a normalized sum:

$$D(h_1,h_2)=\frac{1}{N}\sum_{i=1}^{N}\frac{(h_1[i]-h_2[i])^2}{f_1[i]^2+f_2[i]^2} \tag{7}$$

where N is a number of histogram bins. This formula is normalized with respect to the bin number and in each bin, the distance between two histogram values is calculated relatively to the average value of both bin content. In the case of sequential analysis of the video stream, consecutive values of (7) form a numerical sequence and a rapid change in these distance values proves a change in the output coal stream qualitative parameters. There are also many other possibilities of histogram distance calculation methods, described in [5][6].

3.2. Analysis of the homogenous coal stream

Analysis of the homogenous stream has been performed using several times the same coal sample (without gangue) but with particles placed in different layouts. Sample images are presented in fig. 3. and their corresponding grayscale intensity histograms in figure 4.

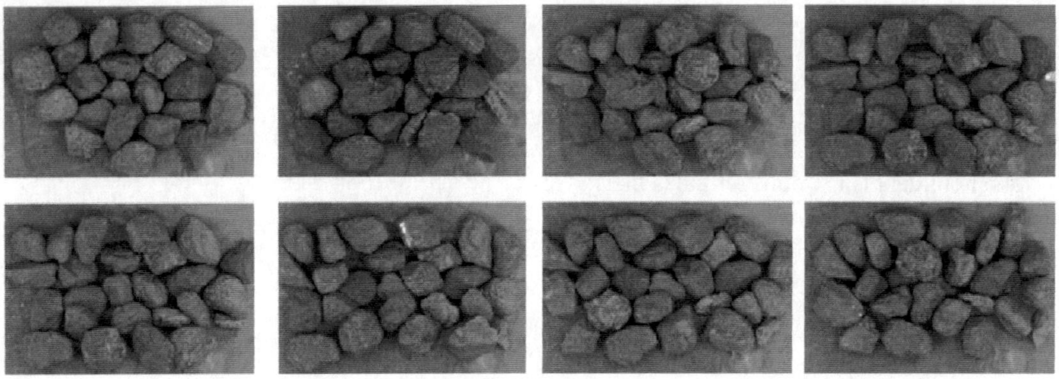

Figure 3. Coal sample images representing a homogenous coal stream.

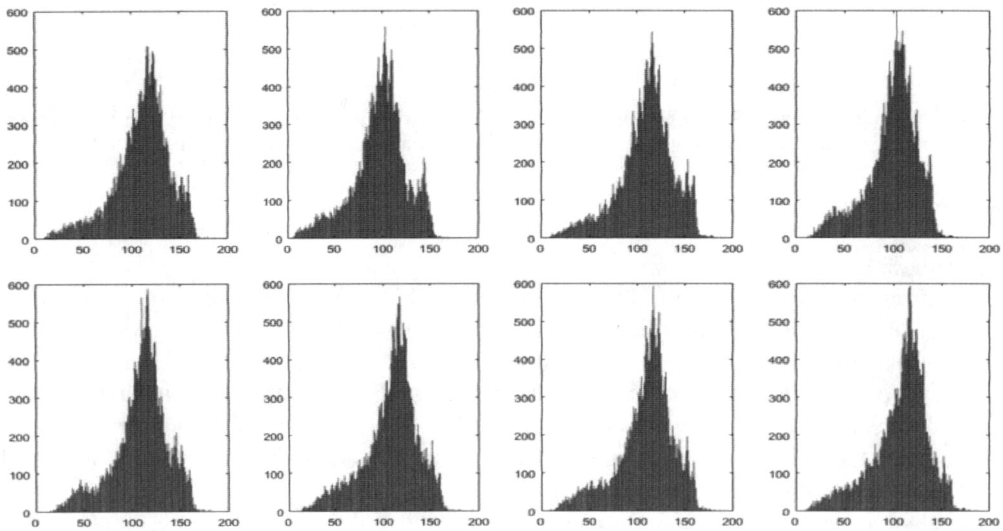

Figure 4. Grayscale intensity histograms of a homogenous coal stream samples from figure 3.

On the basis of these several histograms of the homogenous coal stream samples, there has been calculated an averaged histogram. The other possibility was to calculate a reference histogram as a median of sample histograms. These averaged and median histogram have been presented in figure 5. as thick lines on the background of thin lines of sample histograms from figure 4.

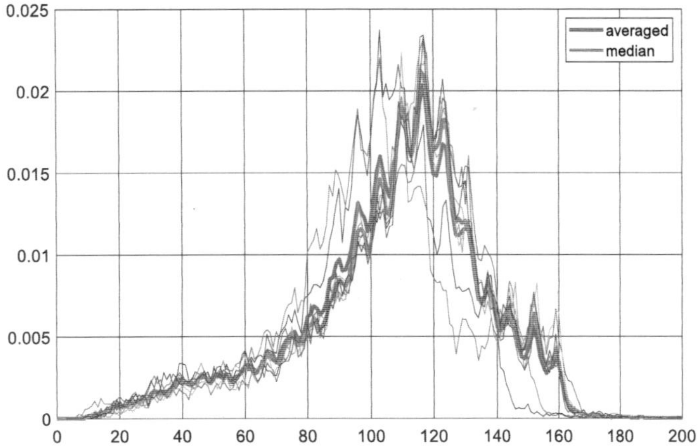

Figure 5. An averaged and median grayscale intensity histograms on the background of sample histograms of the homogenous coal stream.

This averaged histogram has been used as a reference vector for the analysis of contaminated coal stream samples, presented in the following subsections.

3.3. Analysis of the coal stream contaminated with rock/gangue

After analysis of the homogenous coal stream samples, there has been performed an analysis of coal samples contaminated with schist and pyrite particles (figure 6). Their histograms show greater share of the lighter bins, as the surface colour of these gangue particles surface is lighter than the colour of coal particle surface.

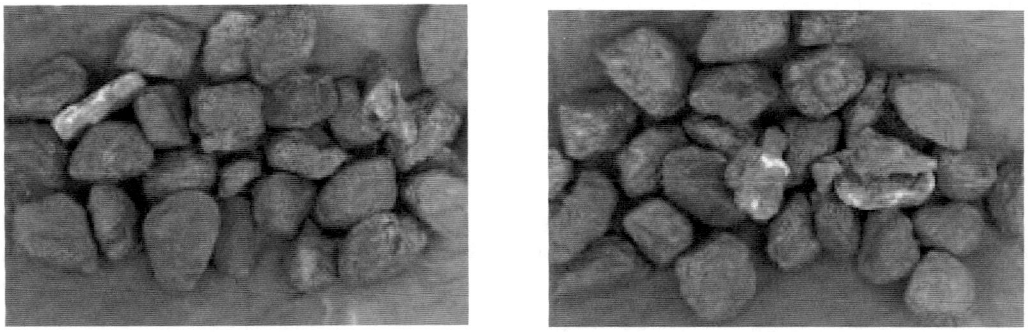

Figure 6. Grayscale images of the coal sample contaminated with schist (left) and pyrite (right).

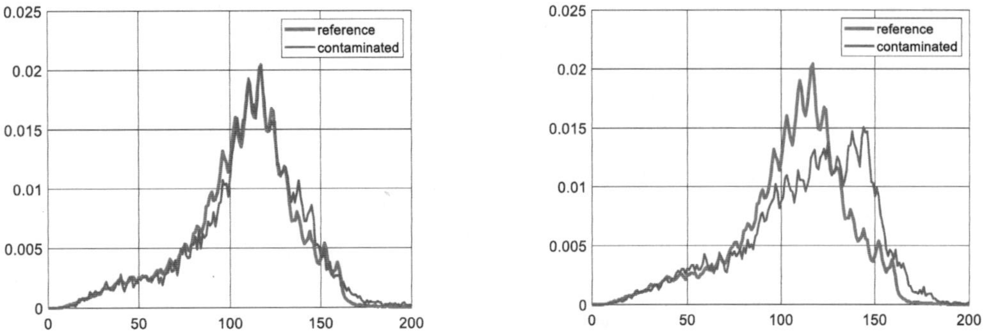

Figure 7. Grayscale intensity histograms of coal samples from fig.6. contaminated with schist (left) and pyrite (right).

3.4. *Analysis of coal stream with wooden blocks*

Accidentally in the coal stream there can also occur some other pollutions –like wooden blocks. They can be also detected by grayscale histogram analysis, as the wood is lighter than coal, but probably in this case even better results can be obtained using analysis in single colour channels. More general method can be based on calculation of hue value for each pixel, as they also form a one-dimensional histogram.

Figure 8. Sample image of a coal with accidentally placed wooden block (left) and its grayscale intensity histogram (right).

3.5. *Influence of the particle size on the coal stream image histogram*

Shape of the grayscale intensity histogram is a function of not only the surface colour (connected to mineralogical properties of the particles) abut also of the surface shape, connected to the particle shape and size. So there has been carried out an analysis of fine coal image (figure 9). It can be seen that the modal value of the histogram is approximately the same (as the surface colour of finer and coarser coal particles is the same) but there is a significantly greater share of darker histogram bins. This fact can be explained by the much more frequent dark inter-particle spaces in the case of finer particles.

IOP Conf. Series: Materials Science and Engineering **427** (2018) 012023 doi:10.1088/1757-899X/427/1/012023

Figure 9. Fine coal sample image (left) and its corresponding grayscale histogram (blue) on the background of the reference histogram of the coarse coal.

3.6. *Statistical signal processing*

In cases where the averaged histogram for clean coal is unknown, it would be useful to apply a chi-square distance and statistical homogeneity test [7][8], where all the recorded N_h histograms (each with N_b bins) form a $N_h \times N_b$ size contingency table with

$$D_F = (N_h - 1) \cdot (N_b - 1) \tag{8}$$

degrees of freedom. The chi-square statistics Q is calculated as a weighted sum (across all the contingency table elements) of squared differences between recorded bin counts f and an expected reference histogram e

$$Q = \sum_{i=1}^{N_h} \sum_{j=1}^{N_b} \frac{(f_i[j] - e_i[j])^2}{e_i[j]} \tag{9}$$

where a reference histogram e_i is calculated for each *i-th* row (in general case e.g. if they have different image sizes and different number of pixels) on the basis of the vectored column-wise total sum of the contingency table as:

$$e_i[j] = e[j] \sum_{k=1}^{N_b} f_i[k] \tag{10}$$

where

$$e[j] = \frac{\sum_{k=1}^{N_h} f_k[j]}{\sum_{k=1}^{N_h} \sum_{i=1}^{N_{Bb}} f_k[i]} \tag{11}$$

Then the homogeneity assessment simplifies to hypothesis testing (null hypothesis H_0 that all the image histograms come from the same grayscale intensity distribution, against the alternative hypothesis H_A) If the calculated for the whole histogram set chi-square statistics is greater than the value of chi-square distributed random variable with D_F degrees of freedom the null hypothesis H_0 should be rejected and the stream probably needs more precise investigation (e.g. more frequent sampling and laboratory sample analysis).

4. Conclusions

Preliminary experiments carried out on the laboratory stand have proved that an image histogram can be a possible tool for evaluating the coal stream homogeneity. There are significant differences in the case of a coal stream contaminated with gangue, or other, e.g. wooden elements. A more reliable analysis requires a much larger number and size of samples, possibly in environmental conditions closer to industrial (e.g. on moving conveyor belt). There will be also tested different formulas for RGB to grayscale intensity conversion (e.g. enhancing lighter or darker parts of the image) and different formulas of histogram distance calculation. Image grayscale histogram analysis cannot be directly compared with an analysis based on sample sieving, because its main application purpose is not a granulometric analysis, but only a stability assessment of a coal blend qualitative parameters. Influence of coal particles edge density (closely related to the particle size) on the histogram shape is only a side effect, secondary to the primary purpose of the coal stream monitoring system.

References

[1] Luo Y Wang Y 2015 *International Conference on Materials, Environmental and Biological Engineering* (Paris: Atlantis Press) pp 847-50

[2] Heyduk A 2017 *2-Dimensional and 3-dimensional image acquisition and processing methods in machine vision granulometric analysis* (Gliwice: SUT Editions).

[3] Ware C 2004 *Information Visualization: Perception for Design* (Burlington: Morgan Kaufmann)

[4] ITU-R BT.709-6: 2015 *Parameter values for the HDTV standards for production and international programme exchange*.

[5] Cha H S 2007 Comprehensive Survey on Distance/Similarity Measures between Probability Density Functions. *Int. J. of Math. Models and Meth. in App. Sci.* **1** pp 300-7

[6] Bityukov S I, Maksimushkina A V and Smirnova V V 2016 Comparison of histograms in physical research. *Nucl. Energy and Techn.* **2** pp 108-13

[7] Greenacre M 2007 *Correspondence analysis in practice* (Boca Raton: Taylor & Francis)

[8] Gagunashvili N D 2010 Chi-square tests for comparing weighted histograms *Nucl Instr Meth. Physics Res A* **614** pp 287-96

IOP Conf. Series: Materials Science and Engineering **427** (2018) 012024 doi:10.1088/1757-899X/427/1/012024

Testing the reduction properties of coke produced from a mixture of hard coal and rubber wastes

P Matusiak[1] and J Kozioł[2]

[1] Institute of Mining Technology, Division of Preparation Systems, 37 Pszczyńska Street, 44-100 Gliwice, Poland

[2] University of Zielona Góra, Faculty of Civil Engineering, Architecture and Environmental Engineering, 15 Prof. Z. Szafrana Street, 65-516 Zielona Góra, Poland

pmatusiak@komag.eu

Abstract. The article presents the method of utilization of rubber waste consisting in the production of coke from a mixture of hard coal and rubber waste from ground tires. In the coke obtained this way, the properties of the coal and rubber mixture were tested, the following indices were determined: reduction properties - CRI and coke strength after the reaction test - CRS.

1. Introduction

Reduction of wastes volume is the main objective of wastes' management. 1991/31/EC Directive, which is in force in EU countries, requires reduction of waste components, which are biodegradable, respectively to the following levels:

- by 25% in 2010 in relation to the level in 1995,
- by 50% in 2013 in relation to the level in 1995,
- by 65% in 2020 in relation to the level in 1995.

IOP Conf. Series: Materials Science and Engineering **427** (2018) 012024 doi:10.1088/1757-899X/427/1/012024

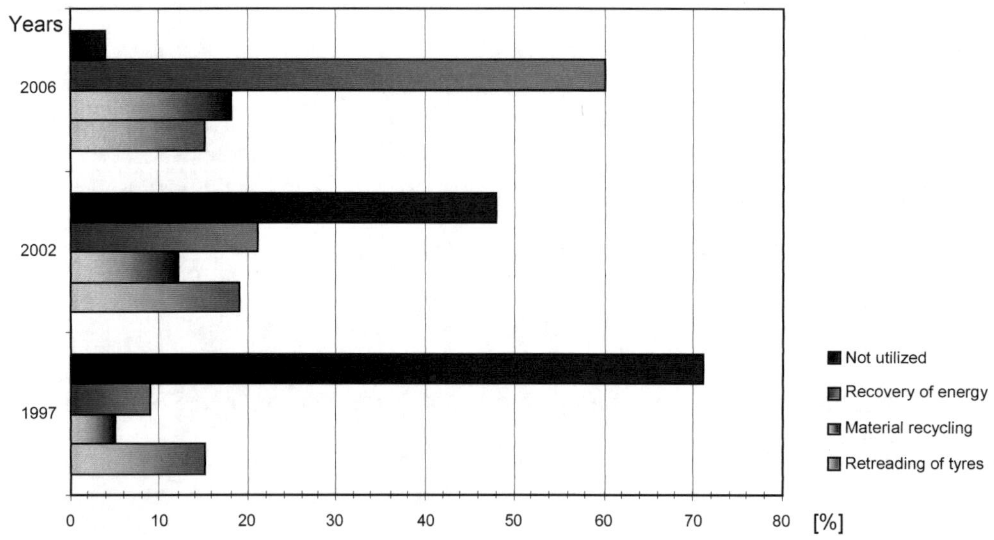

Figure 1. Utilization of rubber wastes from used tyres in EU countries [1].

The diagram (figure 1) presents the utilization of rubber wastes from used car tyres in EU countries in 1997, 2002 and 2006. It clearly shows that within last ten years not utilized wastes (1997 - 71%) were used in much greater amount for power purposes (2006 - 60%).

Recently a series of trials solving a problem, important for the present economy, i.e. rubber wastes' management by their pyrolytic degradation, have been undertaken.

Rubber wastes from car tyres can be used in technological processes, such as production of electric power and heat, or in processes of cement production and metallurgical processes.

It is also known that rubber has reduction properties [2, 3] and can be used for technological processes as a substitute for coke.

The main tests objective was to determine an optimal percentage of coke or coal substitution by rubber for technological use.

Table 1. Production of hard coal and coke in Poland within 1995÷2005 [4].

Specification	1995	1996	1997	1998	1999	2000	2001	2002	2003	2004	2005
Hard coal production, mio ton	137	138	137	116	112	103	104	104	103	100	97.9
Coke production, mio ton	11.6	10.3	10.5	9.8	8.5	9.1	9.0	8.8	10.2	10.2	8.5

In Poland about 10 mio. tons of coke (table 1) is produced per year. Almost 12 mio. tons of coal must be coked in coking plants to produce such amount of coke. The process for rubber wastes utilization of by using them in a form of admixtures (of up to 5% of weight) in a coked coal enabled to eliminate up to 600 thousand tons of wastes per year.

The objects in which such processes take place have suitable technical equipment that enables to carry out the processes according to the requirements of environment protection law.

Diminution of degradation of natural environment by reducing the production of minerals and minimization of nuisance of wastes as regards natural environment, by limiting the surface area of their disposal, is a benefit, which results from combustion of coke with rubber wastes. Cheap and simple process of grinding, separation and coking technology is a condition for the process profitability.

Coke is commonly used in metallurgical, cement and power processes.

Coke is a fuel, which is obtained by keeping the hard coal in a temperature 600-1200°C in a specially designed coke furnace, which uses off gases (without a presence of oxygen). It has higher calorific value than typical fossil coal because it includes at least 90-95% of pure coal element. However, special sorts of coal, at a suitable technology, can give coke of purity even up to 98%. Gases, liquids, inflammable substances (e.g. sulphur) and other substances (mainly organic substances), which are degradable in such temperatures, are removed from raw coal in a coking process [5].

2. Laboratory tests

The properties of coked coal and rubber mixture, which are the subject of this paper, were tested at the Measurement-and-Research Central Laboratory in Jastrzębie Zdrój, Poland.

CRI and CSR indices, applied for assessment of quality of industrial cokes used in metallurgical industry, were used to determine the properties of coked mixture (NSC test).

2.1. CRI (coke reactivity index)

CRI characterizes reactivity of coke towards carbon dioxide, i.e. its reduction properties. It can be determined on the basis of coke mass loss during its reaction with carbon dioxide in a temperature 1100 °C within 2 hours (figure 2).

Figure 2. Retort fed with coke just after determination of CRI [6].

2.2. CSR (coke strength after reaction) index

This index characterizes changes of coke strength after reaction. Its value is determined after determination of percentage mass content of grains of size bigger than 10 mm, which remain after mechanical working of sample of gasified coke in a device for barrelling (figure 3).

Figure 3. Device for barrelling of coke sample after testing the reactivity [6].

2.3. Test stand

RF-33/KK device (figure 4), for determination of coke reactivity index (CRI) towards carbon dioxide, consists of three main components: electric furnace consisting of two symmetrical halves, which touch themselves in a vertical plane, together with a set of retorts, of a precise scale as well as of a computer station, which controls all the system (figure 5).

Figure 4. Test stand for determination of coke reactivity index (CRI) towards carbon dioxide [6].

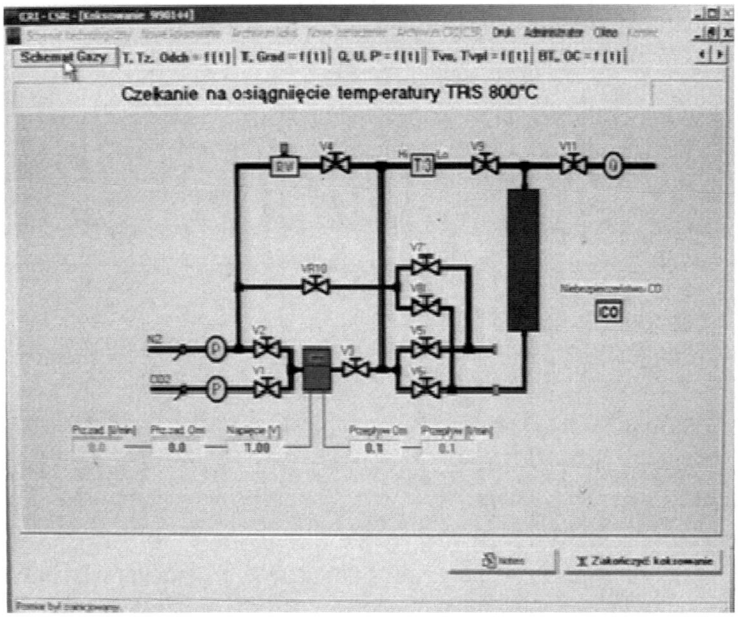

Figure 5. Technological layout of RF-33/KK device for determination of coke reactivity index (CRI) towards carbon dioxide [6].

3. Preparation of samples

Coal was pre-ground in a worm mill and then sieved on a mechanical screen to the grain size of 3-0 mm. Broken rubber from the tyres was sieved to the grain size of 2.5-0.5 mm.

Then the mixtures of hard coal and rubber were prepared according to table 2 and ten 5 kg-samples were prepared from the mixture.

The samples, after dividing into five parts, were poured to the retort in portions; a special bar rammed the layer to obtain approximately the same density. A density of layer was assessed by monitoring the depth of rammer bar indentation in a retort.

3.1. Coking of mixture (feed)

Retort with feed was put into a cold furnace and after bringing the walls closer to retort their intensive warming up started. After about 20 minutes the furnace walls reached a temperature of about 600°C, and after 80 minutes a temperature of 850°C. The temperature inside the feed was constantly controlled during coking. When a formed coke solid reached in its middle part a temperature of 950°C, what happened after 160-180 minutes from the beginning of warming up of retort, a coking process was stopped and the retort was taken out of the furnace. Then the nitrogen passed through the retort at a flow rate 10 l/min till the moment when coke reached a room temperature.

3.2. Testing method

Preparation of analytical coke sample for tests and determination of CRI and CSR indices were conducted according to a standardized method [7].

3.3. Determination of measuring points

Samples of content given in table 2, which meet the requirements of experiment and guarantee the best representativeness of results of conducted analysis, were prepared for tests. Below measuring points correspond to roots of Chebyshev's polynomial.

5 measuring points were initially assumed, and maximal and minimal values were additionally measured.

Masses of coal and rubber for each measuring point were presented in table 1. Measurement No. 3, as a mean value, was repeated 4 times for statistical analysis.

Table 2. Distribution of measuring points with the use of Chebyshev's polynomials and assumed minimal and maximal values [6].

Measuring point	$x_k = \cos\left(\frac{2 \cdot j - 1}{2 \cdot k} \cdot \pi\right)$	Rubber content Z_g(%)	Coal (kg)	Rubber (kg)
max	1.00000	10.00	4.5000	0.5000
1	0.95106	9.75	4.5122	0.4878
2	0.58779	7.93	4.6031	0.3969
3	0.00000	5.00	4.7500	0.2500
4	-0.58779	2.06	4.8969	0.1031
5	-0.95106	0.24	4.9878	0.0122
min	-1.00000	0.00	5.0000	0.0000

4. Results

The first tests for power coal were not successful as the mixture of power coal and rubber gave loose coke, which was not suitable for testing the reduction properties.

Figure 6. Sample, which was not suitable for determination of reactivity index [6].

Thus, coke coal of 35 type from "Zofiówka" Colliery of the following parameters: ash content $A_d = 7.07\%$, total water content $W_{tr} = 10.0\%$, volatile compounds $V^{daf} = 24.4\%$, was used in the next tests.

This coal enabled coking of coal-and-rubber mixture and in the whole range from 0 to 10% of rubber admixture it did not show a tendency to crushing (figure 6).

IOP Conf. Series: Materials Science and Engineering **427** (2018) 012024 doi:10.1088/1757-899X/427/1/012024

Figure 7. Coke with 5% rubber content [6].

A sample of coke, which was formed of coal with 5% of rubber admixture, is presented in figure 7. Darker spots are the coked rubber.

Table 3. Results for measuring points with the use of Chebyshev's polynomials and assumed minimal and maximal values [6].

Measurement	rubber content (%)	CRI (%)	CRS (%)
1	0	31.8	56.8
2	0.24	31.6	58.3
3	2.06	33.4	58.2
4	5	34.1	54.6
5	7.93	35.7	48.5
6	9.75	37.3	41.1
7	10	37.9	40.5

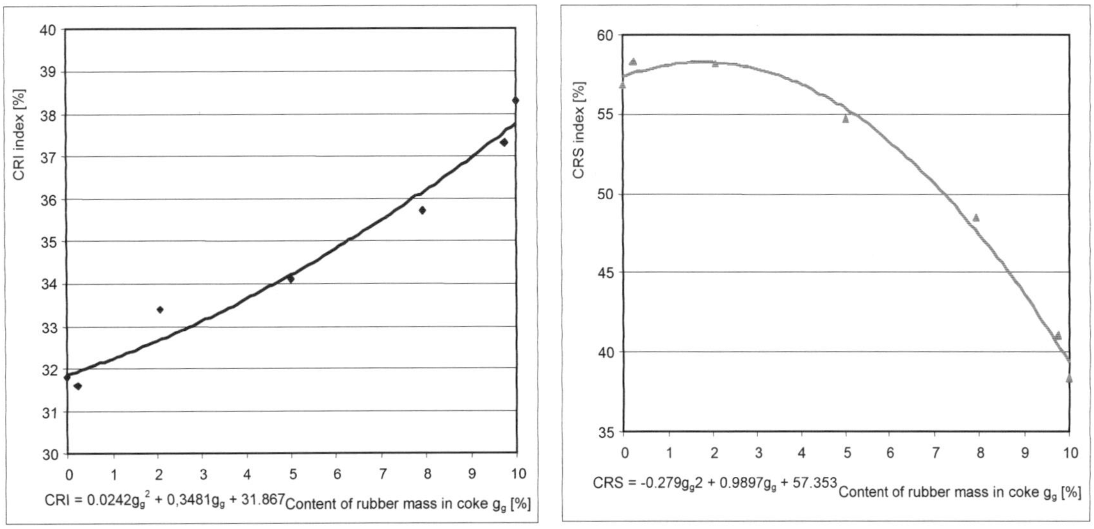

Figure 8. Results of CRI and CRS indices for mixtures of hard coal with rubber admixture from 0 to 10% [6].

IOP Conf. Series: Materials Science and Engineering **427** (2018) 012024 doi:10.1088/1757-899X/427/1/012024

5. Conclusions

Results of measurements show that reduction properties of coke increase with increase of rubber admixture, but coke strength decreases. This phenomenon is described by the following equations [6]:

- reduction properties of coke: $CRI = 0.0242g_g^2 + 0.3481g_g + 31.867$
- coke strength $CRS = -0.279g_g^2 + 0.9897g_g + 57.353$

The test on superposition could not be conducted due to difficulties in CRI and CRS results for pure rubber, because it could lead to the failure of RF-33/KK device.

The use of rubber waste as fuel for production of coke limits degradation of natural environment by a decrease of output of minerals.

During combustion of coke, which contains rubber, i.e. increased amount of chlorine (action of microcomponents), it is emitted again (reemitted) in the amount, which mainly depends on conditions of combustion process.

References

[1] Srogi K and Koszorek A 2004 Methods for recycling of rubber wastes *Fuels and Power Management* **4** pp 15-22

[2] Koszorek A and Mianowski A 1999 Possibilities and limitations of the use of classical coking process as a method for utilization of industrial wastes (Możliwości i ograniczenia w wykorzystaniu klasycznego procesu koksowania jako metody utylizacji odpadów przemysłowych), *Annual Set The Environment Protection* **1** pp 27-56

[3] Owczarek A 2004 Impact of mineral substance of hard coals on co-incineration of car tyres with coal fines in industrial stoker fired boilers *Fuels and Power Management* **2** pp 22-4

[4] *Statistical Yearbooks* Central Statistical Office (1998-2006)

[5] www.wikipedia.org

[6] Kozioł J and Matusiak P 2010 Power and ecological effects of hard coal burning with rubber wastes *Monograph* **32** (ITG Komag: Gliwice)

[7] PN-C-04312 Coke from hard coal *Determination of reactivity as regards carbon dioxide and strength after reaction*

IOP Conf. Series: Materials Science and Engineering **427** (2018) 012025 doi:10.1088/1757-899X/427/1/012025

Technology of fine coal grains recovery from the mining waste deposits

D Kowol and P Matusiak

Institute of Mining Technology, Division of Preparation Systems, 37 Pszczyńska Street, 44-100 Gliwice, Poland

dkowol@komag.eu

Abstract. Mining waste deposits (heaps) consist mainly of the waste rock, however, they may contain significant amount of coal, which was left after the coal beneficiation process. The laboratory tests, that verified possibilities of applying the technology used in the process of gravitational separation of raw minerals for recovery of grain class 6-0 mm are presented. The concept of recovering the combustible materials from mine waste by integration of the jig beneficiation technology with autogenous suspension bed technology is discussed.

1. Introduction

Mining waste deposits, created over many years, contain mainly mining wastes from the mechanical coal processing in coal mines. They often contain large amount of useful coal grains, which in majority can be recovered after applying the appropriate technology.

In contrast to the specificity of easily separated, large-sized grain material, processed with a good result in a pulsating jig [1,2,3], the fine-grain material creates much more problems in the industrial beneficiation process of [4].

The selection of the most profitable technology for the beneficiation of fine coal material (including coal waste), requires analysis of available methods for material separation, determination of quantitative and qualitative feed material parameters and wanted beneficiation products, as well as the process tests to verify the technological assumptions and the process parameters.

Unlike raw coal beneficiation, where the share of grains of waste fractions amounts to 20-30%, beneficiation of mining waste, containing mostly heavy grains with high ash content, is much more difficult due to impact of feed quality parameters on the ash content in the concentrate.

The selection of an appropriate technology for beneficiation of fine-grain mining waste should also take into account the impact of material graining range on the efficiency of separation process, carried out in with changes in the quality parameters of the feed. In the narrow grain class, the difference in grain density determines the grain separation effectiveness. At constant conditions, the flow through the separator with a wide graining range, separation effectiveness depends on the grains size and density, the grains size may have a greater impact than their density.

In order to obtain high separation effectiveness, beneficiation of 6-0 mm mining waste should be carried out in a multistage process for the previously separated grain classes.

2. Laboratory tests

2.1. Testing methodology

Based on the analysis of state of the art in coal feed beneficiation process of grain class 6-0 mm, the developed testing methodology was aimed at verification of effectiveness of the various separation technologies of each grain classes in mining wastes smaller than 6 mm [5].

Effectiveness analyses of the material separation of grain size 6-2 mm were made, on the basis of two-product jig beneficiation technology, in a single-chamber experimental pulsating jig, using periodic flow of material through the working chamber and discharging the separation products from it

Grain size 2-1(0.5) mm material separation effectiveness analysis was carried out on the basis of:

- technology of hydraulic two-product separation at the stand of a two-chamber suspension separator under turbulent flow of rising water stream,
- technology of an autogenous suspension bed in a fluidised bed of the beneficiation chamber under conditions of laminar flow of rising water stream.

After each test, the separation products were subjected to quantitative and qualitative analyses to determine the separation effectiveness in each grain class.

Density analyses included separation of material in heavy liquids of the following densities: 1.3; 1.4; 1.5; 1.6; 1.7; 1.8 and 2.0 g/cm^3. The ash content was determined in each of the obtained size and density fraction, and then, based on the results of laboratory analyses, the qualitative parameters of the beneficiation products in form of d_{50} separation density, probable error E_p and imperfection I were determined. Figure 1 schematically shows the algorithm of the conducted tests.

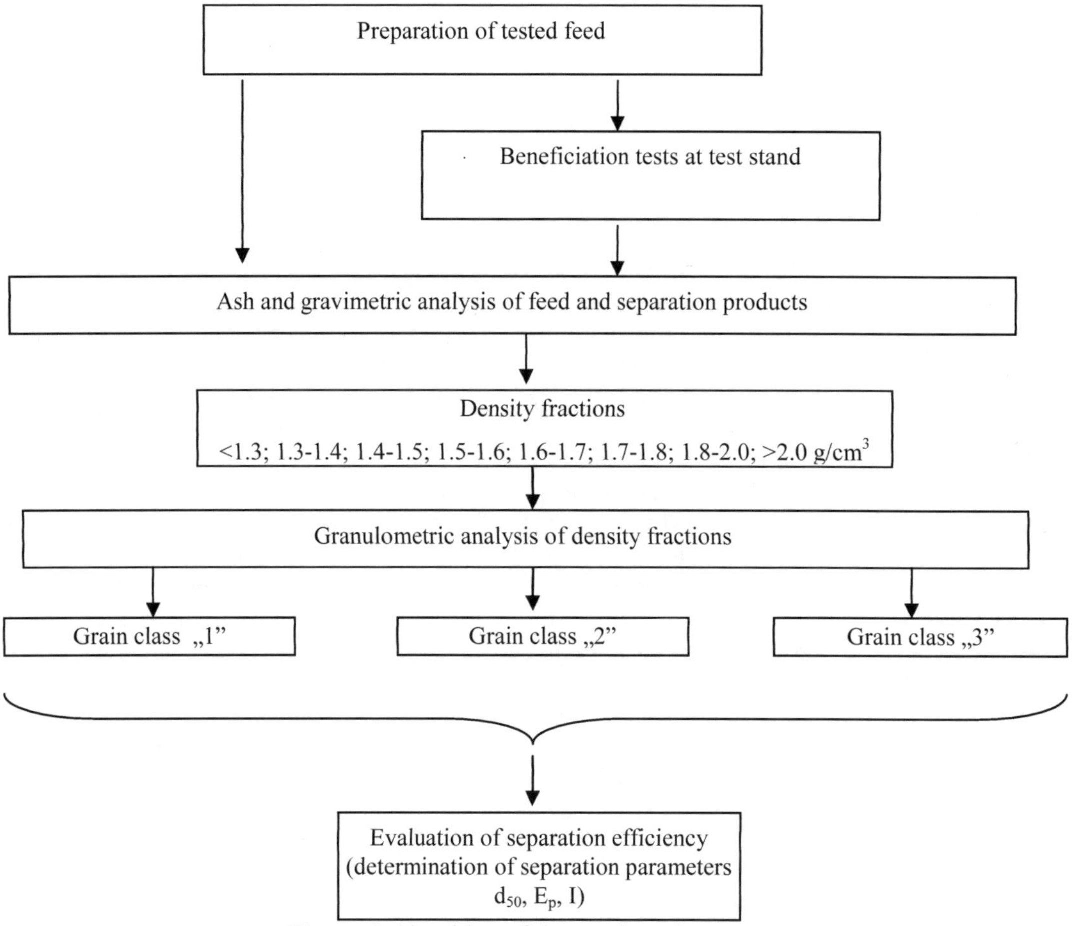

Figure 1 Algorithm of the conducted tests.

2.2. Tests of separation effectiveness for material in class 6-2 mm using the water pulsating movement
Testing the possibility of combustible substance recovery from coal wastes of the grain class 6-2 mm, using a gravitational separation process, was carried out at the laboratory jig stand, equipped with electronically controlled air pulse valves (figure 2). In pulsating jig the screen deck was equipped with a polyurethane slotted screen of 2 mm slot size.

Figure 2. Laboratory jig stand [6].

As a result of preliminary technological tests, airflow and bottom water controlled parameters in the jig were selected and adjusted to grain size of the separated material. The obtained parameters are given in table 1.

Table 1. Controlled parameters of experimental jig [5].

Parameter	Value
Frequency, min^{-1}	70
Single cycle time, ms	860
Inlet valve opening time, ms	200
Inter-signal interval time1, ms	310
Outlet valve opening time, ms	200
Inter-signal interval time2, ms	150
Rate of bottom water inflow, m^3/h	3
Water pulsation in a working chamber, m	0.03
Unit load, $t/h/m^2$	3.43

During separation process, the material was separated into two products: "lightweight" one and "heavy" one.

The "lightweight" product - concentrate, after discharging through the working chamber overflow, along with the process water, was dewatered on sieves with a 2 mm slots and 0.5 mm holes, successively.

The "heavy" product - waste product was continuously discharged in form of a material passed through the sieve openings of the jig's working chamber and periodically through a vertical slot with an articulated flap.

Separated material was subjected to quantitative and qualitative analyses to determine the separation effectiveness in the 6-2 mm and 6-4 mm and 4-2 mm grain size fractions.

In addition, comparative analyses of the parameters of the waste product discharged from the jig's working chamber through the sieve holes and through the vertical receiving slot of the heavy product were carried out.

The results of the beneficiation process of the 6-2 mm grain class are presented in table 2.

Table 2. List of parameters of separation process of the grain class 6-2 mm [5].

Parameter		**Grain class, mm**		
		6-2	6-4	4-2
Separation density, d_{50}, g/cm^3		1.705	1.691	1.705
Probable error E_p, g/cm^3		0.122	0.123	1.121
Imperfection I		0.173	0.178	0.171
Feed	<1.5	9.09	10.31	8.36
	1.5-1.8	4.23	4.96	3.79
	>1.8	86.68	84.73	87.85
	Share	100.00	37.50	62.50
Concentrate	Output	11.67	12.41	11.21
	<1.5	67.10	72.47	63.68
	1.5-1.8	21.46	22.73	20.48
	>1.8	11.44	4.80	15.84
Tailings	Output	88.33	87.59	88.79
	<1.5	1.43	1.50	1.37
	1.5-1.8	1.95	2.45	1.69
	>1.8	96.62	96.05	96.94
Partition number	<1.5	13.83	12.78	14.59
	1.5-1.8	40.80	43.17	39.42
	>1.8	98.46	99.30	97.98

Analysis of the separation test results for coal waste of 6-2 mm grain size class showed that for separation density d_{50} = 1.705 g/cm^3 separation efficiency indicators were as follows: the probable error E_p = 0.122 g/cm^3, and imperfection I = 0.173.

Tests of two-product separation process for fine coal waste – size 6-2 mm, conducted in a pulsating jig, had small share of carbon grains and middlings, enabled obtaining a concentrate with an output of 11.7%, ash content 21.7% and a calorific value of 25.5 MJ/kg.

A comparative analysis of beneficiation results for grain sizes of 6-4 mm and 4-2 mm with shares equal to 37.5% and 62.5% showed that with similar values of separation accuracy indicators, separation of 6-4 mm grains was more effective due to the larger yield of coal fractions of density <1.5 g/cm^3 by 1.8% in a concentrate product.

Concentrate obtained from the beneficiated material of 6-4 mm grain size had a higher quality and had an ash content of 17.73% and a calorific value of 27.02 MJ/kg, while in the concentrate from feed material of class 4-2 mm the ash content was 23.05% and the calorific value was equal to 24.52 MJ/kg.

The smaller share of coal incorrect grains in the waste product, discharged through the screen (0.4%), in comparison to the share of these grains in the waste product discharged through the receiving slot (1.6%) indicates the possibility of intensification of the heavy product discharge process from the jig's working surface through the screen openings. This will increase the effectiveness of 4-2 mm grain separation and the quality of the concentrate product.

Figure 3 presents in a graphic form the relationship between the separation parameters and the material grain size.

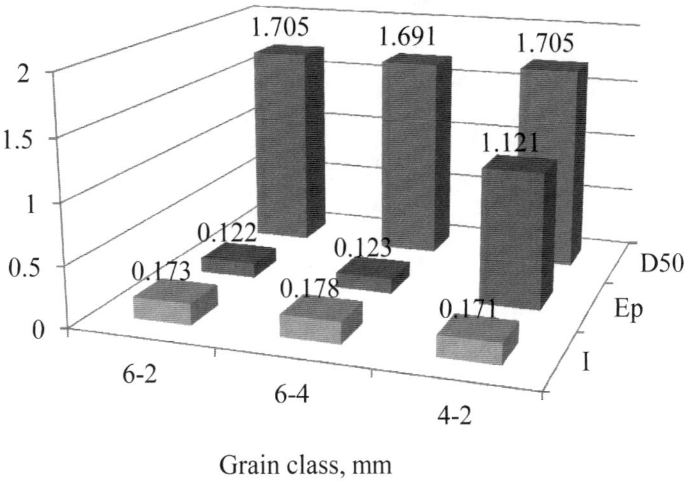

Figure 3. Impact of grain size on the distribution parameters [7].

2.3. Tests of material separation effectiveness for class 2(1)-1(0.5) mm using rising water stream

In the tests of separating the 2-0.5 mm grain size feed, a hydraulic method of material beneficiation was used. The method uses the turbulent flow of rising water stream in the stream-and-suspension separator's working chamber (figure 4).

Figure 4. Stream and suspension separator test stand [7].

Separation of the feed in the separator's chamber into lightweight (concentrate) and heavy (waste) products was carried out using the process (top) water supplied together with the feed to the bottom of the working chamber and the bottom water flowing through the heavy product slot.

The results of the gravitational process of the fine material depended on the flowrate of the rising water stream in the separator's working chamber as well as on the separation density.

The selection of the top water inflow rate Q = 0.95 m^3/h and the bottom water Q = 0.25 m^3/h allowed obtaining, in the working chamber with a cross-section area of 0.0108 m^2, the average water flow velocity of 0.031 m/s. The flow velocity of the rising water stream through the heavy product slot, with a cross-section area of 0.0008 m^2, was equal to 0.084 m/s.

The separation products were subjected to quantitative and qualitative analyses to determine separation effectiveness in 2-0.5 mm, 2-1 mm and 1-0.5 mm grain size classes.

The results of analyses enabled determining the separation density for the whole material in grain sizes 2-1 mm and 1-0.5 mm and determining the separation process effectiveness for the same grain size classes basing on the probable error values E_p.

The results of conducted tests are presented in table 3.

During beneficiation tests of 2-0.5 mm grain class material with a significant 72.5% ash content, the concentrate with 16.3% yield and of 25% ash content was obtained, the waste product had 82.4% ash content.

The obtained separation process effectiveness in the grain size 2-0.5 mm had the probable error E_p, amounted to 0.094 g/cm^3 and separation density D_r=1.54 g/cm^3.

The results of analysis of the material distribution parameters in size and density fractions showed a significant impact of the grain size and gravimetric composition of the feed on the yield and the quality parameters of the beneficiation products.

Comparison of the separation results for the feed of the grain class 2-0.5 mm as well as in the 2-1 mm and 1-0.5 mm classes showed better beneficiation results (E_p=0.069 g/cm^3 with D_r=1.5 g/cm^3) for the size range of 2-1 mm grains share of which in the feed was 69.5%.

The concentrate of grain size 2-1 mm had the smallest ash content (20.54%), while this parameter in the 2-0.5 mm and 1-0.5 mm grain classes was 25.0% and 29.9% respectively.

IOP Conf. Series: Materials Science and Engineering **427** (2018) 012025 doi:10.1088/1757-899X/427/1/012025

Table 3. List of beneficiation process parameters of the 2-0.5 mm grain size class [5].

Parameter		Grain class, mm		
		2-0.5	2-1	1-0.5
Separation density, d_{50}, g/cm^3		1.537	1.501	1.671
Probable error E_p, g/cm^3		0.094	0.069	0.137
Imperfection I		0.175	0.137	0.203
Feed	<1.5	12.93	11.03	17.27
	1.5-1.8	4.28	3.84	5.22
	>1.8	82.79	85.12	77.51
	Share	100.00	69.50	30.50
Concentrate	Output	16.26	12.22	25.47
	<1.5	73.02	81.20	64.09
	1.5-1.8	7.50	4.66	10.61
	>1.8	19.48	14.14	25.30
Tailings	Output	83.74	87.78	74.53
	<1.5	1.26	1.26	1.27
	1.5-1.8	3.65	3.73	3.38
	>1.8	95.09	95.01	95.35
Partition number	<1.5	8.16	10.03	5.48
	1.5-1.8	71.48	85.18	48.25
	>1.8	96.17	97.97	91.69

Two times smaller separation effectiveness in the grain class 1-0.5 mm (E_p=0.137 g/cm^3), in comparison to the results of the class 2-1 mm, with a much higher (by 0.13 g/cm^3) separation density, indicates:

- dependence of the separation process in the counter-current classifier mainly on the grain settling velocity,
- the necessity to take into account the share of 1-0.5 mm grains in a 2-0.5 mm feed in forecasts of beneficiation results.

Figure 5 presents in a graphic form the relationship between separation parameters and the grain size distribution of the material.

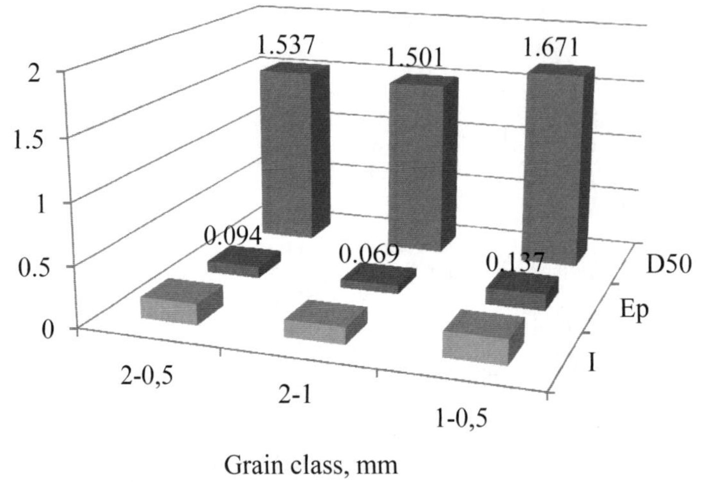

Figure 5. Impact of grain size on the distribution parameters [7].

2.4. Tests of the material separation effectiveness in class 1 (0.5) -0 mm using an autogenous suspension bed

The tests of the material beneficiation process in the 2-0.5 mm grain class, obtained from the coal waste were carried out in a laboratory using the suspension part of the hydraulic fluid suspension separator model (figure. 3).

The analysis used the method of suspension beneficiation of the material using an autogenous fluidised bed, obtained in a result of action of the rising water stream on the material in the upper part of the beneficiation chamber.

Separation of the material in the suspension chamber of the separator into lightweight (concentrate) and heavy (waste) products was carried out using the stream of water supplied by the flat diffusion plate, which is at the same time was the settling bottom of the working chamber.

The results of the gravitational separation process of the fine material depended on flow velocity of the rising water stream in the working chamber and the material density as well as on the parameters of the separation layer in the autogenous suspension bed and the height of its position in the chamber.

In a result of selection of the inflow intensity of the bottom water, which amounted to Q=4.5 m^3/h, in the working chamber with the area of 0.075 m^2, the flow velocity of the rising water stream was equal to 0.017 m/s.

The material of the gravitational beneficiation products was subjected to the same analyses as in the case of tests using the rising water stream.

The results of the conducted tests are presented in table 4.

During beneficiation tests of 2-0.5 mm material with a significant 71.81% ash content, a concentrate product with 12.8% yield and 16.92% ash content and a waste product with 80.68% ash content were obtained.

In comparison to the test results using a counter-current classifier, separation of the material using an autogenous suspension bed and separation density lower by 0.1 g/cm^3, higher quality concentrate with ash content lower by 8.1% was obtained.

The effectiveness of beneficiation process in the 2-0.5 mm grain class was characterized by the probable error E_p, was equal to 0.075 g/cm^3 and imperfection I was equal to 0.1171.

Table 4. List of beneficiation process parameters of the 2-0.5 mm grain class [5].

Parameter		**Grain class, mm**		
		2-0.5	2-1	1-0.5
Separation density, d_{50}, g/cm^3		1.441	1.392	1.505
Probable error E_p, g/cm^3		0.075	0.042	0.079
Imperfection I		0.171	0.107	0.155
Feed	<1.5	13.75	11.03	17.27
	1.5-1.8	4.45	3.86	5.22
	>1.8	81.80	85.12	77.51
	Share	100.00	56.30	43.70
Concentrate	Output	12.84	7.89	19.22
	<1.5	86.29	99.03	79.69
	1.5-1.8	3.77	0.97	5.24
	>1.8	9.94	0.00	15.07
Tailings	Output	87.16	92.11	80.78
	<1.5	3.06	3.49	2.42
	1.5-1.8	4.55	4.10	5.21
	>1.8	92.39	92.41	92.37
Partition number	<1.5	19.40	29.14	11.32
	1.5-1.8	89.12	98.01	80.69
	>1.8	98.44	100.00	96.26

Analysis of the separation results for the fed material of grain class 2-0.5 mm as well as of grain classes 2-1 mm and 1-0.5 mm showed that the highest beneficiation efficiency (E_p=0.04 g/cm^3 at D_r = 1.39 g/cm^3) was obtained for the class 2-1 mm, share of which in the feed was 56.3%.

The concentrate of grain class of 2-1 mm had the smallest, 8.34% ash content, while the value of this parameter in the 2-0.5 mm and 1-0.5 mm grain classes was 16.92% and 21.37, respectively.

In comparison to the results of 2-1 mm class separation, the efficiency of separating the grain class 1÷0.5 mm (E_p = 0.079 g/cm^3) was two times lower, and the separation density was higher by 0.11 g/cm^3. The difference in density of material separation was significantly lower (more favourable) than the one obtained in the counter-current classifier tests, where it reached 0.17 g/cm^3.

The above difference in the separation density of the size and density-fraction in the suspension separator indicates for:

- dependence of the material separation process on both, the grain settling velocity and the density of the autogenous suspension bed,
- the necessity to take into account the share of 2-1 mm and 1-0.5 mm grain classes and the differences in their separation parameters, while forecasting the beneficiation results of feed 2-0.5 mm.

Figure 6 in the graphic form presents the relationship between the separation parameters and the grain size of the material.

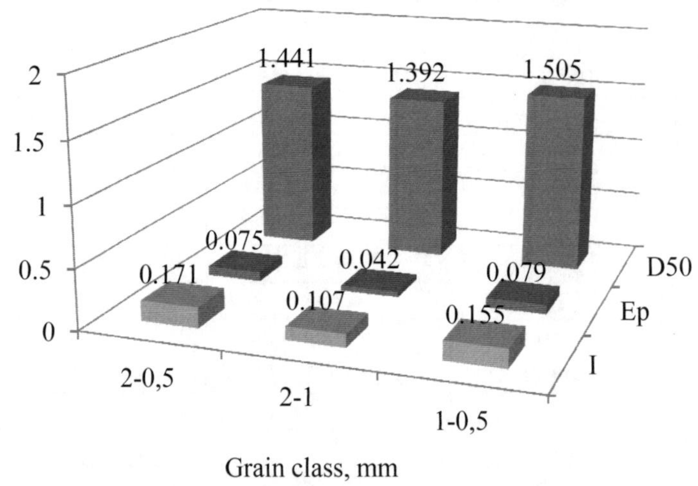

Figure 6. Impact of grain size on the separation parameters [7].

Figure 7 presents a comparison of the separation effectiveness for class 2-0.5 mm obtained by the hydraulic method using the rising water stream and by the method using an autogenous suspension bed.

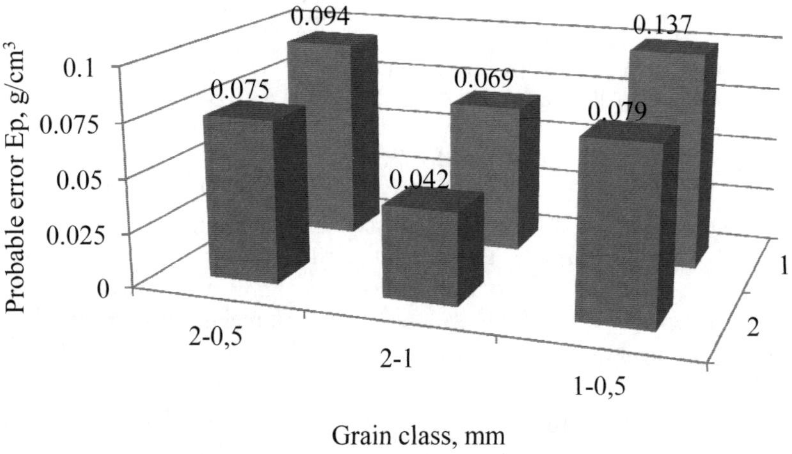

Figure 7. Comparison of separation effectiveness of 2-0.5 mm grain class by hydraulic (1) and suspension (2) methods on the example of probable error [7].

3. The concept of beneficiation technologies for fine coal feed

Basing on the results of laboratory tests, a concept for the recovery of combustible material from coal waste was developed, integrating the technology of sedimentary beneficiation with technology using an autogenous suspension bed [5].

In order to obtain high separation effectiveness, the beneficiation of 6-0 mm coal waste will be carried out in a multi-stage manner, separately for the previously classified grain classes.

Grain class 6-2 mm

Material of grain class higher than 6-2 mm can be alternatively beneficiated in a one-compartment pulsating jig together with a 30-6 mm grain class (pulsating classifier) or after broadening the grain size range to 8-2 mm in a pulsating jig with an internal slot for heavy product discharge and the system for removing the sludge from the lower densifying box.

In comparison to the beneficiation results of the 30-2 mm grain class in a single device, the above solution, consisting in beneficiation of coal waste in 30-8 mm and 8-2 mm grain classes in separate jigs (classifiers), will ensure greater separation effectiveness due to a smaller difference between the largest and smallest grains.

Grain class 2-0 mm

For the beneficiation of the material in the 2-0 mm grain class, it is planned to use a device based on the flow of the rising water stream in the working chamber enabling the separation of grains according to their density, using an autogenous suspension bed as well as one and two-stage separation process.

In the case of beneficiation of coal waste product, which share of heavy fractions may exceed 80%, the two-stage material separation process in the 2-0 mm grain class allows obtaining a concentrate product with lower ash content, compared to one-stage beneficiation, with the same separation density.

The density separation in the narrow grain classes will be more efficient and allows to obtain a concentrate product with a lower ash content compared to the material separation with a larger grain class range.

Separation of 2-0 mm waste grains from the feed of grain class 2-1 mm, during the first material separation stage, will have a positive effect on the separation effectiveness of the second stage in a separate working chamber or in another beneficiation unit due to uniformity of density of the layers forming the autogenous bed.

As the grain size of the heavy grains increases in the feed, the size range of the grains which can be effectively separated in the suspension bed will reduce. This is due to the need to increase the rising water velocity to maintain the homogeneity of the bed layers.

Separation in one, single-chamber separator, with an autogenous suspension bed is an alternative method for the 2-0 mm grain class material beneficiation. With the application of the above technology, the significance and impact of the process of dewatering and classification of the concentrate product on its quality will increase. Separation of the smallest grains from the concentrate product, with the highest ash content, and then directing them for beneficiation in a hydro-cyclone may be a beneficial solution allowing to obtain a sufficiently low ash content in the final product concentrates.

4. Conclusions

Mine waste deposits often contain significant amount of coal grains, which can be recovered after using proper beneficiation technology.

The beneficiation of mine waste containing significant, often exceeding 90% of waste fraction grains is characterized by a considerable difficulty in obtaining a stable quality concentrate product during the material separation process.

The above problem appears especially in the case of fine material beneficiation.

Basing on the analysis of the solutions using the technologies for the separation of fine grain feed according to grain density the assumptions for testing methodology in which the grain size range of beneficiated material using different technologies was determined, were elaborated.

The results of laboratory tests were the basis for the development of recovery technologies for the 6-0 mm class fine coal grains recovered from the mine heaps, in which the integration of the jig separation technology with the technology using an autogenous suspension bed was suggested.

IOP Conf. Series: Materials Science and Engineering **427** (2018) 012025 doi:10.1088/1757-899X/427/1/012025

References

[1] Kowol D and Matusiak P 2016 Possibilities of using the pulsating classifier for revitalisation of mine landfills Innowacyjne rozwiązania rewitalizacji terenów zdegradowanych *Instytut Ekologii Terenów Uprzemysłowionych, Centrum Badań i Dozoru Górnictwa Podziemnego sp. z o.o.* **8** pp 163-72
[2] Matusiak P and Kowol D 2016 Application of the jig beneficiation operation for minerals recovery from mining waste deposits *Min. Sci., Miner. Aggreg* **23** pp 15-25
[3] Matusiak P and Kowol D 2017 Technology of recycling coal from mining waste through implementation of KOMAG device *Masz. Gór.* **150(2)** pp 42-52
[4] Różański Z, Suponik T, Matusiak P, Kowol D, Szpyrka J, Mazurek M and Wrona P 2016 Coal recovery from a coal waste dump *E3S Web of Conferences* **8** p 01052
[5] Łagódka M et al. 2016 Development of technology for recovery of fine coal grains in the class 4-0 mm form mine dump sites basing on gravitational beneficiation in narrow-size classes, (not published) Gliwice: ITG KOMAG
[6] Kowol D, Matusiak P and Łagódka M 2017 Laboratory tests for efficiency of jig beneficiation of coal wastes of grain classes 35-3 mm and 35-0 mm *Masz. Gór.* **150(2)** pp 3-15
[7] Not published materials ITG KOMAG, 2017-2018

IOP Conf. Series: Materials Science and Engineering **427** (2018) 012026 doi:10.1088/1757-899X/427/1/012026

Development of research work in the air-water spraying area for reduction of methane and coal dust explosion hazard as well as for dust control in the Polish mining industry

D Prostański

KOMAG Institute of Mining Technology, Gliwice, Poland

dprostanski@komag.eu

Abstract. The article presents a method for equipment designing and techniques to reduce methane ignition hazard, excessive dust in the place of operation and in conveyor transportation lines as well as methods to eliminate the risk of coal dust explosion. It describes technical solutions, the effects of these solutions and tests' results related to effectiveness of the described air-water solutions as well as water spraying devices. The spraying systems used in longwall shearers and roadheaders, on conveyor belt transfer points as well as the spraying curtains in the roadways are described. A new approach to the assessment of the safety level in the zones protecting against coal dust explosion hazard is also given. Intensity of dust accretion along the explosion protecting zones is analyzed. The spraying device is suggested, the operation of which depends on dust concentration in the ventilation air. The article presents the results of tests on the effectiveness in dust reduction and suppression of methane ignition. The tests were carried out in the laboratory, on testing facilities and in-situ.

1. Introduction

Methane and coal dust ignition hazards are the reason of many mining disasters. Within dozen or so years, five disasters, which killed 64 people took place [1, 3]. Presence of methane and coal dust at the place of mining and its ignition by sparks generated by tools contacting with mined rock or by other ignition sources are the main reason of the said disasters. Contamination of mine air by coal dust generated in the result of rock cutting and grinding or in transportation processes is another serious hazard. Airborne dust in mine is the reason of about 500 cases per year of pneumoconiosis among workers [1, 2].

Finding the possibilities of elimination or reduction of methane and/or coal dust explosion hazard as well as improvement of air quality by air-water spraying systems and development of other design solutions, eliminating such hazards, was the objective of KOMAG Institute research projects. The practical character of the developed solutions was kept in the research work.

2. Assumptions for the method of designing the air-water spraying installations

Innovative approach to reduction of impact of methane ignition and explosion hazard as well as reduction of dust concentration by using the air-water aerosols were suggested as result of the research work.

Available research results indicated that use of air-water spraying system can be efficient method for reduction of methane and/or coal dust ignition hazard due to high energy of ejecting the water drops evenly atomized in a stream of compressed air.

The scope of research and design work, enabled to develop the method for designing the air-water spraying installation for longwall shearers. Its general description is presented in figure 1. The method combines computer modelling, virtual prototyping, functionality tests on mock-ups, laboratory stand tests as well as verification of the installation in real mine conditions.

The developed method is of general character, what was confirmed by the results of further design and research work for other mining machines. According to this method, air-water spraying installation for roadheaders as well as the installations reducing dust concentration in roadway were developed and tested.

The method was developed to obtain smart solutions. For this purpose the models for determination of dust settlement intensity in a function of dust concentration were elaborated. The models were the basis for development of stationary and mobile spraying installations. The applied predictive algorithms allow to control the spraying process in changing dust generation intensity in mine roadways. These solutions enabled a creation of zones protecting against propagation of coal dust explosion and increased work safety in mining faces. Also the spraying installation acting selectively at the place of the dust generating mining machine operation, was included in this group of solutions.

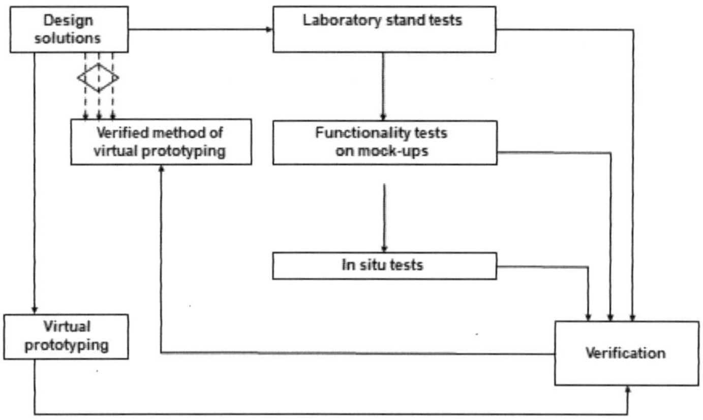

Figure 1. General description of the method for designing air-water spraying installations.

Each solution of the spraying installation, irrespectively of its application undergoes designing and testing. After application of solution, it is further modified as the result of in-situ observations and tests. Continuous development of air-water spraying installations is associated with solving the problems and meeting the requirements of increasing operational safety in mines as well as with improvement of work safety in underground mines. In the consequence of this development, the original technical solutions, improving functionality and automation level, are also elaborated.

3. Control of methane ignition possibility and reduction of dust concentration in a longwall panel

First projects on air-water spraying installations were undertaken in the result of many accidents including mining disasters caused by methane ignition during coal cutting by shearers and roadheaders [1, 3]. The disasters in the longwall face may be influenced by the type of cutting knives [24, 25] and by the effectiveness of spraying systems for the suppression of methane ignition. The concept of spraying installation, intended for a longwall shearer in a form of 3D model, was developed under projects. It was used in the simulation tests. The expected parameters of spraying streams were specified: range and the stream angle as well as arrangement and number of spraying nozzles, which enable a creation of water curtain around the shearer's cutting drum. The shape of the stream aims at preventing against propagation of methane explosion out of the contact zone of cutting drum with cut rock. An analysis of available designs of two-media spraying nozzles with the stream parameters, flow rate and pressure of spraying media, was performed. Then the laboratory tests of two-media spraying nozzles were carried out. Basing on this research methodology, the tests schedule was planned and the test stand was constructed. Within these tests degree of stream atomization, its range and shape were assessed. The designs of two-media nozzles of even distribution and properly atomized spraying stream and a range of the stream not less than 2 meters were selected. The assumption was that nozzles should have flow rate not less than 1 dm^3/min and pressure not exceeding 0.6 MPa and air flow rate 150 dm^3/min at pressure below 0.45 MPa. Further step included a development of original solutions with increased mechanical strength. The laboratory tests, conducted with use of Sprytec drops analyzer, showed that median drops diameter does not exceed 25 μm [6]. The spraying installation put on wooden mock-up of shearer's arm with the cutting drum was designed and tested. It enabled to assess the functionality of air-water spraying installation. According to the assumptions the atomized water ejected in a form of stream from nozzles, deployed on the cutting drum, made a tight water curtain around it. It prevented propagation of methane flame out of the cutting zone [4]. The installation of nozzles dipped in a longwall shearer's arm with determined directions of spraying streams was suggested and ratio of water and compressed air flow rate as well as supply and mixing of water and air in external spraying system were determined. After completion of tests and development of the technical documentation, an air-water spraying installation for arms R-200N of KSW-460NE shearer was designed [4]. After completing the installation, the testing stand was designed and methodology was developed for testing the installation regarding assessment of its effectiveness in extinguishing the methane flames and in controlling its ignition. During the tests, the gas ignition was initiated at the same time by two sparks of energy 10 MJ each (figure 2a). The tests were conducted in Barbara Experimental Colliery (GIG). After the tests full effectiveness of the designed installation for the assumed scope of its use was proven [4]. After certification procedures realized by the manufacturer, the pilot shearer with the spraying installation was tested in-situ in Pniówek mine in longwall W-9 [4]. During tests the functionality of air-water spraying installation in underground conditions, assessment of its reliability and effectiveness in dust control was evaluated. After the tests full functionality of air-water spraying installation was assessed and proven (figure 2b). [5]. Next tests of the shearers showed full usability in reduction of methane ignition hazard as well as in efficient dust control especially, regarding the respirable fraction [5].

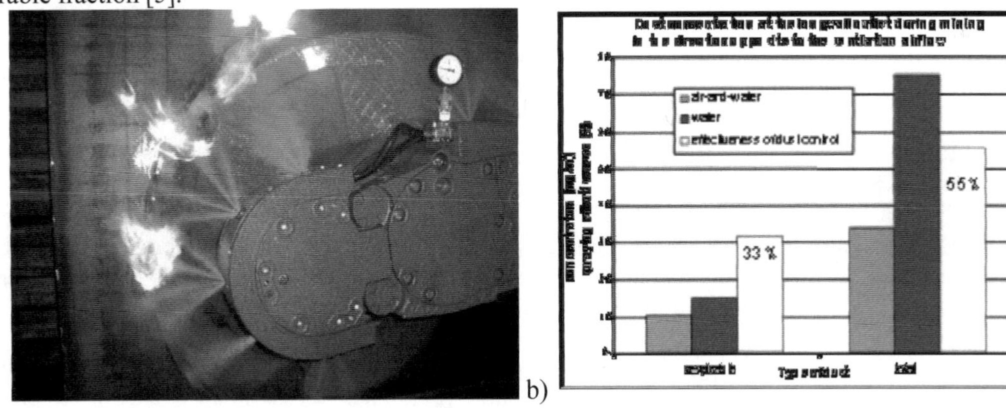

a) b)

Figure 2. KSW-880EU longwall shearer a) extinguishing of flames by air-water spraying installation, b) efficiency of dust control with air -water and water spraying system [5].

Validity of selection of this problem for research work as well as realization of research projects were confirmed by successful implementation of the spraying installation, especially that it was the first such application in the longwall shearer in the world.

Positive effects of the implementation of air-water spraying systems in the KSW-460NE longwall shearer caused the producers' interest in applying this solution in other types of longwall shearers.

Process of implementation the air-water spraying installation in longwall shearers caused undertaking the research work on increasing the automation level of operation of these installations. In longwall shearers with air-water spraying installation there is a separate high-pressure water installation, which is an alternative solution. Change of spraying type requires restarting the shearer and causes breakdowns, which reduce shearer's availability and coal production losses. As a part of the design work another solution was suggested, in which pressure of compressed air decreases, water supply to air-water nozzles on the shearer's arm is cut off. Such a solution limits water flow rate, which increases by tens dm^3/min if there is no compressed air. Also original solution was suggested, which enabled automatic selection of spraying variant operation mode depending on compressed air presence or absence, by switching the check valves incorporated with pressure sensors. The research work on development of compact set of check valves opening alternately depending on pressure of compressed air, was started. This solution was used in KSW-880EU shearer.

Within the next steps research and design work, associated with air-water spraying installation in KSW-800NE longwall shearer was realized. The project included the research work oriented onto development of new design of air-water spraying installation installed not only in the shearer's arms, but also in the cutting drum. Preparation of air and water mixture before feeding it to the cutting drum, where its rotational movement caused mixing water with air was the solution idea. The research work showed full effectiveness of the spraying installation in extinguishing methane flames [7]. Then tests were carried out in the underground mine in the scope of assessment of dust control efficiency with the applied solution of the spraying installation [7]. To provide stability of spraying installation parameters irrespective to the stability of feeding parameters, a device balancing media pressure on output irrespectively to the media pressure on input was designed. Balancing water pressure and compressed air pressure is mechanical by using the pressure scale.

Together with application of other types of longwall shearers with air-water spraying installation in-situ tests were conducted in coal longwalls in scope of effectiveness of dust control. The measurements were taken at longwall outlet and at the miner's work-place using the CIP-10 gravimetric dust meters. Analysis of test results indicated that in each case degree of dust reduction in the case of using the air-water spraying was up to 77% of total dust and respirable fraction up to 93%, in comparison with a typical water spraying system [5, 9].

4. Systems for reduction of methane ignition in roadheaders

Recognizing the problem of exceeding the dust concentration and methane ignition hazard in roadway faces, it was suggested to transfer developed designing method for using air-water spraying installations in roadheaders. In roadheaders there are many solutions of the systems protecting against methane ignition, where internal sectoral behind-tool spraying systems, installed in cutter heads, are highly efficient in reduction of methane ignition hazard [10, 23]. However, these solutions are complicated and damage prone devices supplied by pumps with pressure in the range from 5 MPa to 10 MPa and water flow rate from 100 to 180 dm^3/min. Efficient air-water spraying system, suggested by other producers [11], requires dozen or so cubic meters of compressed air per minute, what often exceeds possibility of delivery of such volume of air in the conditions of the Polish mines.

Conceptual, designing and research projects concerned external air-water spraying installations in roadheaders. In this solution experience gained from the installations designed longwall shearers was used, what significantly simplified and shortened the designing process. The assumption was to supply this installation with water and compressed air of pressure from 0.35 MPa to 0.5 MPa, water flow rate about 30 dm^3/min and air from 1.5 to 3.0 m^3/min. In the result of tests it was suggested to apply spraying pack and spraying nozzles in a shape of reversed letter C, placed just after the cutter head on connection of a gear with a motor, enabling covering entire cutting zone by the spraying curtain. Each type of roadheader requires individual approach due to design differences. Spraying pack was assumed to be of universal character. After manufacturing the spraying installation and mounting it on the roadheader as well as checking its operation, it was tested on the test stand and tests confirmed proper effectiveness of extinguishing the gas flames (figure 3) [12]. The tested installations were installed in different types of roadheaders. First air-water spraying installations for roadheaders were implemented in Murcki, Jas-Mos and Borynia coal mines.

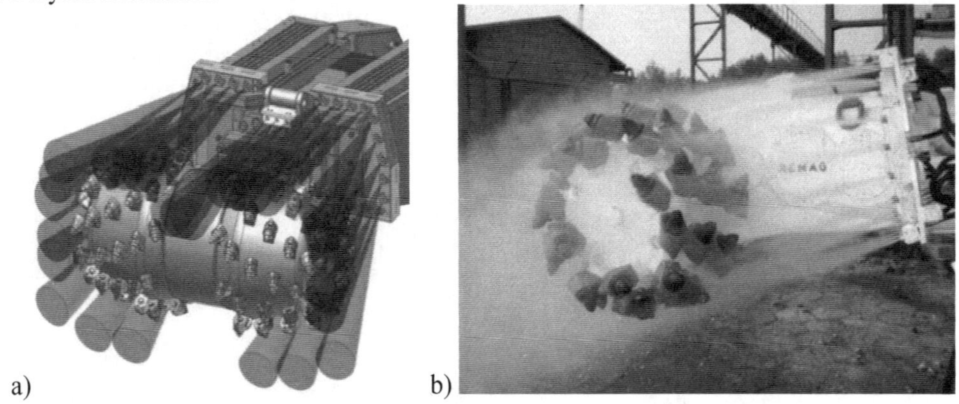

Figure 3. Air and water spraying installation in the roadheader [12].

Despite a very good extinguishing characteristics, mining regulations do not allow using the external spraying installations as the main spraying system in the atmosphere where methane concentration exceeds 0.5%. Within the concentration ranging from 0.5% to 1.0% only roadheaders equipped with internal sectoral spraying system can operate. Through elimination of main disadvantages of known designs of sectoral spraying and replacing them with new approach solutions without these disadvantages, the internal sectoral spraying installation was suggested, where water is delivered out of the cutter head gear under pressure not exceeding 1 MPa with flow rate below 30 dm^3/min. Variant of this installation is also provided for air-water version. After development of internal sectoral spraying system (figure 4a) and simulation tests carried out as well as design work, a sectoral spraying installation for KPD roadheader was made (figure 4b). The stand tests of operational safety in relation to methane ignition hazard confirmed the effectiveness of this installation.

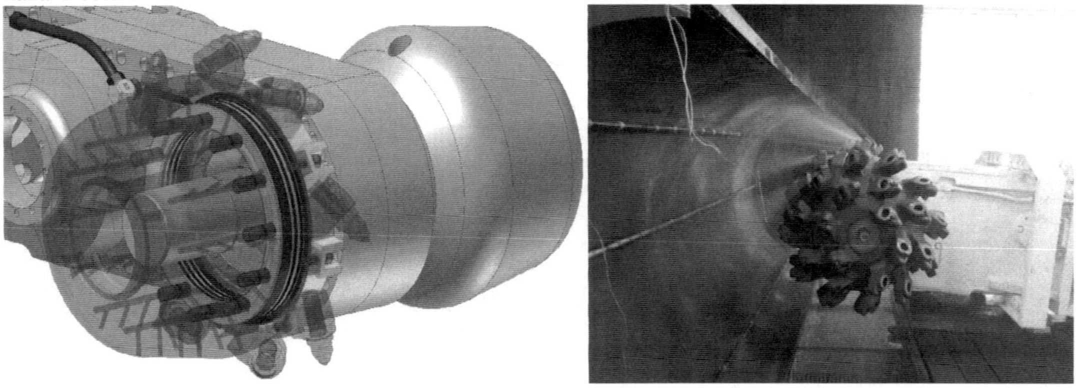

Figure 4. Installation of internal sector spraying system before trying to prevent ignition of methane [13].

KPD roadheader equipped with internal and external spraying installations was used in Jas-Mos mine and next in Marcel mine.

5. Application of air-water spraying in reduction of airborne dust concentration in roadways

Experience gained in realization of research and design projects on spraying installations intended for longwall shearers and roadheaders was used in development of air-water spraying installations to be applied in other places in the mine as dust concentration is one of the important problems that so far have not been solved comprehensively. Dust problem in roadways, besides mining operations and faces is important as dust is almost everywhere in coal mines. The problem is most important in roadways with the stream of used air and in the area of conveyors.

Undertaking design work on air-water spraying installation CZP BRYZA for dust control in mine roadways was one of the projects. The work assumed that airborne dust would be captured by water atomized by compressed air and after its watering, the dust particles would settle on louvre-type grate of special design (figure 5) [6]. The flow rate of the sprayed water does not exceed 3 dm^3/min. Such inclination of louvres (crossbars) was suggested not to obstruct free flow of air and to collect as much of wetted dust as possible.

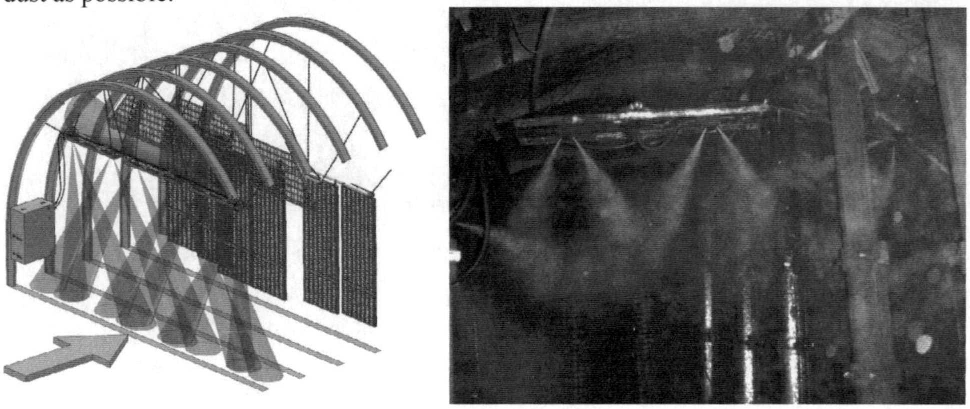

Figure 5. CZP BRYZA roadway dust barrier [6].

Spraying installation in operation should not negatively affect the working personnel, what was realized with the method for switching on/off the curtain on both sides. After manufacturing the first unit of CZP BRYZA the tests were conducted in Budryk mine in the scope of dust control efficiency in the zone protecting against propagation of coal dust explosion. The test results showed a double reduction of dust concentration in air flowing out of the longwall panel (figure 6). During observation it was noticed that CZP BRYZA device removed dust settled on entire circumference of the working on a distance of 100 m. Dust settled on the floor increased its water content, extending time of protecting against dust explosion by 50% [6].

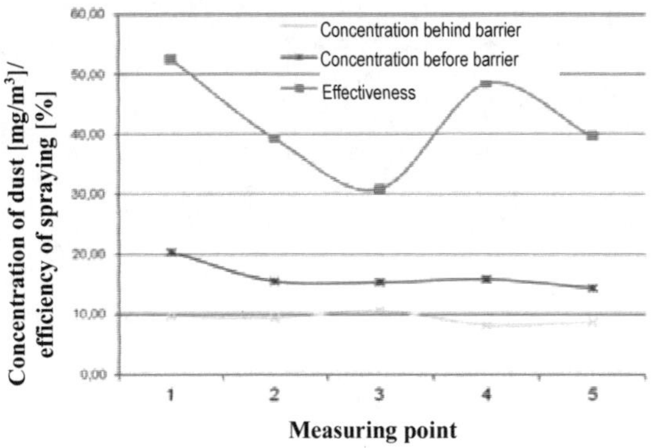

Figure 6. Efficiency of reducing the airborne dust concentration behind CZP BRYZA roadway dust barrier measured along the working width [6].

Positive test results of the device were installed increased work safety in this working. Ten CZP BRYZA anti-dust barriers were installed in eight coal mines. Regularly dust control effectiveness of newly implemented devices was tested, which showed a double reduction of total dust as well as of its respirable fraction irrespectively of the place of device installation. The roadway anti-dust barriers were also manufactured only in a water version.

6. Reduction of dust on coal haulage equipment

Transfer points of coal haulage conveyors are another field, where the air-water spraying devices were applied. Adaptation of this method seemed to be easy and natural. Conveyor transfer points required installation of spraying devices just over these points, however, the most important issue was to direct precisely the spraying stream towards the places of most intensive dust generation during loading the coal from one conveyor to another. It was noticed that dust starts generation on external surface of coal at the area of contact with the receiving conveyor. Propagation of dust during its transfer between conveyors was another problem. These problems were reduced by directing the air-water streams spraying installation to the edge of transferred run-of-mine and placing the drainage grate over the conveyor, which enabled catching the watered dust. After required conceptual work, simulation tests and designing the spraying installation VIRGA was installed in the underground workings of Brzeszcze mine. This solution consisted of a frame with nozzles evenly installed in two rows along entire frame width. In-situ tests, which were conducted, covered measurements of dust concentration at the area of conveyor operation, behind the transfer point, about 7 meters from the sprinkler and 7 meters ahead of the transfer point. The conducted measurements in the selected measuring points for the case without spraying, with water and with air-water spraying installations showed that efficiency of dust control for water spraying was 50% for total dust and only 4,4% for removal of respirable dust fraction. When using the air-water spraying installation, efficiency of dust control was 80% for total dust and even 90% for respirable dust fraction [6]. In the result of test, four VIRGA spraying installations implemented in Brzeszcze mine, and they were installed on the transportation route from the longwall panel. Water flow rate was reduced to 2 dm^3/min. The tests confirmed high efficiency of dust reduction up to 80% (figure 7) [14].

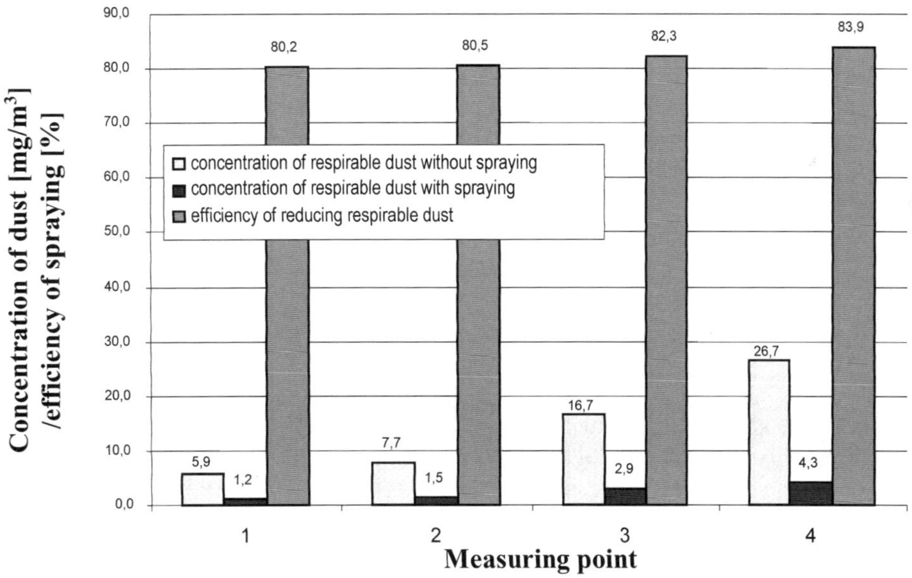

Figure 7. Measurement of respirable dust concentration and efficiency in reducing the respirable dust fraction using the VIRGA air-and-water spraying system [5].

Application of drainage grate cooperating with BRYZA-1200 spraying device was planned in another concept, where it was assumed that the water flow rate would not exceed 0.5 dm^3/min (figure 8a). Such a result was obtained by changing the design of spraying nozzles i.e. by reduction of inlet and outlet opening diameter. After designing and manufacturing of BRYZA-1200 prototype, it was tested in in-situ

IOP Conf. Series: Materials Science and Engineering **427** (2018) 012026 doi:10.1088/1757-899X/427/1/012026

conditions of LW Bogdanka mine. The analysis of these results showed that almost 83% of dust control effectiveness was achieved (figure 8b) [5]. Tests of BRYZA-1200 transfer points' spraying installations in other mines showed at least 70% efficiency of dust removal from mine air.

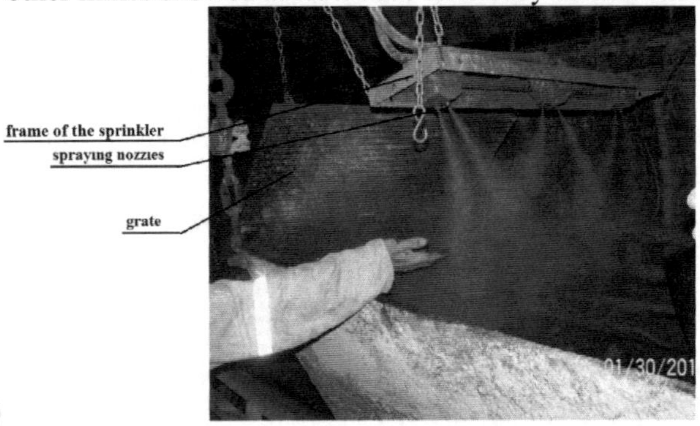

a)

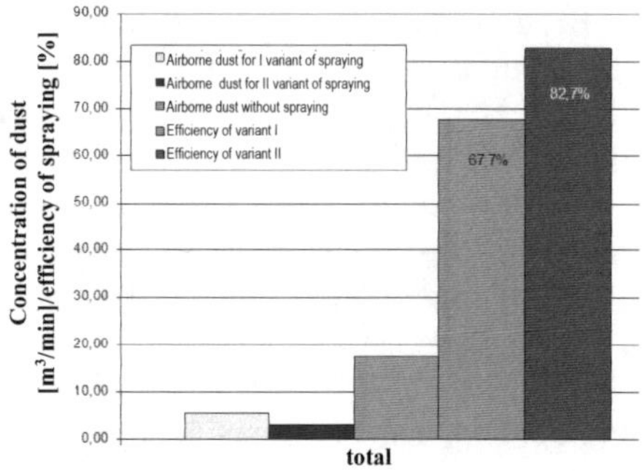

b)

Fig. 8. Bryza-1200 transfer point sprinkler a) first application of Bryza-1200 transfer point sprinkler b) measurements of total airborne dust concentration and efficiency of dust control using Bryza-1200 sprinkler [5].

Both types of conveyor sprinklers were initially designed as manually started devices. It was suggested to start up the spraying installation by belt movement aided by mechanical roller water distributor with the roller coupled with the conveyor belt and in another application using an optical dust sensor. In one of spraying installations it was suggested to start-up with proximity sensor detecting height of feed on the conveyor. These solutions enabled to increase the automation level of spraying installations, additionally limiting water consumption. Mine water used for spraying systems, not always has parameters that favour its combination with coal or rock dust particles. That is why a device, dosing the liquid agents reducing the water surface tension driven by flow of spraying water, was developed. Connection of intrinsically safe pilot electro-valve with check valve was the solution increasing the automation level of spraying installation. It enabled failure-free start-up of spraying installation by any electric impulse. In more complicated systems of spraying installations, which required protection against flow in wrong direction, the solution was suggested, which was a combination of solenoid valve with a pilot operated check valve as well as with check valve operating in the opposite direction. At the places, where installation of greater number of spraying systems is planned, a solution of the control valve of high water flow intensity was developed. All these devices found practical applications and contributed to increasing efficient dust control in coal mine air as well as in improving work safety and comfort.

7. Longwall system for dust control

The method for designing the air-water spraying installation was developed within the spraying installation working selectively at the place of mining machine generating dust during operation in longwall panels. This idea consists of installation of a single pack of spraying installations on the selected powered roof supports along the longwall and production of air-water mixture in the spraying packs in the area of generated airborne dust [8]. It was suggested to start only those spraying installations fixed to powered roof supports which are between the operating shearer and longwall end at the side of stream of used dust-laden air (figure 9). Additionally, it was assumed that each sprinkler should have a possibility of periodical switching off to enable passing the personnel through the spraying zone or in the case of required work in this spraying zone. The observations and measurements taken in the longwall showed that the highest dust concentration was in the shearer's mining path (over the AFC) and significantly lower concentration in the zone of personnel movement.In this zone dust concentration increased proportionally to a distance from the longwall shearer and was increased due to mixing of the ventilation stream and taking the dust from above of AFC to other parts of the working cross-section.

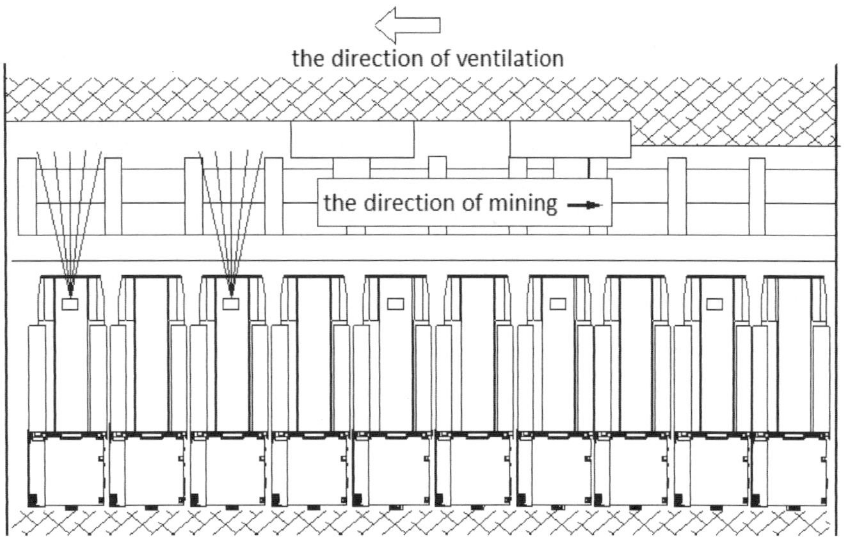

Figure 9. Operating zone of the KOMAG spraying system [15].

It was observed that the highest efficiency of dust removal from the longwall face can be achieved by directing the spraying streams towards the mining path to the zone between the longwall shearer and the ventilation roadway, which takes off airborne dust from the longwall. Such an approach enabled efficient reduction of dust by tens percent with water consumption of the whole system from 0.5 to 3.0 dm^3/min depending on the shearer position in the longwall face. To increase efficiency of dust control in low longwalls, below 2 m, sprinklers were developed, which had a function of articulating spraying streams (figure 10a). The tests of spraying system in Budryk mine longwall panel, with use of two sprinklers in the entire longwall, confirmed 80% efficiency in dust reduction in the distance up to 27 m from the sprinkler and about 63% in the distance of 72 m from the sprinkler (figure 10b). Similar tests were conducted in Sośnica mine, where one sprinkler was installed in the entire longwall and pressure of compressed air was 0.05 MPa. Despite the significantly lower spraying parameters, use of the sprinkler enabled dust reduction by over 92% in the distance of 6 m from the sprinkler and over 63% in the distance of 12 m from the sprinkler [15]. Positive test results of the longwall spraying system contributed to industrial implementation of this solution in longwall panels in other coal mines.

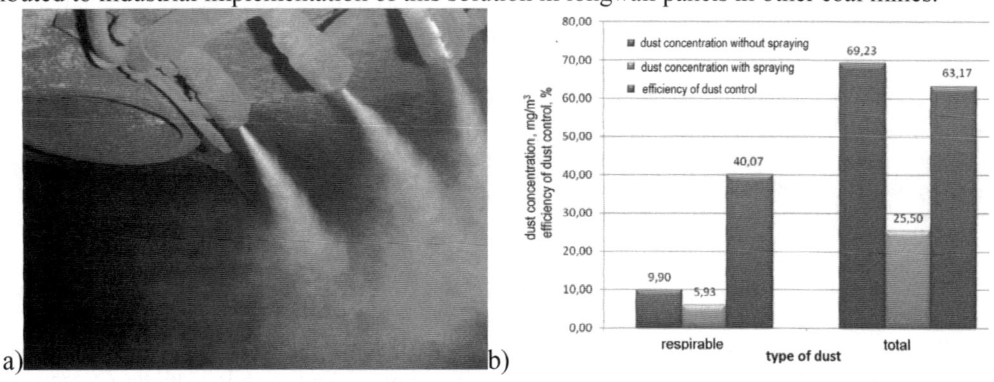

Figure 10. a) KOMAG-n spraying system in the longwall of Budryk mine, efficiency in dust reduction in the distance up to 72 m [15].

8. Improvement of safety in the zones protecting against coal dust explosion

During research work on air-water spraying systems, it was found that the zones protecting against propagation of coal dust explosion were made using the methods which did not enable their automation. Arrangement of such zones involves manual application of rock dust or watering it in such an amount to eliminate flammability and volatility of coal dust. Preparation of these zones is governed by the relevant regulations [16] and national standards [17, 18]. Effectiveness of the protecting zones preparation is controlled by tests conducted, among others, by Barbara Experimental Colliery. Proper quality of the protecting zone is usually assured by adding surplus amount of agent neutralizing the explosive coal dust. The project provided possibilities process automation of protecting zones preparation with proper quality of neutralizing the explosive coal dust as well as with limitation of penetration of coal dust to the protecting zone by use of air-water spraying systems. There has not been any such research work suggesting such an approach before. This subject was proposed in the framework of cooperation between the KOMAG, GIG (KD Barbara) and Brzeszcze mine as a part of the MEZAP research project [19].

The project included the following two variants of supporting the preparation of protecting zones:

- installation of air-water spraying system on a suspended rail (figure 11a),
- stationary installation of air-water spraying installation coupled with dust meter and the control device (figure 11b).

a) 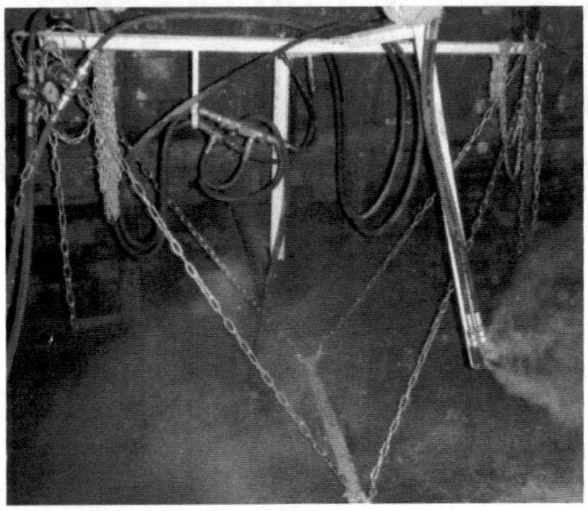 b) 

Fig. 11. Air-water sprinkling installation a) mobile installation b) stationary installation [19].

Only after detailed in-situ tests on measuring dust concentration in the zones protecting against coal dust explosion, which were planned in a few mine workings, it was possible to use the suggested solutions in mines. The following measurements were taken within the tests: rate of dust settlement and presence of the settled dust, moisture, dust concentration with division into the fractions from nanometers to 1 millimeter as well as measurements of air flow rate, its temperature and water content.
Empirical models of relationship between intensity of dust settlement and changes in its concentration were developed [20]. The results showed that while walking away from dust source, rate of dust sedimentation and its concentration decreased what can be described by the exponential functions (figure 12). Depending on type of the roadway, its character, airflow direction and its rate, slope of the function is different [20, 21]. The developed mathematical models of dust sedimentation rate enabled a development of algorithm for intelligent control module coupled with an optical dust meter, controlling stationary air-water spraying installation. Controlling the spraying process in the function of dust concentration in mine workings was possible due an installation of predictive algorithm [22]. The solution paved the way to aiding a preparation protecting zones against propagation of coal dust explosion. In-situ tests of the stationary protective zone showed an extension of the protecting zone life between its consecutive renewals [20].

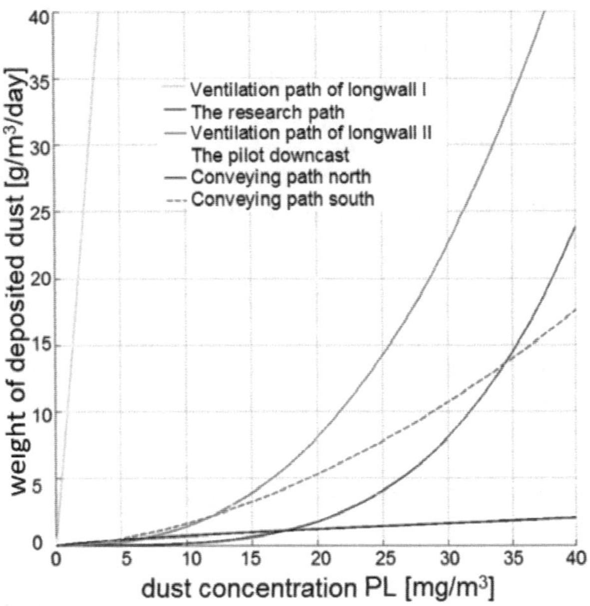

Fig. 12. Equivalent characteristics of dust deposition depending on dust concentration in the air in paths under analysis [20].

The second version of aiding the preparation of protecting zones was based on the solution of mobile protecting zone. The design of spraying device was developed and its testing was planned, which predicted assessment of dust sedimentation and dust transient moisture as well as a movement assessment of dust settled on the floor to other parts of the working [20].

Both variants of aiding the preparation of protecting zones showed positive results with lower sedimentation rate and also higher content of transient water. Variant with a stationary protective zone is technically easier in implementation and common use as it does not require movement along the protective zone as well as dust concentration control with use of optical dust meter and adaptive control of spraying efficiency, depending on change in dust concentration such a solution gave a better image of safety conditions in the protective zone as one of defending barriers against propagation of coal dust explosion.

Research work, related to maintaining the protecting zones against propagation of coal dust explosion, is continued within ROCD European project. It is planned to develop a system for prediction of dust propagation and its concentration in the selected part of the mine, basing on developed empirical models and sensors installed in the key places of ventilation system.

9. Summary

The proposed method for designing air-water spraying installations enabled a development of effective solutions to reduce dust concentration and methane ignition risk as well as coal dust explosion. The solutions are widely used in the coal mining industry. Development of a simple, reliable in operation and effective solutions allowed to extend installations' scope of use in almost all the areas of mines. Other implementations that are not the subject of this article have been applied in preparation plants and aggregate mines. Up-grading of spraying installations has led to an increase in the level of automation, improving the safety and optimizing operation of these installations. In the design and implementation process, particular attention was paid to improvement of miners' work comfort in the area of operation of these installations. During an operation of cutting drums, when machines were equipped with air-water spraying systems no methane ignition event occured and all the air-water spraying installations contributed significantly to the reduction of coal dust in mines.

References

[1] State Mining Authority 2017 Assessment of work safety, mining rescue and general safety in connection with mining and geological activities in 2016 *Report WUG* 2 7-28

[2] Kabiesz J et al 2015 Annual report (2010) on main natural and technical hazards in hard coal mining industry *Central Mining Institute*

[3] https://pl.wikipedia.org/wiki/Katastrofy_górnicze_w_Polsce

[4] Prostański D and Rojek P Designing, testing and operational tests of air-water spraying installation for dust and methane control installed in KSW-460 longwall shearer *Masz. Gór.* **24** 37-44

[5] Prostański D 2012 Dust Control with Use of Air-Water Spraying System *Arch. Min. Sci.* **4** 975-990

[6] Prostański D 2013 Use of air-and-water spraying systems for improving dust control in mines *J. Sust. Min.* **2** 29-34

[7] Prostański D et al 2013 Testing the utilitarian features of air-and-water spraying installations on KSW-800NE shearer *KOMTECH 2013 Innovative Techniques and Technologies for Mining Industry* pp 211-24

[8] Bałaga D et all 2016 Innovative spraying system for longwall panels *Masz. Gór.* **3** pp 14-22

[9] Prostański D 2011 Assessments of efficiency of air-water and water spraying systems in reduction of airborne dust concentration. *Gór. Śr., Pr. Nauk. GIG* **4** pp 398-408

[10] Kalukiewicz A 1998 High pressure assistance systems in roadheaders. *Prz. Górn.* **3** pp 12-18

[11] Krackhecke M and Finkenbusch R 2002 Wassernebelbedüsungen für Querschneidköpfe und wettertechnische Optimierungen in TSM–Votrieben *Glückauf* **5**

[12] Libera K and al 2010 System of air-water spraying installation in roadheaders manufactured by REMAG *Masz. Gór.* **2** pp 19-24

[13] Bałaga D and Siegmund M 2016 Stand tests of effectiveness of spraying system of KPD roadheader in extinguishing the explosion and preventing against methane ignition *Masz. Gór.* **2** p 312

[14] Prostański D and Ryszka L 2012 Assessment of effectiveness of dust control with use of air-and-water spraying on run-of-mine transportation routes *KOMTECH 2012 Innovative Techniques and Technologies for Mining* pp 455-64

[15] Prostański D 2013 Air-and-water spraying installation reducing airborne dust concentration in longwall panels *KOMTECH 2013 Innovative Techniques and Technologies for Mining Industry* pp 189-99

[16] Regulation of the Minister of Energy of November 23, 2016 on detailed requirements for the operation of underground mining facilities *Journal of Laws of 2017* item 1118

[17] PN-G-04036 1997 Explosion protection of mining plants - Protection against coal dust explosion - Determination of the intensity of settling dust

[18] PN-G-04037 1998 Explosion protection of mining plants - Protection against coal dust explosion - Determination of the content of non-combustible parts in mine dust

[19] Project MEZAP Modelling mechanism of combustible coal dust deposition in the vicinity of exploitation fronts to identify, assess and eliminate possibility of its explosion PBS1/B2/4/2012

[20] Prostański D 2017 Empirical models of zones protecting against coal dust explosion *Arch. Min. Sci.* **3** pp 611-19

[21] Prostański D 2015 Experimental study of coal dust deposition in mine workings with the use of empirical models *J. Sust. Min.* **2** pp 108-14

[22] Prostański D et al 2016 Measurements of dust concentration in protective zones *Masz. Gór.* **1** pp 310

[23] Kotwica K 2011 The influence of water assistance on the character and degree of wear of cutting tools applied in roadheaders *Arch. Min. Sci.* **3** pp 353-74

[24] Krauze K et al 2015 Parametric Factors for the Tangential-Rotary Picks Quality Assessment *Arch. Min. Sci.* **1** pp 265-81

[25] Kotwica K and Krauze K 2007 Selection and underground tests of the rotary tangential cutting picks

used in cutting heads of the longwall and roadway miners *Arch. Min. Sci.* **2** pp 195-217

IOP Conf. Series: Materials Science and Engineering **427** (2018) 012027 doi:10.1088/1757-899X/427/1/012027

Preliminary assessment of new approach to scale-up the laboratory results of flotation reagents tests based on mineralogical analysis.

M Krzemińska and P Kurzydło

KGHM Polska Miedź S.A., Oddział Zakłady Wzbogacania Rud

pawel.kurzydlo@kghm.com

Abstract. The question of introducing new collectors into copper concentrator plants referring to KGHM experience has been presented in this paper. Description of present methodology of flotation reagents testing has been indicated followed by explanation of crucial problems and limitations with proper qualification of reagents. Some attempts to apply scale-up methodology have been reported in technical literature. In this article assumptions of new methodology based on advanced mineralogy tests have been suggested. This methodology should allow the more accurate forecast of flotation performance of industrial plant after the laboratory tests.

1. Introduction

The right selection of flotation collectors is an essential issue for each flotation process optimisation. As reagents have main influence on process performance, scientist' efforts are focused on continuous development of their efficiency. In KGHM the research investigations on flotation reagents has been still continuing since commissioning ore processing plants in KGHM structures (1968-1974).

Testing of reagents were aimed to introduce various mixtures of collectors to optimise technology efficiency and reagents consumption rates. Works have been performed by KGHM own forces as well as with cooperation with outside research and development centres. Consequently, some tens of reagents with different properties have been examined. They were delivered by well-known global scale manufacturers as well as by domestic local producers. Reagents have been tested as separate collectors or as additives to regularly used collectors. Their efficiency has been tested together with optimal consumption rate and synergy with other reagent(s) but also the influence of different points of dosing in flotation and grinding circuits on process efficiency has been examined [1].

As a result of long-term research the proper mixtures of collectors have been chosen and approved for each KGHM concentrator plant. Nowadays mixtures with fixed quantities of sodium xanthates and mixtures of sodium xanthates with sodium diethylditiophosphate are used as a regular collectors chosen for different qualities of flotation feed. Selected reagent species are the most popular collector applied worldwide not only for copper sulphide ore but also for Cu-Mo and Cu-Au ores. They are recognized as presently most efficient reagents for polymetallic sulphide ores [1].

2. Some aspects of development the appropriate method of flotation reagents testing

In spite of satisfactory results of collector tests there is a continuous effort to develop new reagents or their mixtures which might provide better process efficiency performed on copper ores of Legnica-

Glogow Copper District (LGOM) deposit. The investigation attempts are often finalized with no positive recommendation. Conclusions made on laboratory scale results are not sufficient to recommend the plant trial. Among the large amount of tested tens of reagents only few of them were approved to be tested in industrial scale. Hence the researchers activity should be still focused on continuously development of methods of new reagents testing which are applied in concentrator plants.

The problem of scale-up results is a complex issue. It is known that in laboratory scale investigation should range many variable parameters, so they should be conducted on small amounts of tested samples [2]. The laboratory scale experiments should be proper planned and developed to reflect the real industrial conditions. Testing parameters should be representative what means they should mirror specific physicochemical phenomena and hydrodynamic conditions present in industrial flotation cells [3]. Laboratory experiments should be replicate at least twice or three times and their results should be statistical evaluated. The method suggested by Dell is based on set of multiple fractional flotation experiments followed by scale-up prognosis. Another method targeted to forecast industrial results is comparison of flotation selectivity observed in laboratory flotation with selectivity so called re-flotation of industrial final concentrates and tailings [4].

Comprehensive study on scale-up methodology proves that complete forecasting industrial results based on laboratory testing has not been developed yet. Even though the aspect of scaling-up reagent tests in KGHM conditions was thoroughly analysed, the reliable, universal and ready-to-use method of laboratory tests still needs to be established. According to current practise, tested reagent is approved to industrial scale trial only on slight differences between lab results for reference and tested reagents. Generally laboratory results do not repeat in industrial scale for majority of conducted tests [5-7]. Subsequently the aspect of scale-up and results reproducing must be better investigated as a crucial factor of testing the new reagents. The aim of this approach is to minimize or eliminate negative results of industrial scale trials and in results to avoid undesirable costs.

3. The error sources of method of laboratory tests and industrial trials

Applied in KGHM new reagents testing procedure includes preliminary laboratory testing stage followed by industrial trial for approved reagents with promising prognosis of their efficiency in plant conditions. The range of laboratory testing is practically limited to fractional flotation in pneumomechanic cell of one litre volume. On the other hand, comparing to industrial conditions, laboratory scale allows to dose precisely reagents and to control air flowrate, to ensure scheduled retention time, froth level and other important process parameters.

It needs to be emphasized that in real, industrial circuits the process efficiency depends not only on controlled flotation cell operating parameters but also on set of various process factors and their influence on process is difficult to predict. Among others, one of the important aspects of industrial conditions is recirculation of slurries resulting with gathering material with low floatability and some amounts of reagents. Moreover, in plant conditions many variable parameters affects process stability as fluctuation of ore throughput, feed flowrate, accuracy of airflow and reagents dosing and important role of industrial circuits inertia after changing operating regime.

Difficulties with prognosis of industrial reagents efficiency are also related with complexity of geological morphology of LGOM deposit. KGHM owned resources, formed as stratoidal type of deposit, constitute sulphides minerals accumulation in layer of sandstone, rotliegend, zechstein shales and carbonates (mainly dolomites) rocks. Based on this stratigraphy the three species of ore lithology have been described: sandstones, shale and dolomites. In large part of deposits they occur common but in different ratio and with various mineralization. Copper orebody has several forms of sulphide minerals occurrence with different grain size. Hence geological conditions strongly influence on lithological and mineralogical constitution of run-of-mine ore.

To summarize all the limitations listed above, it is difficult to conclude about forecast reagents efficiency in real environment avoiding important uncertainty of calculated result. It is necessary to develop more accurate scale-up model which is dedicated strongly to KGHM conditions.

IOP Conf. Series: Materials Science and Engineering **427** (2018) 012027 doi:10.1088/1757-899X/427/1/012027

4. Guidelines for developing the new procedure of reagents testing

Unsuccessful effects of present reagents testing are the background to resume effective model of scaling-up laboratory results to industrial reality. The more specialized and specific approach is needed, which can deliver more reliable laboratory results on which decision of industrial trial is considered.

This task should be based on experience and knowledge of researchers who investigated procedures of application flotation reagents to KGHM concentrator plants. As a new concept it is suggested to support industrial trails with parallel (simultaneous) laboratory tasting. Results of laboratory tests achieved during industrial trial will be compared with previous laboratory data delivered by tests conducted to decide on industrial trial. These data together with industrial results will be base to develop algorithm which allows to forecast industrial performance of reagents only on laboratory tests data. It is a new approach but with reasonable background both theoretical and practical. The details of such methodology is based on reports presented in scientific literature but also derives from realized testing of reagents conducted during years of KGHM activity. This is why it is promising direction of future research with a great potential. This approach will allow include a range of characteristic parameters of laboratory flotation in scaling-up its results. These parameters are difficult to consider separately and can be exemplified by slurry hydrodynamic phenomena occurring in different conditions in laboratory, pilot or industrial flotation cells.

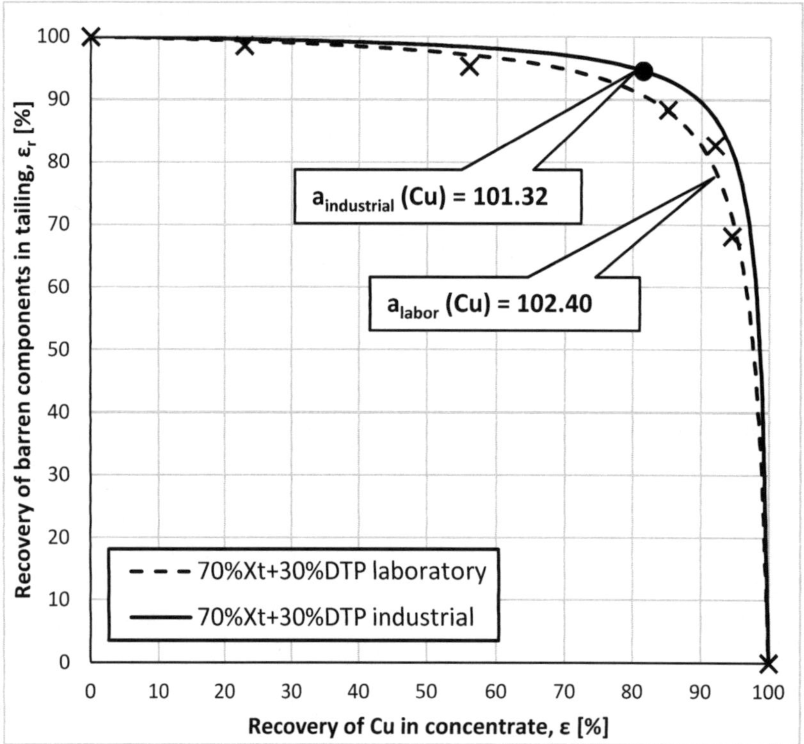

Figure 1. Comparison of Fuerstenau upgrading curves for copper obtained during reagent test in laboratory and industrial scale for one of the technology line of KGHM concentrator plant.

Present attempts to develop the reliable method of reagents testing are based mainly on chemical assay to determine and balance Cu content, sometimes complemented by Ag and other elements, in analysed samples. Nowadays the more detailed and specific information may be obtained by mineralogy investigation of analysed samples. Referring to geological complexity of LGOM deposit described in previous paragraph, it seems to be reasonable to determine and compare upgrading susceptibility of specific copper sulphide minerals in laboratory and industrial scale.

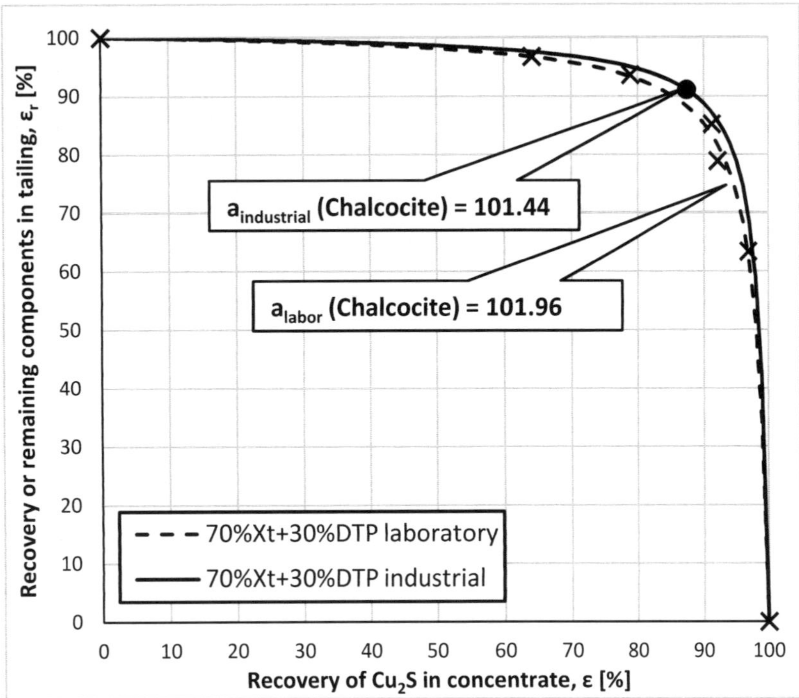

Figure 2. Comparison of Fuerstenau upgrading curves of chalcocite obtained during reagent test in laboratory and industrial scale for one of the technology line of KGHM concentrator plant.

During some research on application of new mixtures of collectors a set of laboratory flotation experiments was executed [1]. The feed samples were collected from one of the technology line of KGHM concentrator plant to examine performance of selected collector mixture. Concentrates and tailing were examined with MLA technic to quantify mineralogical species. The results of comparison laboratory and industrial upgrading of copper and major copper bearing copper sulphide minerals are shown in figures 1 – figures. 4. The comparison of laboratory and industrial data was possible due to laboratory research with collector mixtures conducted simultaneously with regular metallurgy assay of production process. Both experiments were conducted with the same collectors mixture (70% mixture of ethyl/isobutyl sodium xanthate and 30% sodium diethylditiophosphate). To present results of reagents efficiency Fuerstenau upgrading curve with corresponding selectivity factors "a" was chosen as a curve not sensitive to α variation. This is necessary to avoid influence of differences between feed grade in laboratory and industrial observation.

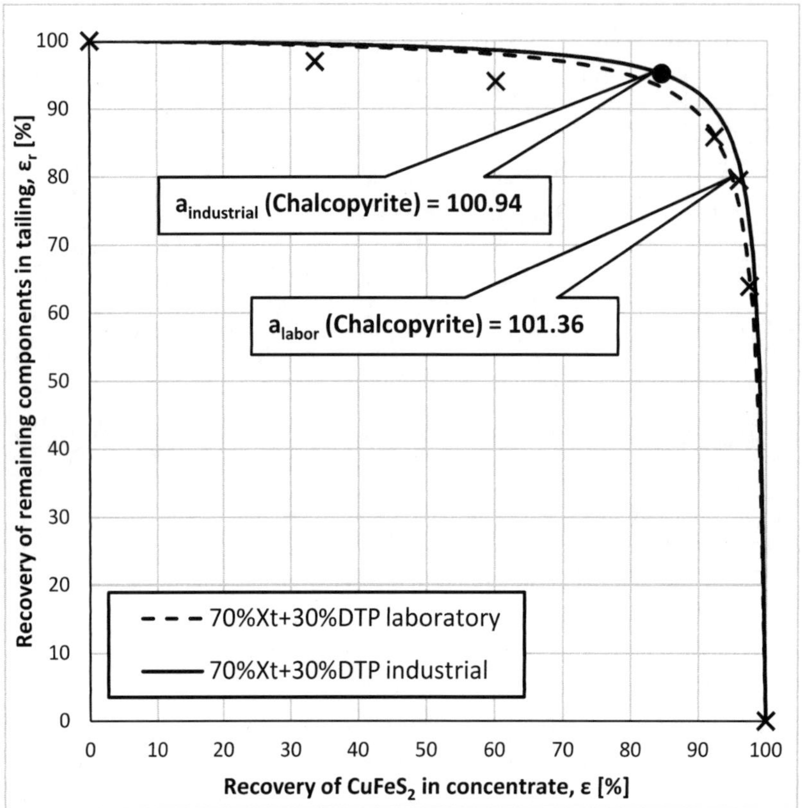

Figure 3. Comparison of Fuerstenau upgrading curves of chalcopyrite obtained during reagent test in laboratory and industrial scale for one of the technology line of KGHM concentrator plant.

Obtained results (figure 1 – figure. 4) show the differences between laboratory and industrial upgrading behaviour of specific mineral components present in feed. Comparison of upgrading curves for Cu and analysed copper sulphide minerals demonstrates that difference between selectivity factors for elemental copper is evident higher than relevant differences for chalcocite, chalcopyrite and bornite. It appears the possibly influence of other copper bearing minerals and barren minerals on total copper upgrade performance. Therefore, minor sulphide minerals may negative affect flotation efficiency. Observed phenomenon should be investigated and verified by broaden analysis of relationships between laboratory and industrial upgrade rates of remaining copper bearing minerals as well as accompanying sulphide minerals present in flotation feed. Extended mineralogical tests may bring crucial information on mutual interactions between specific minerals during scaling-up process conducted in order to implement new flotation reagents.

IOP Conf. Series: Materials Science and Engineering **427** (2018) 012027 doi:10.1088/1757-899X/427/1/012027

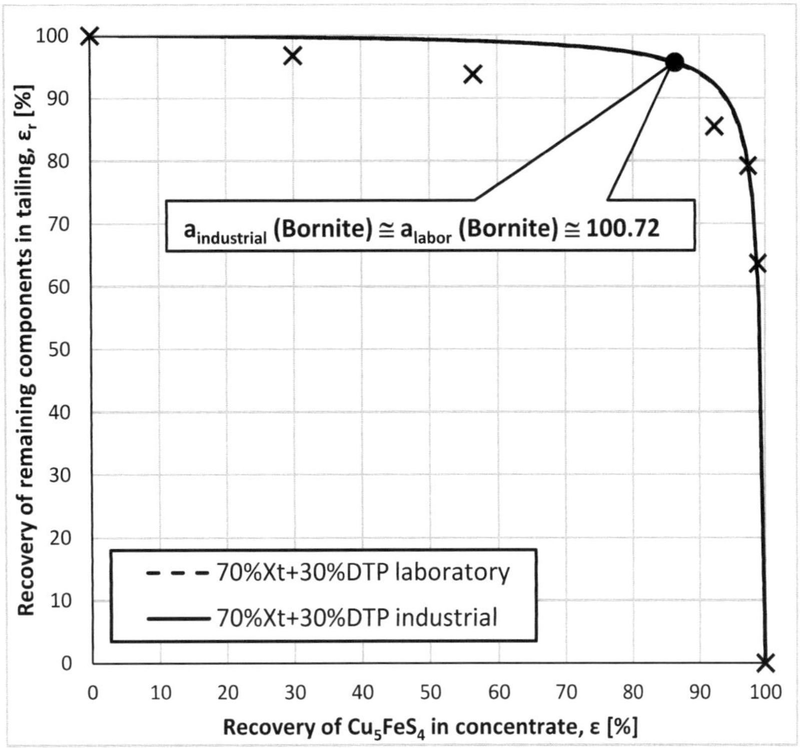

Figure 4. Comparison of Fuerstenau upgrading curves of bornite obtained during reagent test in laboratory and industrial scale for one of the technology line of KGHM concentrator plant.

5. Summary

Latest research on selection flotation reagents was generally based on assessment of upgrading properties of coppers and other valuable metals in laboratory and industrial conditions. Collected experience proves that positive results of laboratory tests do not guarantee satisfactory effects in concentrators plant when new reagents is applied. Many attempts have been taken to form scale-up model for test results from laboratory to industrial environment. However, no reliable, universal and easy-to-use laboratory reagent testing method to predict their industrial performance has been developed yet. The accuracy of scale-up models is affected by many factors influencing on process stability besides geological complexity of LGOM deposit. In industrial conditions research and testing policy and operational procedures of new collectors implementation have the key impact on process economy. Since the present knowledge on practical aspects of scaling-up is unsatisfactory, discussed subject desires complex approach which should be develop in further studies. As preliminary tests, upgrading behaviour of main copper sulphide minerals has been investigated, simultaneously in laboratory and industrial scale. Results reveals complexity of scaling-up models for specific minerals. Therefore the scope of future investigation should include possibilities delivering by contemporary mineralogical analysis methods.

References

[1] Krzemińska M 2017 *Wpływ mieszaniny kolektorów flotacyjnych na efektywność wzbogacania rud miedzi* Praca doktorska (Wrocław: not published)

[2] Woźniakowski B and Grotowski A 1975 *Powiększanie skali – programowanie badań*

http://www.minproc.pwr.wroc.pl/journal/pdf/1975/2%20Wozniakowski.pdf

[3] Laskowski J 1975 Problemy skali badań w fizykochemicznych metodach wzbogacania *Physicochemical Problems of Mineral Processing* **9(1)** 33–49 (Wrocław: Wrocław University of Technology)

[4] Drzymala J, Luszczkiewicz A and Foszcz D 2010 Application of upgrading curves for evaluation of past, present and future performance of a separation plant *Mineral Processing and Extractive Metallurgy Review* **31(3)** 165–175 (Abingdon: Taylor and Francis Ltd)

[5] Tumidajski T, Łuszczkiewicz A, Drzymała J, Trybalski K, Foszcz D, Muszer A, Niedoba T, Saramak D and Henc T 2007-2008 Określenie wpływu wzbogacalności przerabianych rud na jakość koncentratów miedziowych dla potrzeba optymalizacji górniczo-hutniczego procesu wytwarzania miedzi, Etap I i uzupełnienie *Sprawozdanie z Pracy AGH 5.5.100.60/2007* (Kraków: not published)

[6] Trybalski K, Łuszczkiewicz A, Drzymała J, Tumidajski T, Foszcz D, Muszer A, Niedoba T and Henc T 2008 Określenie wpływu wzbogacalności przerabianych rud na jakość koncentratów miedziowych dla potrzeba optymalizacji górniczo-hutniczego procesu wytwarzania miedzi, Etap II *Sprawozdanie z Pracy AGH 5.5.100.60/2007* (Kraków: not published)

[7] Foszcz D, Łuszczkiewicz A, Drzymała J, Tumidajski T, Trybalski K, Muszer A, Niedoba T and Henc T 2009 Określenie wpływu wzbogacalności przerabianych rud na jakość koncentratów miedziowych dla potrzeba optymalizacji górniczo-hutniczego procesu wytwarzania miedzi, Etap III *Sprawozdanie z pracy AGH 5.5.100.390 /2009* (Kraków: not published)

IOP Conf. Series: Materials Science and Engineering **427** (2018) 012028 doi:10.1088/1757-899X/427/1/012028

Hydrometallurgical processing of serpentine ore

E A Abdel-Aal[1], D El-Sayed[1], A K Ismail[1] and F I El-Hosieny[2]

1 Central Metallurgical Research and Development Institute, P.O. Box 87, Helwan, Cairo, Egypt
2 Ain Shams University, Faculty of Science, Cairo, Egypt

Abstract. Serpentine ore from the Eastern Desert of Egypt was leached with aqueous hydrochloric acid. Experimental statistical design program (ESD) was applied to study the influence of different variables on leaching efficiency (reaction temperature, leaching time and Hydrochloric acid (HCl) to Magnesium oxide (MgO) molar ratio (stoichiometry) where the liquid/solid ratio of 5:1 ml/g was maintained constant along the experiments. The mathematical model and the different contours (curves generated from ESD) were discussed. The results showed that, all the studied parameters are effective. The results have been collected at a 3-D cube showed that at high levels of parameters (4 hrs , 95 °C and 1.6 S), the Magnesium oxide recovery was about 96.5%, while M agnesium oxide recovery was 67.1% at the low levels (2 hrs, 65 °C and 1.4 Stoichiometry (S).

1. Introduction

Serpentine, [$Mg_3Si_2O_5(OH)_4$], a hydrated magnesium silicate, is abundant mineral all over the world. Egypt has serpentine resources. The serpentine sample in this study is characterized by high contents of magnesium and silica (36.8% SiO_2) as well as low calcium and aluminum contents. Moreover, important quantities of nickel (0.4 % NiO) and chromium (0.4 % Cr_2O_3) are also present in the ore.

This mineral has been utilized in various applications, such as fertilizer production, soil amelioration, extraction of amorphous silicate, carbon sequestration, extraction of pure magnesium compounds and silica production, refractories [1-9]. Magnesium metal has wide range of applications in our life. It is manufactured by electrolysis of molten anhydrous magnesium chloride [10, 11]. So, production of magnesium chloride salts and other magnesium compounds from serpentine have great interest due to various applications in different industries. Magnesium has high chemical reactivity and always present as compounds in nature e.g. minerals and salts [12].

Production of magnesium compounds depends mainly on extraction of magnesium minerals using different mineral acids from serpentine or other ores such as magnesite [13-17] and dolomite [18]. Serpentine ore is widespread all over the world. Serpentine minerals include three minerals namely chrysotile, lizardite, and antigorite, all of which have similar compositions based on the presence of the main nucleus $Mg_3Si_2O_5(OH)_4$, and differ slightly in the degree of substitution of Mg by Fe and Al [19].

The main commercially occurrences of serpentine ores in the world are in Canada, U.S.S.R and Africa [20]. In addition, serpentine ores are widely spread at different regions in Egypt, namely, in the Eastern Desert (Barrameya, along the Quift-Quseir road, Wadi Gabel El-Rubshi) [21-23].

2. Experimental

2.1. Raw materials and apparatus

A sample of serpentine ore from the Eastern Desert (Egypt) and 36% pure hydrochloric acid were used. The reaction was performed in a 500 ml round bottom flask fixed in water bath that adjusted to the required temperature. Reaction mixture was agitated at a rate of 500 r.p.m. and then filtered under vacuum in a Buchner-type filter using polypropylene filter cloth of 200 mesh aperture size.

2.2. Characterization

Powdered samples (100 % -200 mesh) was analyzed using X-ray powder diffraction (XRD), Bruker AXS diffractometer (D8-ADVANCE) with Cu Kα radiation, operating at 40 kV and 10 mA. The diffraction data were recorded for 2θ values between 4° and 70° and the scanning rate was 0.5° min^{-1}. Types of the phases in the samples were identified using the X-ray powder data file, published by the American Standard for Testing Material (ASTM). Scanning Electron Microscope, JEOL instrument (Japan) model JSM-5410 was used for sample imaging.

Philips X–ray fluorescence (XRF) Spectrometer was applied for determining the chemical composition of the ore.

2.3. Procedure

Calculated amount of serpentine was added gradually into agitated hydrochloric acid solution in the reaction flask. After the desired reaction time, slurry was filtered and washed with warm water from silica and other insoluble materials. The filtrate was analyzed for MgO and CaO contents and then evaporated. Resultant crystals were analyzed for total MgO content. The residue was washed two times with warm water.

An experimental Box–Behnken design was used to study the interaction of the following variables (leaching time, reaction temperature and HCl:MgO stoichiometry) that affects the leaching of serpentine ore.

The applied variables are: reaction time in hours (2, 3, 4), reaction temperature, °C (65, 80, 95) and stoichiometry (1.4, 1.5, 1.6).

The filtration fundamentals are obtained by applying Darcy's equation. The filtration characteristics of magnesium chloride solution from un dissolved silica and other components are expressed by the filtration rate, the specific cake resistance and the specific filtering medium resistance. High efficiency of the filtration operation is achieved when the filtration rate is maximum whereas cake resistance and filtering medium resistance are minimum. The basis of the filtration study is the correlation derived from Darcy's equation which is given as follows:

$$\frac{dV}{d\theta} = \frac{G_e\, \Delta P\, A^2}{\mu\, \alpha\, \omega} \cdot \frac{1}{V + V_e} \qquad (1)$$

where:

V - Volume of filtrate at the time used, cm^3

θ - Filtration time, sec

V_e - Equivalent volume of filtering medium which is defined as the volume of filtration corresponding to fictitious weight of cake that causes resistance equal to the effective filter medium resistance.

Specific Cake Resistance (α), cm/g

Equivalent Volume of Filtering Medium (Ve), cm^3

Pressure difference in g/cm^2 (ΔP cm Hg x 2.54 x13.32)

Viscosity of Filtrate at 90 °C (μ), Poise

Weight of Solids per Unit Volume of Slurry (ω), g/cm^3
Gravitational constant (G_c)
The reciprocal of the above equation gives:

$$\frac{d\theta}{dV} = \frac{\alpha\,\mu\,\omega}{A^2\,\Delta P\,G_c}(V + V_e) \tag{2}$$

Which represents an equation of a straight line when $\Delta\theta/\Delta V$ is plotted against V. The calculations of the filtration data comprise the following:

Experimental data of V and θ,

Viscosity, calculated from the rate of flow of the filtrate compared with that of water at the same temperature (poise), Pressure difference in g/cm^2 (ΔP inch Hg x 2.54 x13.32), Plotting V against Δθ/ΔV and calculated the slope (sec/cm^3), Specific cake resistance, α from the conversion:

$$\alpha = \text{slope} * A^2 * G_c * \Delta P * \frac{1}{W} * \frac{1}{\mu} \text{ cm/g} \tag{3}$$

Equivalent volume of filtering medium, V_e:

$$V_e = \frac{\text{intersecion}}{\text{slpoe}} \text{ cm}^2$$

Filtering medium resistance, R from the conversion:

$$R = \frac{\alpha W}{A} * V_e \text{ cm}^{-1} \tag{4}$$

Filtration rate at V = α cm^3 . $\frac{dV_x}{d\theta}$ from the conversion:

$$\frac{dV_x}{d\theta} = \frac{G_c\,\Delta P\,A^2}{\mu\,\alpha\,W} \cdot \frac{1}{\alpha + V_e A} \quad cm^3/sec \tag{5}$$

The following filtration conditions were applied:
Type of filter: Laboratory Buchner Type Funnel
Diameter of filter: 9.5 cm
Filter area: 70.9 cm^2
Pressure difference: 550 mm .Hg
Type of filter cloth: Single layer polypropylene
Temperature of slurry: 90 °C
Density of slurry: 1.2 g/ml

For calculating the filtration rate, the slurry is poured in the Buchner Funnel while suction is applied and the rate is determined from the volume accumulating in the receiver in the time required to all the liquid to be sucked and the surface of cake becomes dry.

IOP Conf. Series: Materials Science and Engineering **427** (2018) 012028 doi:10.1088/1757-899X/427/1/012028

3. Results and discussion

3.1. Characterisation of the ore

Mineralogical composition

As stated earlier this was accomplished using XRD. Figure 1 shows the pattern obtained. It shows that the major mineral is antigorite [$Mg_3Si_2O_5(OH)_4$].

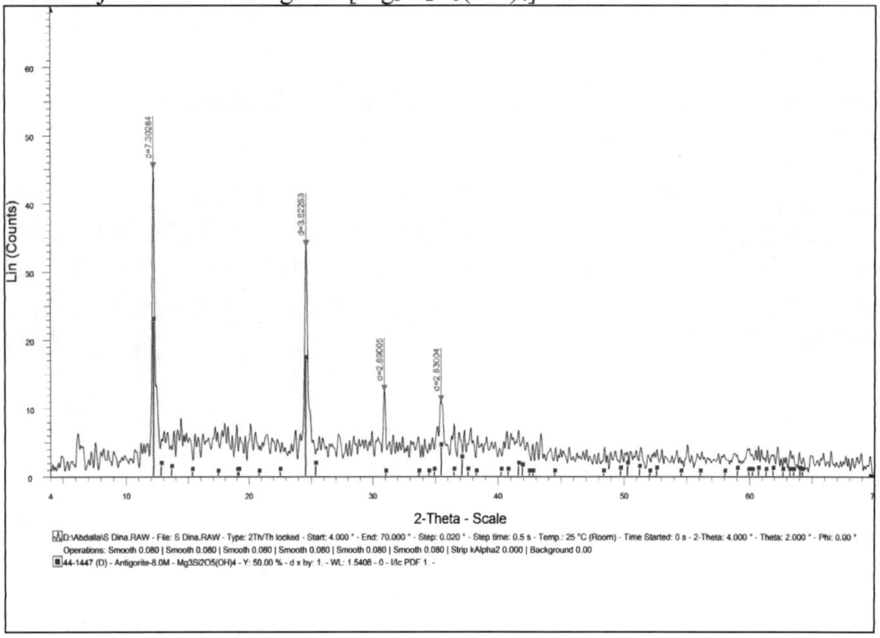

Figure 1. XRD patterns of serpentine ore (*Antigorite).

Chemical composition

As subjected to XRF analysis, the chemical composition of the ore was shown in table 1. The serpentine sample in this study is characterized by high contents of Magnesium (35.8 % MgO) and Silica (36.8% SiO_2), low calcium content (3.8 %) and low aluminum content (0.4 % Al_2O_3). Moreover, an important quantity of nickel (0.4 % NiO) and chromium (0.4 % Cr_2O_3) is also present in Egyptian serpentine ore.

Table 1. Chemical composition of serpentine ore.

Component	Concentration, %	Component	Concentration, %
SiO_2	36.80	**ZnO**	0.091
MgO	35.80	**SO_3**	0.031
Fe_2O_3	7.64	**Co_3O_4**	0.026
CaO	3.81	**SrO**	0.015
NiO	0.441	**P_2O_5**	0.008
Al_2O_3	0.381	**CuO**	0.007
Cr_2O_3	0.361	**L.O.I.***	14.40
MnO	0.163		

L.O.I.*: Loss on Ignition at 1000 °C.

3.2. Statistical experimental design

An experimental Box–Behnken design [24] was used to study the effect of some variables (reaction time, reaction temperature and stoichiometry) affects the leaching of serpentine ore (% recovery of MgO). The design-matrix of different runs, 17 experiments, as well as the levels of each factor are shown in table 2. The applied variables were: reaction time in hours (2, 3, 4), reaction temperature °C (65, 80, 95) and stoichiometry (1.4, 1.5, 1.6).

Table 2. Experimental Box Behnken Design with 3 levels and 3 variables

Run	Factor 1 A: Time, hr	Factor 2 B: Temperature, °C	Factor 3 C: Stoichiometry
1	-1	+1	0
2	0	-1	-1
3	0	+1	+1
4	-1	0	-1
5	0	0	0
6	0	-1	+1
7	+1	0	+1
8	+1	+1	0
9	-1	0	+1
10	-1	-1	0
11	+1	0	-1
12	+1	-1	0
13	0	+1	-1
14	0	0	0
15	0	0	0
16	0	0	0
17	0	0	0

Levels

Variables	-1	0	+1
Reaction Time, hr	2	3	4
Reaction Temperature, °C	65	80	95
Stoichiometry	1.4	1.5	1.6

Magnesium oxide recovery was the response used for evaluation and comparing experimental results. Box–Behnken experimental design for the variables is shown in table 3. Plots of the response (% recovery of MgO) surface contours and the best predictive models for estimate of the response variables were developed. The Box–Behnken design in table 3 can fit the following model [24] following a second order polynomial function by which correlation between studied factors and response was generated.

$$E(y) = \beta_0 + \sum_{i=1}^{3} \beta_i X_i + \sum_{i=1}^{3} \sum_{j=1}^{3} \beta_{ij} X_i X_j \quad (6)$$

where y is the estimate of the response variable and Xi's are the values of independent variables [reaction time, reaction temperature and concentration] those are known for each experimental run. The parameters β_0, β_i and β_{ij} are the regression parameters.

Software package, Design-Expert 6.1, Stat-Ease, Inc., Minneapolis, USA, was used for regression analysis of experimental data and to plot response surface. Analysis of variance (ANOVA) was used to estimate the statistical parameters. The extent of fitting the experimental results to the polynomial model equation was expressed by the determination coefficient, R^2. F-test was used to estimate the significance of all terms in the polynomial equation within 95% confidence interval.

3.3. Leaching

Leaching is based on the decomposition of the serpentine with hydrochloric acid according to the following equation:

$$6\,HCl + Mg_3Si_2O_5(OH)_4 \rightarrow 3MgCl_2 + 2SiO_2 + 5H_2O \qquad (7)$$

Moreover, hydrochloric acid reacts with some impurities e.g. $CaCO_3$, Fe_2O_3, ... etc.

The obtained results of magnesium oxide recovery for each run are represented in table 3. It can be seen from this table that all the studied factors are significant regarding the magnesium oxide recovery. The statistical design data showed that the standard deviation was 4.73. The good predictability of the models can be indicated by correlation coefficient R^2 of 0.9032.

The correlation between the magnesium oxide recovery and the process parameters (reaction time, reaction temperature and stoichiometry) is shown in the following model that is generated from experimental data in table 3:

$$\text{Magnesium Oxide Recovery, \%} = 24.61 + 1.98A + 1.46B + 0.11C - 9.67A^2 - 0.05C^2 - 0.04AB - 5.17B^2 - 0.04BC \qquad (8)$$

where A is the reaction time (hour), B is the reaction temperature (°C) and C is the stoichiometry. The estimated values of magnesium oxide recovery using the above model are calculated. The experimental results are presented in table 3. These values are confirming the agreement of the generated previous model with the experimental results.

IOP Conf. Series: Materials Science and Engineering **427** (2018) 012028 doi:10.1088/1757-899X/427/1/012028

Table 3. Magnesium Oxide Recovery according to the experimental statistical design conditions.

Run	Factor 1 A: Time (hours)	Factor 2 B: Temperature (°C)	Factor 3 C: Stoichiometry	Response 1 MgO Recovery %
1	2	95	1.5	87.1
2	3	65	1.4	67.1
3	3	95	1.6	94.5
4	2	80	1.4	79.4
5	3	80	1.5	88.2
6	3	65	1.6	71.8
7	4	80	1.6	90.7
8	4	95	1.5	96.5
9	2	80	1.6	82.3
10	2	65	1.5	69.3
11	4	80	1.4	86.3
12	4	65	1.5	78.9
13	3	95	1.4	89.2
14	3	80	1.5	88.3
15	3	80	1.5	88.5
16	3	80	1.5	88.2
17	3	80	1.5	88.7

Figure 2 shows the relation between the effect of reaction time and stoichiometry on magnesium oxide recovery at different temperatures: figure 2.a at 65°C, figure 2.b at 80°C and figure 2.c at 95°C. These results showed that, the MgO recovery increases with increasing the reaction temperature. The optimum temperature is 95 °C which gives 96.5% MgO recovery. Further increase of the reaction temperature gave no increase in MgO recovery. At low reaction temperature, the viscosity of liquid phase is higher that leads to a decrease in the mobility of reacting ions and consequently lowers the reaction rate. On the other hand, higher temperatures result in excessive foam formation together with increasing corrosively of hydrochloric acid.

Figure 3 shows the effect of Temperature and stoichiometry at different leaching times (figure 3.a for 2 hr, figure 3.b for 3 hr and figure 3.c for 4 hr). Results reveal that the reaction of serpentine with hydrochloric acid is a spontaneous reaction of a high rate. It is obvious that recovery of MgO increases with the increasing reaction time. The optimum reaction time is 4 hours where 96.5% of MgO recovery is achieved. Increase of reaction time over 4 hours gives no appreciable increase in MgO recovery.

IOP Conf. Series: Materials Science and Engineering **427** (2018) 012028 doi:10.1088/1757-899X/427/1/012028

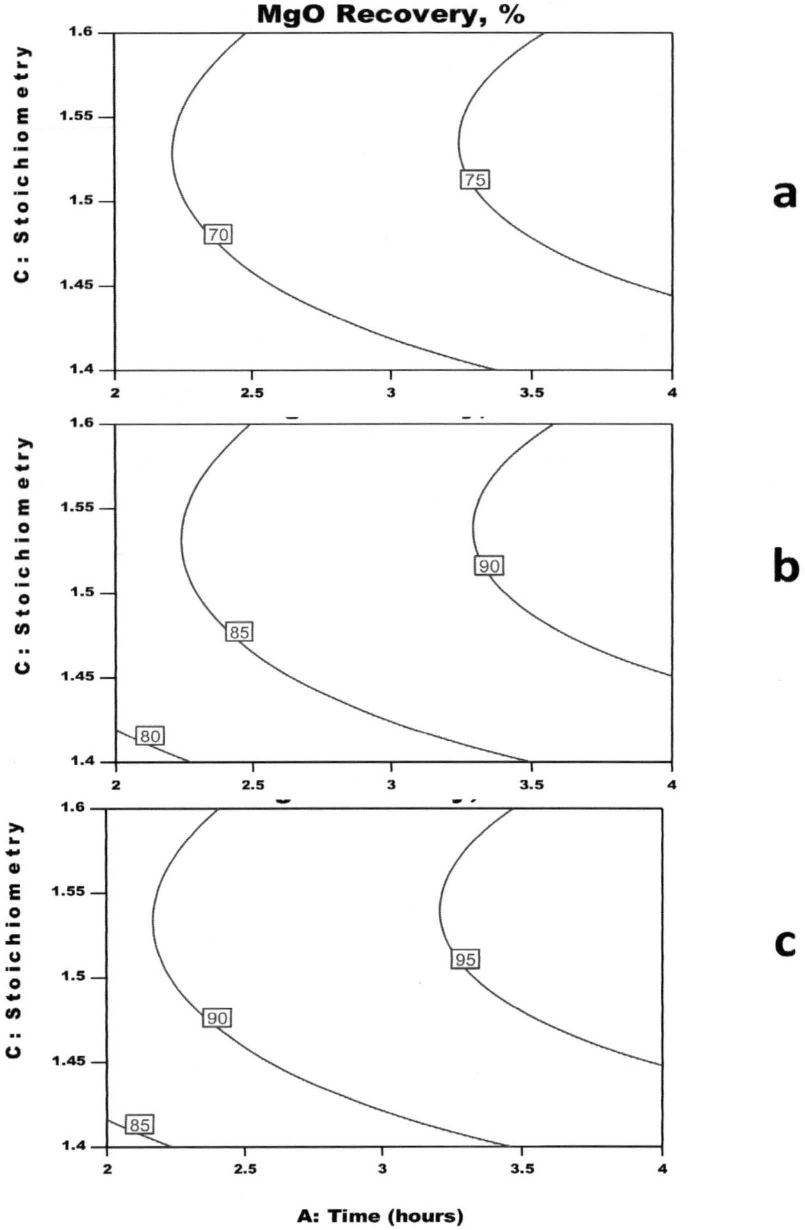

Figure 2. Effect of reaction time (hour) and stoichiometry on magnesium oxide recovery at different temperatures: figure 2.a at 65°C, figure 2.b at 80°C and figure 2.c at 95°C.

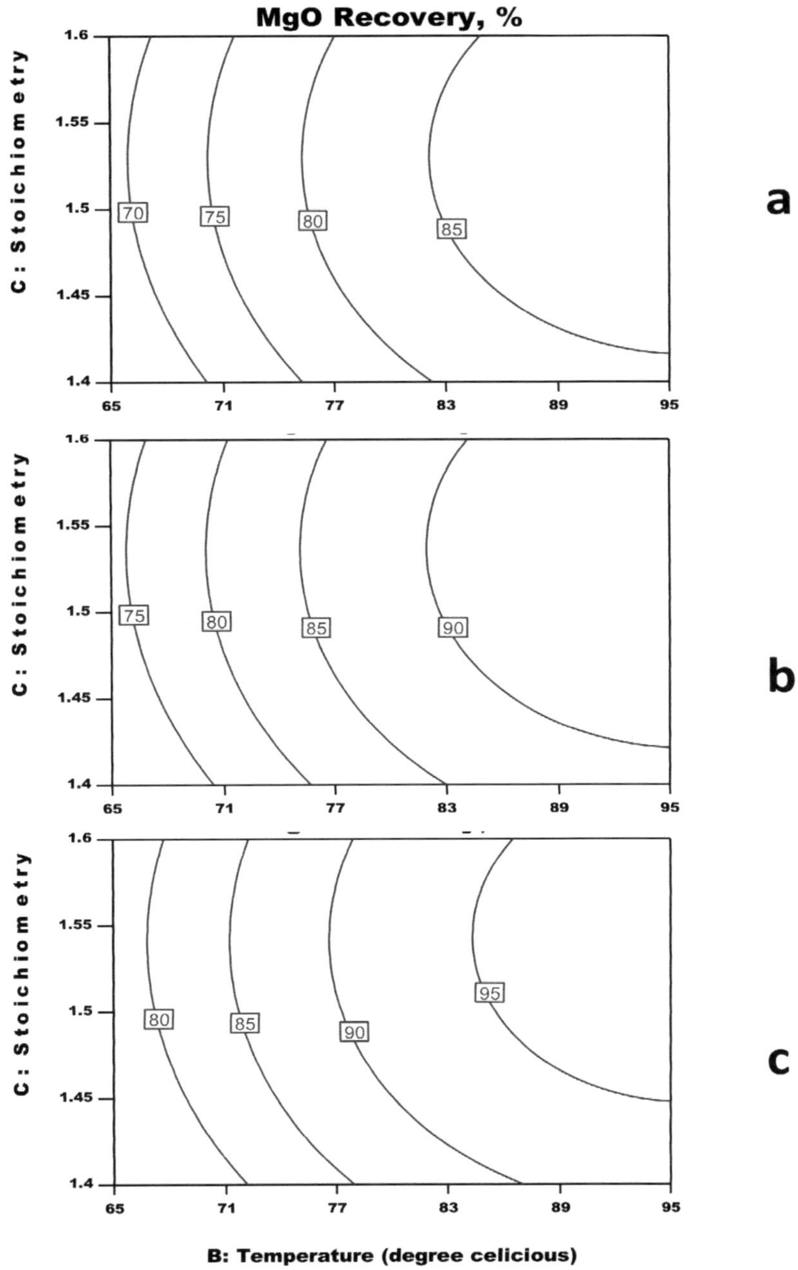

Figure 3. Effect of temperature and stoichiometry for different leaching times (fgure 3.a: 2 hr, figure 3.b: 3 hr and figure 3.c: 4 hr).

Figure 4 shows the effect of leaching time and temperature at different stoichiometry. The achieved results reveal that, with increase in HCl : MgO molar ratio leads to MgO recovery increase. MgO recovery at molar ratio of 1.6 goes up to 98.5 %. Further increase in the molar ratio produces an increase of free HCl content in magnesium chloride solution and does not lead to MgO recovery increase.

IOP Conf. Series: Materials Science and Engineering **427** (2018) 012028 doi:10.1088/1757-899X/427/1/012028

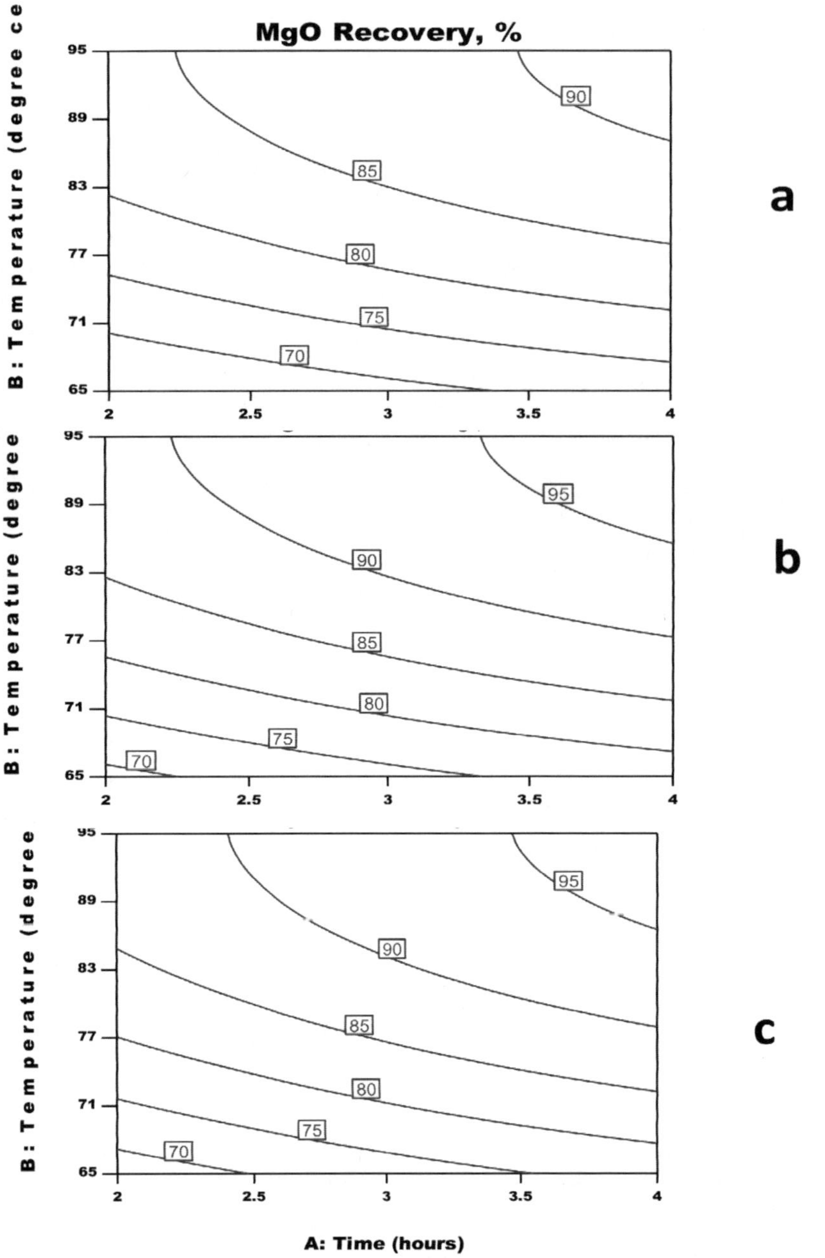

Figure 4. Effect of leaching times and temperatures at different stoichiometry (figure 4.a: 1.4, figure 4.b: 1.5 and figure 4.c: 1.6).

All the experimental data has been collected at the 3-D cube as shown in figure 5. This cube show that the highest magnesium oxide recovery of 96.5% can be obtained at the highest levels of parameters. The smallest magnesium oxide recovery of about 63.9% was obtained at the lowest levels of parameters. Decreasing any of the main variables lead to the decrease of magnesium oxide recovery.

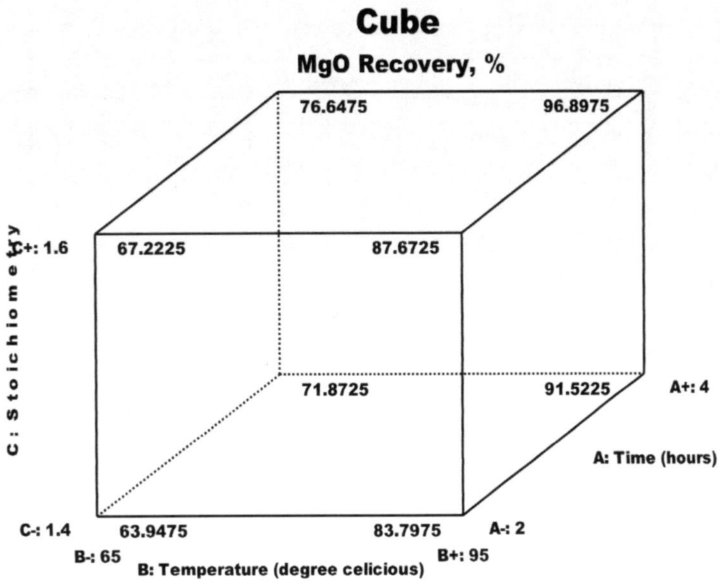

Figure 5. 3-D Plot for all the experimental data for leaching of serpentine ore.

3.4. Optimum conditions for the leaching

The optimum conditions and results of decomposition of the serpentine ore with HCl are summarized in table 4. The maximum MgO recovery achieved is 96.5%.

Table 4. Optimum conditions and results of serpentine leaching.

Conditions	Values, %	Results	Values, %
Particle size	86 % -200 Mesh	MgO Recovery	96.5
Temperature, °C	95		
Reaction time, hr	4		
Liquid / Solid Ratio, ml/g	4 : 1		
HCl Concentration, %	12.07		
Stoichiometry	1.6		

3.5. Solid liquid separation

The filtration data are tabulated in table 5. The filtration characteristics of magnesium chloride slurry are determined by deriving the values of slope, intersection and equivalent volume from the curve shown in figure 6. The slope and intersection were calculated. The viscosity, specific gravity of slurry and weight of solids per unit volume of slurry were determined experimentally. From these data in the afore-mentioned substituting equations, the filtration characteristics were obtained and given in table 6. The main factors affecting he filtration rate is the viscosity of slurry, viscosity of the solution, insoluble residue thickness as well the applied pressure difference.

Table 5. Filtration Data of Magnesium Chloride solution from Precipitated Silica.

Volume (V), cm^3	Time (θ), sec	▲V	▲θ	▲θ/▲V
5	3	5	3	0.6
10	7	5	4	0.8
15	12	5	5	1.0
20	18	5	6	1.2
25	26	5	8	1.6
30	35	5	9	1.8
35	45	5	10	2.0

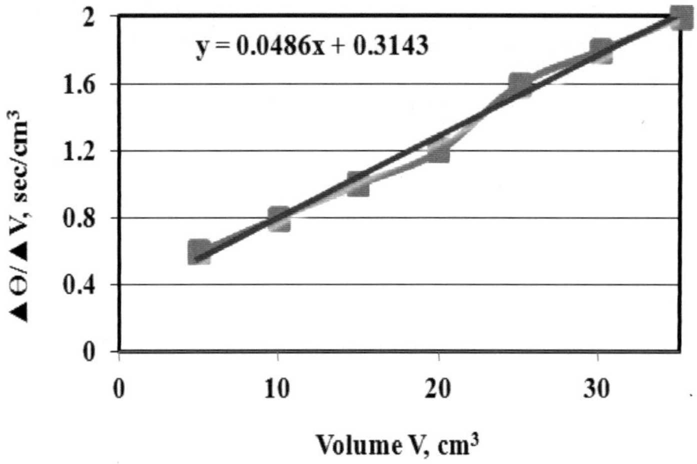

Figure 6. ▲θ/▲V Versus Volume.

Table 6. Filtration Characteristics of Magnesium Chloride solution from Precipitated Silica.

Item	Value
Slope, sec/cm^3	0.0486
Intersection, sec/cm^3	0.3143
Specific Cake Resistance (α), cm/g	$2.9X10^9$
Equivalent Volume of Filtering Medium (V_e), cm^3	6.5
Filtering Medium Resistance (R), cm^{-1}	$2.3X10^8$
Filtration Rate $dV_{max}/d\theta$, ml/min.	36
Filtration Time, sec	35
Viscosity of Filtrate at 90 °C, Poise	0.02
Specific Gravity of Slurry, g/cm^3	1.2
Weight of Solids per Unit Volume of Slurry (W), g/cm^3	0.2

4. Conclusions

Egyptian Serpentine ore from the Eastern Desert Area was leached with aqueous hydrochloric acid. Box–Behnken experimental statistical design is applied to study the influence of different parameters on leaching process. The liquid/solid ratio of 5:1 ml/g was applied. The mathematical model was generated. The results showed that, all the studied parameters of reaction temperature, reaction time and stoichiometry are effective. The results are presented as a 3-D cube. These results reveal that, high magnesium oxide recovery of about 96.5% was achieved at reaction time of 4 hours, reaction temperature of 95 °C and HCl : MgO stoichiometry of 1.5. On the other hand, the recovery of magnesium oxide was 67.1% at low parameters levels (3 hours reaction time, 65 °c reaction temperature and 1.4 HCl : MgO stoichiometry). Magnesium chloride slurry was filtered to separate the clear solution from insoluble residues using vacuum filtration technique (horizontal pan filter). The achieved filtration rate of magnesium chloride solution is 36 mL/min under pressure difference of 550 cm^3.Hg.

5. References

[1] Lackner K S, Butt D P and Wendt C H 1997 Magnesite disposal of carbon dioxide *In Proceedings of the 22nd International Technical Conference on Coal Utilization and Fuel Systems*, Clearwater, Florida pp 419–30
[2] Auzende A L, Daniel I, Reynard B, Lemaire C and Guyot F 2004 *Phys. Chem. Miner.* **31** pp 269–77
[3] Cheng T W, Ding Y C and Chiu J P 2002 *Miner. Eng.* **15** pp 271–5
[4] Pietrikova A, Bugel M and Golja M 2004 *Metalurg.* **43** pp 299–304
[5] Kanari N, Menad N and Gaballag I 1998 *Thermochim. Acta* **319** pp 97–104
[6] Cheng T W and Hsu C W 2006 *Chemosph.* **64** pp 510–4
[7] Wang X and Maroto-Valer M M 2011 *Fuel* **90** pp 1229–37
[8] Wang D G and Li Z B 2011 *AIChE. J.* **57** pp 1595–606
[9] Gao W C and Li Z B 2013 *AIChE. J.* **59** pp 1855–67
[10] Zhang Z, Lu X, Yang S and Pan F 2013 *Ind. Eng. Chem. Res.* **51** pp 9713–8
[11] Li Z B and Wang J F 2010 CN Patent 201010108988
[12] Gary L M, Donald A. Tarr 1991 *Inorg chem* prentice Hall Int Inc.
[13] Takeuchi S, Endo T, Kurumada Y and Otani T 1972 Japan Inst. Min. Metal., Hokkaido Chap 21-46, November 1972
[14] Apostolidis C I and Distin P A 1978 *Hydrometallurgy* **3** pp 181–96
[15] Abdel-Aal E A, Ibrahim I A, Rashad M M and Ismail A K 1994 *Physico. Chem. Probl. of Min. Process. J,* **28** pp 165-75
[16] Abdel-Aal E A, Ibrahim I A, Rashad M M and Ismail A K 1996 *Physico. Chem. Probl. of Min. Process. J.* **30** pp 207-16
[17] Abdel-Aal E A and Rashad M M 1997 *Physico. Chem. Probl. of Min. Process. J.* **31** pp 7-17
[18] Abdel-Aal E A 1994 *Physico. Chem. Probl. of Min. Process. J.* **29** pp 165-75
[19] Dietrich R V and Skinner B J 1979 *Rocks and rock minerals* (John Wiley and Sons Inc.; New York 51)
[20] Michaels L and Chissick S S 1979 *Asbestos* vol. **I,** properties, applications and hazards; (John Wiley and Sons Inc. New York 57)
[21] Abou Sekkina M M, Ghoneim M F, Aly S M and El-Kersh M J 1985 *J. Theral. Anal.* **29 pp** 1309-17
[22] Sultan M, Arvidson H E and Sturchio N C 1986 *Geology* **14** pp 995-9
[23] Basta E Z and Abdel Kader Z 1969 *Mineral. Magaz.* **37** p 287
[24] Box G E P and Behnken D W 1960 *Technometrics* **2** pp 455–75

IOP Conf. Series: Materials Science and Engineering **427** (2018) 012029 doi:10.1088/1757-899X/427/1/012029

A new methodology of floatability tests on Polish copper ores based on grain size-density fractions

A Krawczykowska

Faculty of Mining and Geoengineering, Department of Mineral Processing and Environmental Engineering, AGH, University of Science and Technology, Al. A. Mickiewicza 30, 30-059, Kraków, Poland

aldona.krawczykowska@agh.edu.pl

Abstract Ores difficult to enrich require the appropriate methodology of tests on enrichment for the preparation of ore for flotation tests and precise mineralogical analytics. A feature which should differentiate the separated ore samples for flotation tests is not only the size of grains, but also their density. These features reflect the characteristics of real industrial streams directed for flotation. The methodology for separating narrow grain classes by means of sieves applied to date did not guarantee meeting these separation conditions. The article presents a new methodology for examining the copper ore enrichment based on, instead of testing samples in grain classes, testing grain size-density fractions. To conduct these tests, the hydraulic classification in the field of centrifugal force which ensures the separation of grain size-density fractions varied in terms of granulation and density was used. Samples prepared in this way were subjected to flotation tests.

1. Introduction

Flotation is the main method for enriching the majority of metal ores in the world. It is applied for fine-grained mixtures. It uses differences in wettability of mineral grains' surface [1, 2]. Many factors affect the flotation effectiveness, the most important ones include: the size and shape of enriched grains, their mineralogical composition, surface properties, a type and amount of dosed flotation reagents, suspension parameters.

One of widely discussed issues related to the flotation process effectiveness is the effect of the grain composition on its results. Numerous scientific studies raise the issue of the granulation of the feed directed for the flotation process [3-7]. Many researchers indicate that the best results are obtained for the granulation of 20-100μm, while below this value, there is a big problem with the flotation selectivity.

In the case of difficult to enrich non-ferrous metal ores, in which useful minerals are very finely sprinkled and dispersed in gangue, one of the most important directions of technology development are therefore new solutions concerning ores' milling processes to the granulation from the range of single micrometers, but predominantly the development of conditions for the effective flotation with such granulation [7].

Polish copper ore is difficult to enrich, polymetallic, lithologically variable where separate lithological types differ in enrichment characteristics, additionally, it contains much organic carbon which limits recoveries and the quality of concentrate; it is also fine crystalline, and impregnations and concretions occur even in micrometer-sized particles [9]. It is these finest grains that significantly contaminate concentrates and generate losses in tailings.

The technique of preparing samples for flotation tests is very important in order to correctly locate grains difficult to float. In industrial conditions, ore for flotation is prepared in hydraulic classification processes in hydrocyclones, not on screens. Therefore, the feature which should differentiate the separated samples of ore for flotation tests is not only the size of grains, but also their density. The methodology for separating narrow grain classes by means of sieves applied to date did not guarantee meeting these separation conditions.

In turn, the precise qualitative and quantitative mineralogical analysis of minerals' recognition, mineral liberation, type of concretions with gangue, useful minerals granulation which provides representative results is indispensable to correctly characterize grains difficult to float. Currently, there are many techniques available which enable to obtain such a detailed characteristics of these grains. These methods include the quantitative electron mineralogical scanner, QEMSCAN, and the mineral liberation analyzer, MLA [9,10].

2. Materials and methods

The new methodology of feeds preparation for flotation tests in the hydraulic classification process conducted in the field of centrifugal force is proposed in the paper. This was done in the way ensuring the separation of particle size-density fractions not only in terms of granulation, but also according to density. So far, the methodology of separating narrow particle fractions on sieves did not guarantee maintaining these two separation arguments.

For the purpose of distinguishing products obtained by means of both methods a new term of particle size-density fraction was introduced. This is a population of particles of certain equivalent Stokes diameter related to both size and density. It is obtained by means of flow or gravity classification methods. The argument of separation of the feed into size-density fractions will therefore be composed of three factors: density, particle size and resistance of the medium.

Samples of copper ore in different lithological types representing: sandstone, dolomite and shale were subjected to tests. After the initial comminution of each type to the granulation <10 mm, products were secured in vacuum in order to limit the process of minerals' surface oxidation. Directly before each cycle of flotation tests, samples secured in this way were prepared according to the diagram presented in figure 1.

Laboratory tests of ore enrichment in size-density fractions were conducted on ground ore to the granulation of about <160 μm (P99% =160 μm). The diagram of samples preparation was identical for every lithological type. Samples were ground in a standard Bond mill filled with 14.7 kg of milling balls in 80 minutes.

Ore size-density fractions differing in granulation and density were prepared for flotation tests in a multihydrocyclone classifier. From every lithological type, four size-density fractions (HCI÷HCIV), which were then subjected to flotation tests, were obtained. The non-classified representative sample (feed) was also floated in a wide grain class 0÷160μm. For the conducted flotations, chemical analyses were conducted for all products: concentrates K1÷K5 and the cell product (tailings).

Flotation tests were conducted in a laboratory machine of the Denver D12 type as 30-minute-long fractionated flotations. The flotation chamber volume was equal to V = 1.5 dm^3, slurry density was D_S= 1180 g/dm^3. A mixture of xanthates in the amount of 100 g/Mg was applied as a collector, while Corflot added in the amount of 30 g/Mg was the foaming reagent. The feed pH value varied at the level of about 7.7-7.9. Conditions of the process (reagent regime, hydrodynamic parameters) were identical for all flotation tests. Flotation concentrates were collected in time of: 1, 3, 6, 12 and 13 minutes (respectively: K1, K2, K3, K4 and K5). In total, 15 flotation tests were conducted.

IOP Conf. Series: Materials Science and Engineering **427** (2018) 012029 doi:10.1088/1757-899X/427/1/012029

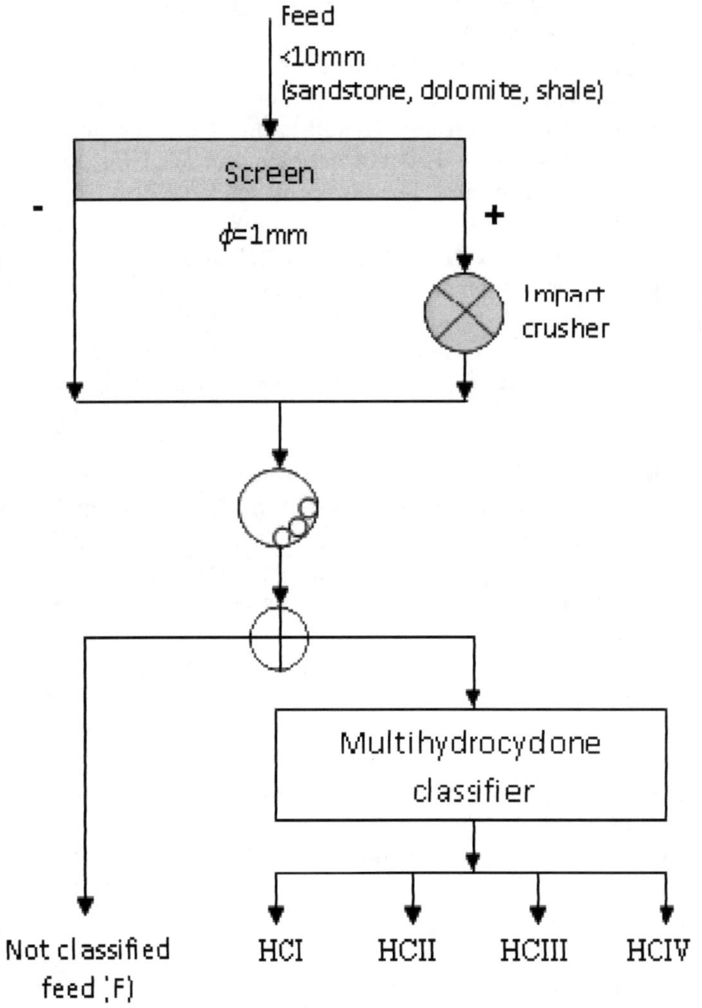

Figure 1. The diagram of samples preparation for flotation tests.

3. Analysis of results

Samples directed for the flotation process: lithological types in a wide grain class and size-density fractions of separate copper ore lithological types were subjected to granulometric analyses by the method of laser diffraction. Analyses were conducted in an Analysette 22 device - a laser particle analyzer.

The detailed results of analyses are presented in table 1 and in a form of curves of the grain composition (figures 2-4). The granulation range of floated feeds: of both the wide class and size-density fraction varies depending on the lithological type, which is caused by the different grindability.

Figures 2÷4 present curves of the grain composition of non-classified lithological types and size-density fractions separated from them.

Table 1. Averaged results of feeds' granulation for the flotation process.

Name of sample	d_{10}, μm	d_{50}, μm	d_{90}, μm
S_F	2.2	27.8	124.5
S_HCI	12.6	90.9	173.6
S_HCII	1.8	21	55.4
S_HCIII	1.2	8.9	27.8
S_HCIV	1.1	5.6	19
D_F	1.3	11.4	37.7
D_HCI	1.6	24.3	68.8
D_HCII	1.3	14.3	39.1
D_HCIII	1.1	8.4	25.8
D_HCIV	1	4.8	17.4
Sh_F	1.8	16.6	53.4
Sh_HCI	2.4	35.5	103.9
Sh_HCII	1.7	17.6	46.6
Sh_HCIII	1.3	10.1	28.9
Sh_HCIV	1.1	6	20.1

where: F - the feed for the multihydrocyclone classifier; S, D, Sh – copper ore lithological type, respectively: sandstone, dolomite, shale; HCI ÷ HCIV – size-density fractions from the multihydrocyclone classifier

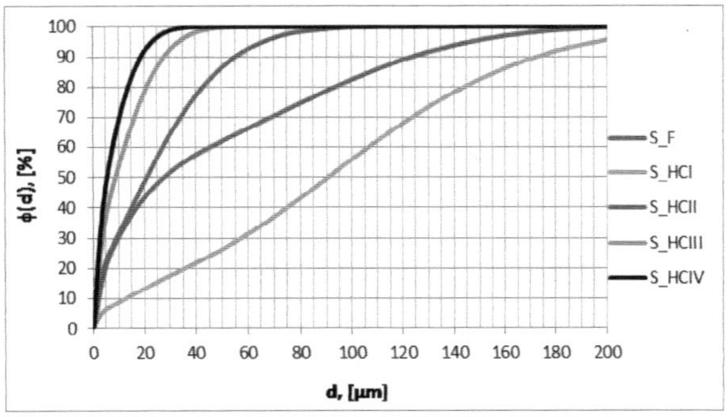

Figure 2. Curves of the grain composition for the feed and separated size-density fractions for sandstone ore.

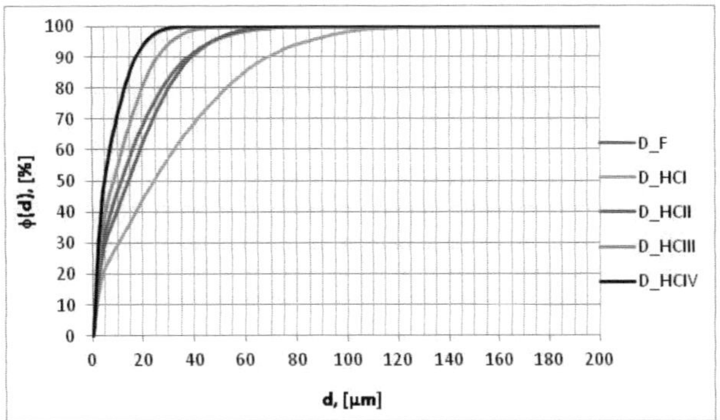

Figure 3. Curves of the grain composition for the feed and separated size-density fractions for dolomite ore.

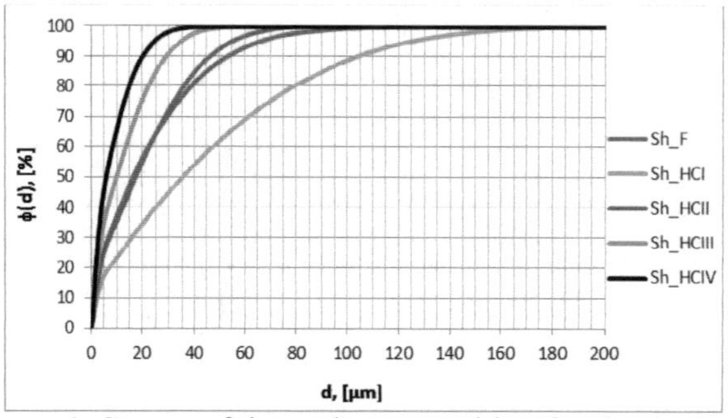

Figure 4. Curves of the grain composition for the feed and separated size-density fractions for shale ore.

Basing on granulometric analyses results, it can be stated that the widest range of granulation is represented by the sandstone ore, while the finest one is the dolomite ore. Grinding conditions were identical in each lithological type; however, individual lithological types differ in grindability and this causes these differences. Products of the multihydrocyclone classifier differ among types in two first size-density fractions: HCI and HCII, which is related to the feed granulation for the classification. Classes: HCIII and HCIV have, on the other hand, characteristic grains which are very close to each other in terms of size, grain d90 in all three types is equal to accordingly 27-29 and 9-10 μm. It is caused by the operation of the classifier which in the first two products separates thicker grains providing products III and IV stable in terms of granulation. Size-density fractions III and IV are ultra-fine classes, separation of which would be extremely difficult in a traditional way, on sieves.

Results of flotation tests of feeds non-classified and size-density fractions for separate copper ores' lithological types in a form of Fuerstenau upgrading curves, kinetics curves (figures 5-7) and in a form of technological indexes are presented in table 2.

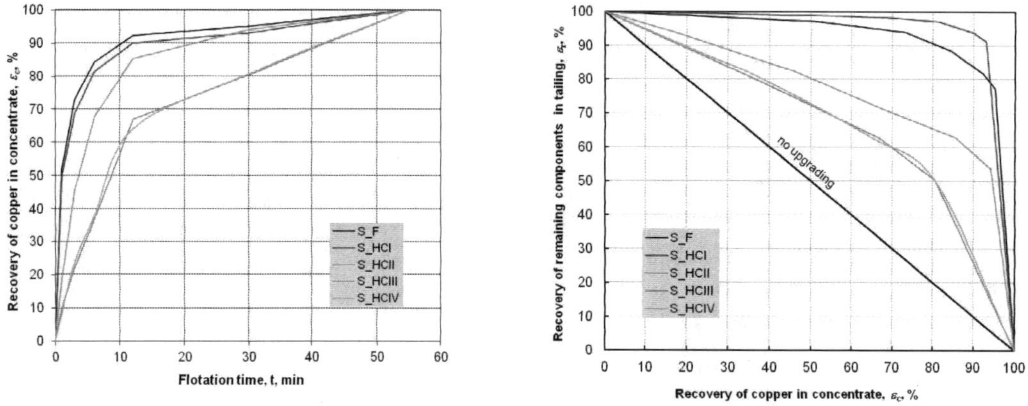

Figure 5. The comparison of results of flotation enrichment of sandstone copper ore for the non-classified and separated size-density fractions.

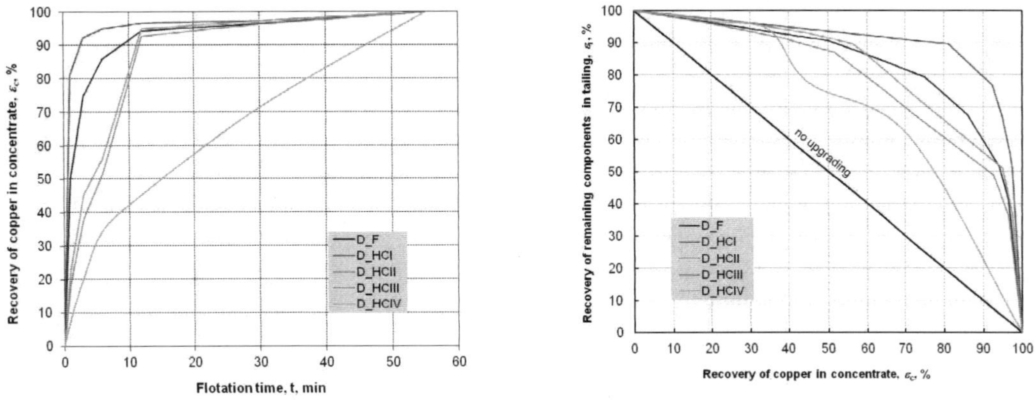

Figure 6. The comparison of results of flotation enrichment of dolomite copper ore for the non-classified feed and separated size-density fractions.

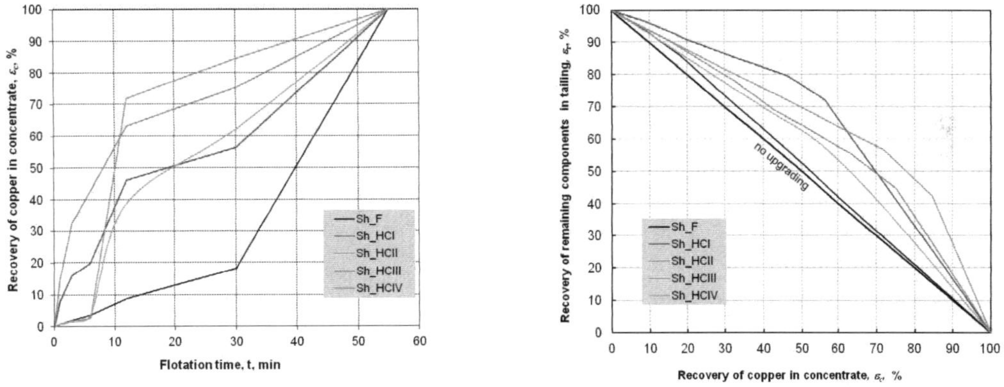

Figure 7. The comparison of results of flotation enrichment of shale copper ore for the non-classified feed and separated size-density fractions.

Table 2. presents the most important qualitative and quantitative indexes obtained in conducted experiments.

Table 2. The comparison of results of conducted flotation tests.

Name of sample	α %	β_1 %	β_c %	υ %	ε_{c1} %	ε_t %	γ_{c1} %	γ_c %	d_{90} μm
S_F	0.81	12.83	3.3	0.051	51.84	4.85	3.26	23.23	124.5
S_HCI	0.49	19.38	4.94	0.037	49.51	6.89	1.25	9.19	173.6
S_HCII	1.24	3.48	2.5	0.135	18.7	5.78	6.69	46.77	55.4
S_HCIII	1.67	3.07	2.68	0.652	9.78	19.55	5.31	50.00	27.8
S_HCIV	1.84	3.56	2.93	0.724	8.32	19.39	4.31	50.66	19
D_F	0.85	4.34	1.39	0.073	49.95	3.49	9.84	59.19	37.7
D_HCI	1.22	8.72	2.38	0.064	80.95	2.63	11.32	49.97	68.8
D_HCII	0.83	6.36	1.27	0.06	22.75	2.63	2.97	63.63	39.1
D_HCIII	0.69	3.21	1.04	0.07	18.68	3.64	4.03	64.05	25.8
D_HCIV	0.29	2.14	0.52	0.138	7.75	28.53	1.03	39.83	17.4
Sh_F	1.22	2.1	1.56	1.16	0.74	81.81	0.43	14.19	53.4
Sh_HCI	1.44	3.57	2.86	0.87	8.1	43.56	3.26	28.38	103.9
Sh_HCII	1.13	1.75	1.66	0.42	0.75	15.41	0.49	57.97	46.6
Sh_HCIII	0.99	1.57	1.34	0.55	14.87	24.78	9.38	55.51	28.9
Sh_HCIV	0.87	1.61	1.1	0.64	0.94	37.67	0.51	49.24	20.1

where:
α, β_1, β_c, υ - content of the Cu, respectively, in the: feed, first fraction of concentrate, whole concentrate, tailings,
ε_{c1} - recovery of the Cu in the first fraction of concentrate,
ε_t - losses of the Cu in tailings,
γ_{c1}, γ_c - yield of the, respectively: first fraction of the concentrate, whole concentrate.

In the case of sandstone ore, the best results were obtained for the coarsest class (S_HCI) and slightly worse for the non-classified feed (S_F). A substantial decrease in floatability is observed in the remaining size-density fractions. despite the increasing content of Cu in the feed.

Results of flotation tests of dolomite ore indicate high cleanness of obtained tailings, low contents of Cu in tailings were obtained in each size-density fraction. Only in the finest size-density fraction (D_HCIV), the value υ was equal to 0.134% (the lowest one from among all types in this class), while in the remaining four classes, this content was at the level of 0.06-0.07%.

Distinctly the weakest results were obtained for shale ore. For size-density fraction (Sh_HCI), losses of the useful component in tailings were equal to over 43%, with the Cu content in tailings at the level of υ=0.87%. Determined qualitative-quantitative parameters for this lithological type of ore indicate large problems with floatability. No satisfactory results were obtained in the conducted tests.

4. Conclusions

The article proposes the new methodology for testing the enrichment of Polish copper ore based on, instead of testing samples in grain classes, testing grain size-density fractions.

The use of a multihydrocyclone classifier enables to prepare and examine the ore enrichment with very narrow ranges of granulation and density. The hydraulic classification in the field of centrifugal force ensures separation of products varied in terms of not only granulation, but also density. These features reflect the characteristics of real industrial streams directed for flotation.

Additionally, thanks to the possibility provided by the device - separation of size-density fractions with specified ranges of granulation and density, it is possible to observe concentration of copper-bearing minerals in separate classification products. Density is related in an obvious way to the mineralogical composition of grains. In the case of carbonate and shale ore, a decrease in the Cu content along with a decrease in granulation is observed; the opposite happens in the case of sandstone ore, in which the copper content in the coarsest size-density fraction is significantly the smallest (0.49%), while the the greatest amount of copper-bearing minerals is accumulated in the finest size-density fraction (1.84% Cu).

The obtained results of tests on size-density fractions confirmed a high effectiveness of flotation for size-density fraction No. 1 (the coarsest grained one and deprived of extremely fine grains) and a gradual decrease in the effectiveness of flotation along with the share of finest grains in size-density fractions. Size-density fractions III and IV (d90 <30μm) were characterized by a sudden decrease in the effectiveness of flotation.

The separation of ore into products with very narrow ranges of granulation and density and the evaluation of density of such size-density fractions provide many opportunities to identify the causes of their difficult enrichment. To that end, further tests on the causes of state of things are necessary.

References

[1] Gaudin A M 1963 *Flotacja* (Katowice: Śląsk)
[2] Wills B A and Napier-Munn T J 2006 *Wills' Mineral Processing Technology. An Introduction to the Practical Aspects of Ore Treatment and Mineral Recovery* Elsevier Science & Technology Books
[3] Gaudin A M, Groh J O and Henderson H B 1931 Effect of particle size on flotation *American Institute of Mining and Metallurgical Engineering. Tech. Publ.* **414**. pp 3-23
[4] Trahar W J 1981 A rational interpretation of the role of particle size in flotation *International Journal of Mineral Processing* **8** 289–327

IOP Conf. Series: Materials Science and Engineering **427** (2018) 012029 doi:10.1088/1757-899X/427/1/012029

[5] Potulska A 2008 *Doctoral thesis: Wpływ drobnego mielenia na flotację krakowych rud miedzi* (Wrocław: Politechnika Wroclawska, Wydzial Geoinzynierii, Gornictwa i Geologii)

[6] Pease J D, Young M F, Curry D and Johnson N W 2004 Improving fines recovery by grinding finer MetPlant *Centenary of Flotation Symposium* (Brisbane: Australia) pp 1-17

[7] Duchnowska M and Bakalarz A 2015 Influence of feed particle size on upgrading selectivity of scavenger stage of industrial copper ore flotation *Minerals & Metallurgical Processing* **32/4** pp 215-21

[8] Piestrzynski A 2007 Okruszcowanie Monografia KGHM Polska Miedź S.A. Piestrzynski A Banaszak A and Zaleska–Kuczmierczyk M eds. (Lubin) pp 167-96

[9] Pascoe R D, Power M R and Simpson B 2007 QEMSCAN analysis as a tool for improved understanding of gravity separator performance *Miner. Eng.* **20** 487–95

[10] Gu Y 2003 Automated scanning electron microscope based liberation analysis *J. Miner. Mater. Character. Eng.* **2(1)** pp 33–41

IOP Conf. Series: Materials Science and Engineering **427** (2018) 012030 doi:10.1088/1757-899X/427/1/012030

Effect of curing time on the dissolution of a secondary copper sulphide ore using alternative water resources

V Quezada[1, 2], L Velásquez3, A Roca[1], O Benavente[2], E Melo[2] and B Keith[4]

[1] Departamento de Ciencia de los Materiales y Química Física, Universidad de Barcelona, Barcelona, España.

[2] Departamento de Ingeniería Metalúrgica y Minas, Universidad Católica del Norte, Antofagasta, Chile.

[3] Departamento de Ingeniería en Minas, Universidad de Santiago, Santiago, Chile.

[4] Departamento de Ingeniería de Sistemas y Computación, Universidad Católica del Norte, Antofagasta, Chile.

vquezare7@alumnes.ub.edu

Abstract. In the north of Chile, due to water shortages, the depletion of oxide ores and the abundance of chalcopyrite ore, mining industry is searching for sustainable hydrometallurgy processes that can use alternative water resources. The leaching process must enhance the dissolution of copper sulphide ore that are refractory to conventional leaching. This paper reports a study on the effect of addition of chloride ion using seawater and discard brine in the agglomeration stage of a secondary copper sulphide ore. The effect of curing time on the same ore also is reported. The leaching tests have been carried out in column irrigated with raffinate under ambient conditions. A size distribution with a P80 of 17 mm is used. A maximum of 72% of copper extraction is obtained using discard brine and 68% using seawater. The use of discard brine and seawater are favorable in all the tests performed. Through an Analysis of Variance (ANOVA), it is determined that the curing time has the highest contribution (92.37%) on the percentage of copper extraction.

1. Introduction

Chile is the largest copper producer in the world [1], exporting 5,552 tons of copper in 2016 [2]. Antofagasta is the main copper producer region in Chile and Chuquicamata and Escondida Mine are located in this region. Because of that Antofagasta Region have supported the economic development of the region to the point where the per capita GDP is the highest in the country [3]. However, the Antofagasta Region is located in the Atacama Desert, the most arid in the world. Antofagasta has the lowest availability of fresh water in Chile [4]. To sustain copper production over time and supply the region's water shortages, copper producing companies have incorporated seawater into their processes (e.g. Michilla, Esperanza, Algorta Norte and Las Luces) [5]. The advantage of seawater is that it has

20 g/L of chloride ion that, according to research data, can enhance the dissolution of copper sulphide ores [6].

1.1. Leaching of copper sulphide ores

Besides to facing the water shortage, some companies go through the depletion of oxidized copper minerals (leachable minerals) facing sulphide deposits that are refractory to conventional leaching [7]. Copper sulphide ores are classified in two zones, secondary and primary. Covellite and chalcocite predominate in the secondary sulphide zone, the latter being the most common [8]. With respect to the primary sulphide zone, the most abundant is the chalcopyrite [9]. Traditionally, the sulfurized zones are processed through concentration by flotation and subsequent treatments at high temperatures. However, this alternative is not economically viable when the copper grade of the deposit fails to cover comminution costs. This becomes a problem when companies already have hydrometallurgical plants and the production of the company is limited to mineralogy. Given this problem is that companies see the continuity of operations in the leaching of copper sulphide ores.

1.2. Pretreatment and chloride leaching

Several leaching media have been studied for the dissolution of copper sulphide ores, of these, the chloride media turns out to be the most efficient due to the reactivity of the chloride ion with sulphide ores [6]. The reactivity of the chloride ion has been declared as beneficial with respect to the dissolution of copper from sulphide ores, since, under certain conditions of pH and solution potential, it is capable of dissolving and reversing the passivating layer that inhibits dissolution [10-11]. Recent research has qualified curing time as a favorable variable in the extraction of copper from sulphide ores [12]. The curing time is a treatment prior to the leaching stage in which the reactivity of the sample with the solution is benefited. According to Dhawan [13], curing time generates a homogeneous distribution of the acid in the mineral bed, increases the kinetics of copper dissolution as well as benefiting the inhibition of aluminum-silicate minerals. However, scarce research has been performed on curing period, which generates limited theoretical background [14]. Thus, this work evaluates the effect of curing time in the extraction of copper from secondary sulphide ores using seawater and discard brine from a desalination company. Leaching experiments were performed in columns, at room temperature and the variables have been analyzed through ANOVA.

2. Experimental

2.1. Ore sample

A secondary sulphide copper ore is used to develop leaching tests. The chemical analysis of the ore indicates a copper grade of 1.19% and an iron presence of 5%. According to the mineralogical analysis presented in table 1, it can easily be seen that the copper present in the system is mainly contributed by secondary sulphide ore such as chalcocite and covellite, both contributions exceed 80% of the total copper in the system. The size distribution, corresponding to the ore, can be seen in figure 1. It is obtained a P80 of 17 mm.

Table 1. Mineralogical composition of the ore sample as determined by QEMSCAN.

	Amount (wt %)	Cu (wt %)	Cu contribution (wt %)
Chalcocite	1.10	0.88	74.14
Chalcopyrite	0.27	0.09	7.76
Covellite	0.11	0.07	6.03
Bornite	0.06	0.04	3.45
Atacamite	0.07	0.04	3.45
Others Cu minerals	-	0.06	5.17
Total		1.19	100

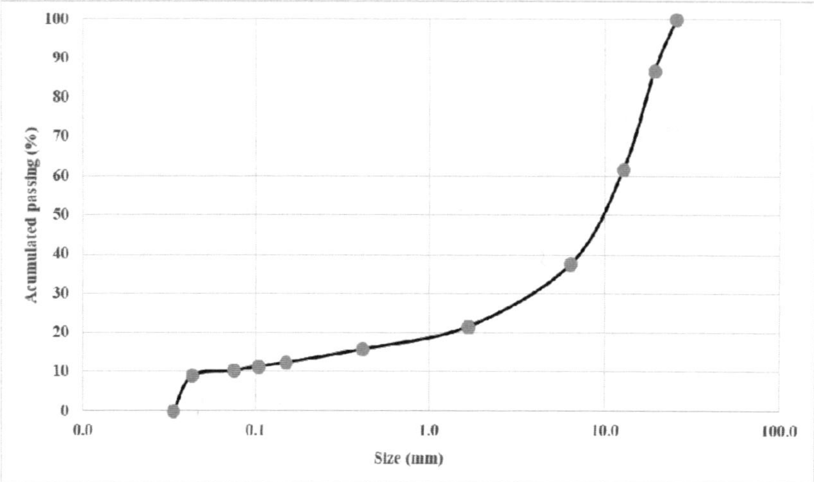

Figure 1. Particle size distribution of the ore.

2.2. Reagents

All tests were performed using sulfuric acid (H_2SO_4) p.a. of 97%. Seawater and discard brine were used as water resources in this study. Seawater was collected from the shore in the southern area of Antofagasta, in a location close to the Universidad Católica del Norte. Discard brine was provided by a desalination company located in Antofagasta. The concentration of chloride in seawater and discard brine is 20 and 32 g / L, respectively.

2.3. Pretreatment to leaching process

Agglomeration tests were conducted in a batch laboratory-scale polypropylene drum through a system of rotating rollers operated at 60 rpm speed. For all agglomeration tests, 4 kg of crushed well-mixed ore with a particles size fraction of 17 mm of P80 were used. Previous results (not shown) indicate that 30 kg of sulfuric acid per ton of mineral and 9.1% moisture are the optimal values to be added in the agglomeration. Curing time studied were 15, 30, 50 and 80 days. All samples were sealed without being exposed to the environment (avoid evaporation).

2.4. Leaching process

Leaching of all agglomerate samples was carried out in acrylic columns, 10.7 cm in diameter and 41 cm in height (figure 2). The leaching solution has 6 g/L of sulfuric acid and the irrigation rate used was 5 L/h/m^2 (0.76 mL/min). All columns leaching test were irrigated with a standard solution (6 g/L of acid). Liquid feed flow rate to the columns was 0.76 mL/min using a peristaltic pump for all the leaching samples, the pH and the redox potential were measured. All pH measurements were performed using a Metrohm 826 pH meter and probe, which was calibrated at pH 2.0 and 7.0, each time before used. The redox potentials were measured using a Metrohm 826 voltmeter with an Ag/AgCl reference electrode and it was converted and reported with respect to the standard hydrogen electrode (SHE). The precision of the measurements was tested using Oxidation-Reduction Potential (ORP) standard solution having a potential of 470 mV at 25°C. During each test, liquid samples were withdrawn periodically and analyzed for Cu and Fe using ASS.

IOP Conf. Series: Materials Science and Engineering **427** (2018) 012030 doi:10.1088/1757-899X/427/1/012030

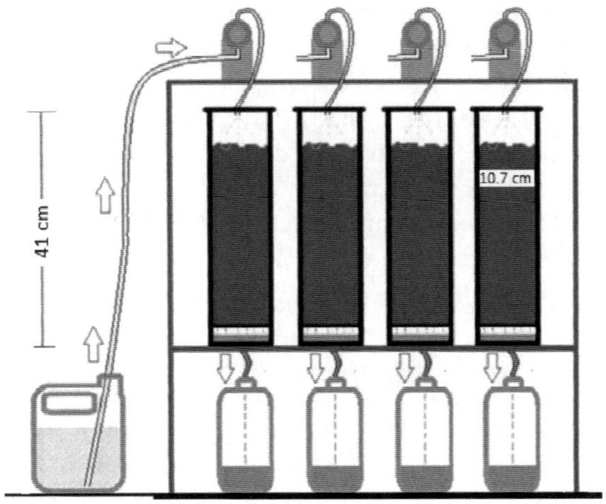

Figure 2. Schematic of column leaching system.

2.5. ANOVA analysis

An ANOVA analysis of two factors without replications was used to compare the effects of curing and the water resource used, with a level of significance of 0.01 on the data [15]. The factors correspond in particular to: curing time (in days) with four levels (3, 10, 30 and 50) and the level of chloride concentration associated with the water resource (in g/L) with two levels (20 and 32, associated with seawater and discard brine, respectively). The response variable corresponds to the final percentage of copper extraction.

Due the number of factors and levels is relatively low, a complete factorial experimental design is used, obtaining a total of 8 samples (8 tests in columns). A summary of the experimental design used is shown in table 2, the curing time has been labeled as A, and the chloride concentration as B.

Table 2. Experimental design used.

A	B	Curing time (days)	Chloride (g/L)
1	1	3	20
2		10	20
3		30	20
4		50	20
1	2	3	32
2		10	32
3		30	32
4		50	32

The data were analyzed using an additive linear model without interaction, under the assumption that this type of model is able to explain the variance of the data in an adequate manner. The effects produced by the interaction of the parameters have not been taken into account in this analysis, since preliminary tests show that they can be disregarded. The predictive results of the model support this assumption, although it could be confirmed by carrying out additional experiments. The additive linear model is shown in equation 1 given by

$$Y_i = \mu + X_i + e_i \qquad (1)$$

where: μ is the general average of the performance value, X_i is the fixed effect of the combination of parameter levels used in the i-th experiment and e_i is the random error of the i-th experiment.

On the other hand, the calculation of the *F* value is performed to evaluate the process parameters and determine if they have significant effects on the performance value. The *F* value of each parameter of the process is the ratio between the average of the squared deviations with respect to the mean of the squared error. In general, the higher the *F* value, the more significant the effect on the yield value due to the change in the parameter. This analysis of variance is followed by Tukey tests at a significance level of $p = 0.01$ for the comparison of means.

To perform the analysis of results, the application of ANOVA and the performance of the Tukey tests, the R language [16] and the DoE.base experiments design library [17] were used.

3. Results and discussion

3.1. Column leaching

In figure 3 it is possible to appreciate that the test with the longest curing time (50 days) is the one that reports the highest extraction, reaching 71.5% after 22 days of irrigation. Opposite case is the test with 3 days of curing, which reaches 58%. As can be seen, there is a directly proportional relationship between curing and final copper extraction. In addition to the above, it is possible to visualize also the slope of the line is greater for those tests that have more curing time. Thus, for the 50-day test, 50% extraction is achieved for the first 48 hours. Regarding the results of the tests using seawater, they can be seen in figure 4. The test with 50 days of curing reaches an extraction close to 68% while the test with 3 days of curing reaches 54.5% of copper dissolution. The same trend is observed regarding the tests with discard brine.

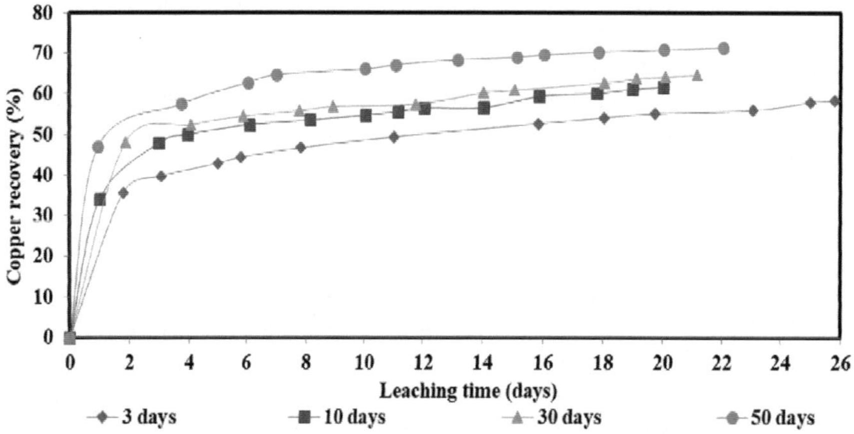

Figure 3. Effect of the curing time on the rate of column leaching of copper sulphide ore in a solution of 6 g/L H_2SO_4 at ambient temperature using discard brine with: 3 curing days (♦), 10 curing days (■), 30 curing days (▲) and 50 curing days (●).

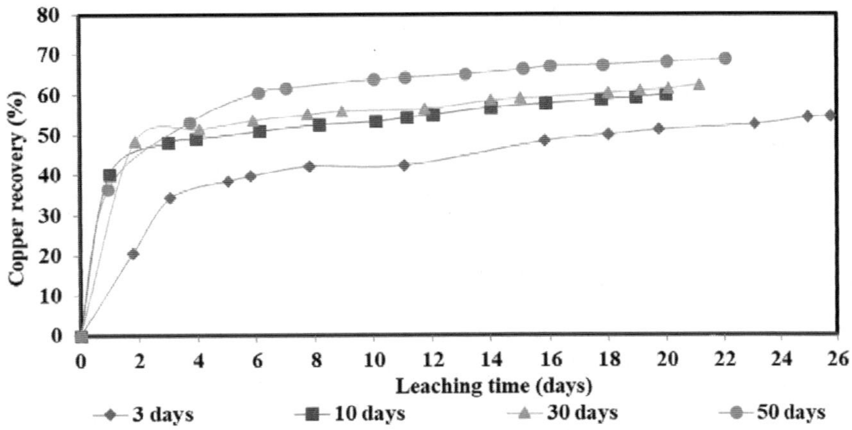

Figure 4. Effect of the curing time on the rate of column leaching of copper sulphide ore in a solution of 6 g/L H_2SO_4 at ambient temperature using seawater with: 3 curing days (♦), 10 curing days (■), 30 curing days (▲) and 50 curing days (●).

In figures 5 and 6 it is possible to appreciate the behavior of the solution potential for the tests with discard brine and seawater, respectively. From both graphs a constant behavior of the potential through time is observed, resulting in a range between 600 and 660 mV for all the tests carried out. This range of variation has been described as beneficial in the dissolution of copper from minerals such as chalcocite [10], but it would tend to be difficult for a potential close to 500 mV.

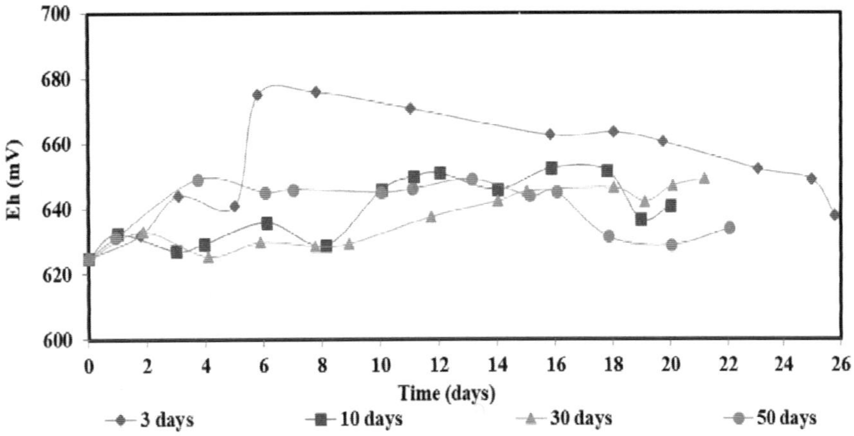

Figure 5. Variation of potential in the leaching solution of copper sulphide ore in a solution of 6 g/L H_2SO_4 at ambient temperature using discard brine with: 3 curing days (♦), 10 curing days (■), 30 curing days (▲) and 50 curing days (●).

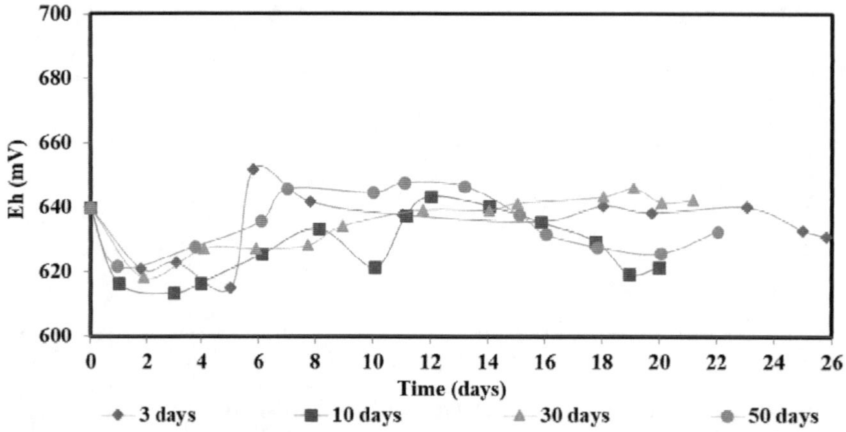

Figure 6. Variation of potential in the leaching solution of copper sulphide ore in a solution of 6 g/L H_2SO_4 at ambient temperature using seawater with: 3 curing days (♦), 10 curing days (■), 30 curing days (▲) and 50 curing days (●).

3.2. ANOVA analysis

The results of the ANOVA performed are shown in table 3. According to the results, parameter *A* has the highest contribution on the percentage of extraction of Cu obtained (92.37%), being able to explain practically all the variance of the data, on the other hand, parameter *B* has a much smaller contribution (7.05%). Furthermore, 99.42% of the variance of the data can be explained between these the two factors, based on these results it can be assumed that the additive linear model is adequate to describe the data. Finally, these contributions are statistically significant according to the obtained *p*-values (all significant at a 99% confidence level). The degree of influence or effect that the parameters have can be visualized in figures 7 and 8.

Table 3. Results of the ANOVA analysis of the parameters for the percentage of Cu extraction (*df* = degrees of freedom, *SSE* = sum of square errors, *MSE* = mean square errors).

	Parameters	*df*	*SSE*	*MSE*	*F* Value	Contribution (%)	*p*-Value
A	Curing time (days)	3	195.264	65.088	160.658	92.37%	0.0008244
B	Cl^- (g/L)	1	14.906	14.906	36.792	7.05%	0.0089928
	Error	3	1.215	0.405			
	Total	7	211.385			99.42%	

From the adjusted additive linear model, it is possible to obtain predictions about the percentages of copper extraction that should be obtained in each configuration. In table 4 these results are detailed and the level of error they present with respect to the empirical results.

In figure 7 it can be seen that the optimum is obtained with level 4, corresponding to 50 days of curing. In general, a positive trend is observed as the days of curing increase, a result that agrees with the theoretical intuition regarding the benefit of curing time in copper extraction.

Table 4. Experimental results and predictions of the model.

A	*B*	Empirical value (% Cu)	Prediction (% Cu)	Error
1	1	54.49	55.08	-0.59
2		59.93	59.46	0.48
3		62.29	62.11	0.19
4		68.63	68.71	-0.08
1	2	58.39	57.81	0.59
2		61.71	62.19	-0.48
3		64.65	64.84	-0.19
4		71.51	71.44	0.08

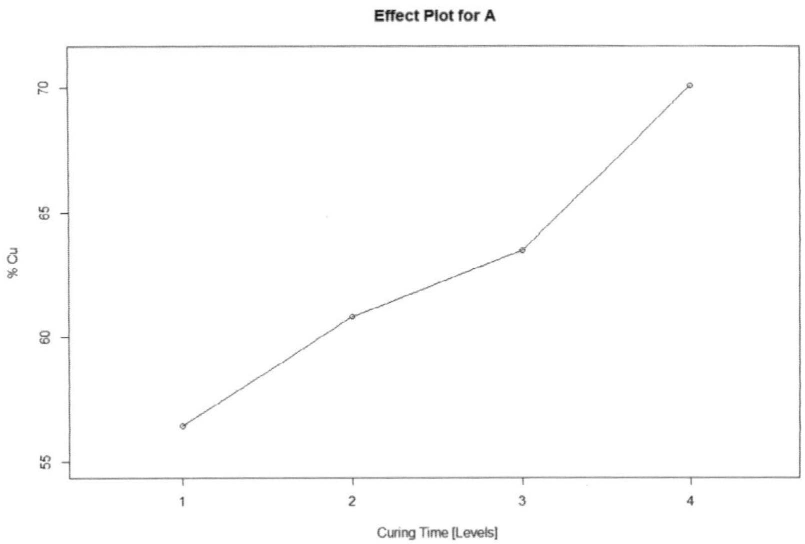

Figure 7. Effect of parameter A (curing time level) on the percentage of Cu extraction.

Figure 8 shows that, at least for these data, a slightly higher yield is obtained when using discard brine (level 2), corresponding to 32 g / L of chloride. It can be seen that the effect is much smaller than that produced by the curing time (note that both graphs are on the same scale on the ordinate axis).

Finally, Tukey's post-hoc test [18] based on the value of Honestly Significant Difference (HSD) is reported that there is a significant difference (at a 0.01 level of significance) between different configurations. In particular, the configuration (4.1) and configurations (1.1), (1.2) and (1.3) show statistically significant differences with adjusted p-values of 0.0007, 0.0022 and 0.0058 respectively, where the difference is in favor of the configuration (4.1). In addition, a statistically significant difference is reported even between the configuration (2.1) and (1.1) with an adjusted p-value of 0.0090, reporting an increase in the percentage of copper extraction. These results correspond to the previous graphic analysis and reinforce the main conclusion of this analysis: the curing time has a statistically significant influence on the percentage of copper extraction.

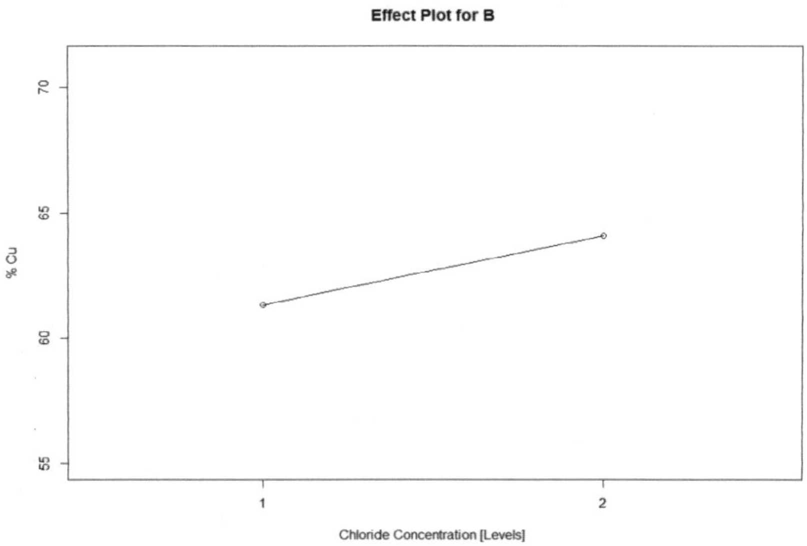

Figure 8. Effect of parameter B (chloride concentration level) o
n the percentage of Cu extraction.

4. Conclusions

- It has been demonstrated that ore agglomerated with acid and chloride ions and cured for long periods of time enhances the dissolution rate of a secondary copper sulphide ore. 72 and 68% of extraction is achieved when the ore is agglomerated and curing for 50 days using discard brine and seawater as the type of water, respectively. In the same medias, 58 and 52% of extraction is obtained with 3 days of curing.
- Despite the difference of the chloride content in each chloride solution type used in the column leach tests at different curing times, similar dissolution was obtained. Thus, it could be concluded that either type of chloride solution in the raffinate for agglomeration and irrigation can be used. Furthermore, the copper sulphide ore dissolution is enhanced at potential between 600 and 660 mV (SHE).
- According to the Analysis of Variance, curing time is the variable that has the greatest contribution in copper extraction (92.37%) versus chloride concentration (7.05%), under the conditions studied.

References

[1] Ghorbani Y, Kuan SH 2016 A review of sustainable development in the Chilean mining sector: past, present and future *Int J Min Reclam Environ* **30** pp 1–29

[2] Chilean Copper Commission (COCHILCO) 2017 Proyección de la producción de cobre en Chile 2017 – 2028. Available at: http://www.cochilco.cl/

[3] Tapia J S, Valdés J, Orrego R, Tchernitchin A, Dorador C, Bolados A and Harrod C 2018 Geologic and anthropogenic sources of contamination in settled dust of a historic mining port city in northern Chile: health risk implications *PeerJ* **6** pp e4699

[4] Quezada V and Velásquez L 2016 Dissolution kinetics of copper ores and concentrate with sea water and discard brine *Hydrometallurgy Conference 2016. Sustainable Hydrometallurgical Extraction of Metals* Cape Town South África pp 101 – 9

[5] Torres C M, Taboada M E, Graber T A, Herreros O O, Ghorbani Y and Watling H R 2015 The effect of seawater based media on copper dissolution from low-grade copper ore *Minerals Engineering* **71** pp 139-45

IOP Conf. Series: Materials Science and Engineering **427** (2018) 012030 doi:10.1088/1757-899X/427/1/012030

[6] Velásquez -Yévenes L 2009 The kinetics of the dissolution of chalcopyrite in chloride media (Doctoral dissertation, Murdoch University)
[7] Ibáñez T and Velásquez L 2013 Lixiviación de la calcopirita en medios clorurados. *Revista de Metalurgia* 49(2) pp 131-44
[8] Niu X, Ruan R, Tan Q, Jia Y and Sun H 2015 Study on the second stage of chalcocite leaching in column with redox potential control and its implications *Hydrometallurgy* **155** pp 141-52
[9] Dutrizac J E 1978 The kinetics of dissolution of chalcopyrite in ferric ion media. *Metallurgical Transactions B* **9(3)** pp 431-9
[10] Miki H, Nicol M J and Velásquez-Yévenes L 2011 The kinetics of dissolution of synthetic covellite, chalcocite and digenite in dilute chloride solutions at ambient temperatures. *Hydrometallurgy* **105 (3-4)** pp 321-7
[11] Nicol M, Miki H and Zhang S 2017 The anodic behaviour of chalcopyrite in chloride solutions: Voltammetry *Hydrometallurgy* **167** pp 72-80
[12] Velásquez-Yévenes L and Quezada-Reyes V 2018 Influence of seawater and discard brine on the dissolution of copper ore and copper concentrate *Hydrometallurgy* **180** pp 88-95
[13] Dhawan N, Safarzadeh M S, Miller J D, Moats M S and Rajamani R K 2013 Crushed ore agglomeration and its control for heap leach operations. *Minerals Engineering* **41** pp 53-70
[14] Lu J, Dreisinger D and West-Sells P 2017 Acid curing and agglomeration for heap leaching.*Hydrometallurgy* **167** pp 30-5
[15] Iversen G R, Wildt A R, Norpoth H and Norpoth H P 1987 *Analysis of variance* (No. 1) Sage
[16] R Core Team 2017 R A language and environment for statistical computing (Vienna: R Foundation for Statistical Computing)
[17] Groemping, U 2018 DoE.base: Full Factorials, Orthogonal Arrays and Base Utilities for DoE Packages. R package version 0.32. https://CRAN.R-project.org/package=DoE.base
[18] Tukey J W 1949 Comparing individual means in the analysis of variance. *Biometrics* pp 99-114

IOP Conf. Series: Materials Science and Engineering **427** (2018) 012031 doi:10.1088/1757-899X/427/1/012031

Analysis of process of grinding efficiency in ball and rod mills with various feed parameters

D Foszcz[1], D Krawczykowski[1], T Gawenda[1], E Kasińska-Pilut[2] and W Pawlos[2]

[1]AGH University of Science and Technology, Faculty of Mining and Geoengineering, Department of Environmental Engineering and Mineral Processing, Krakow

[2] KGHM Polska Miedź S.A.

Abstract. The article presents the results of laboratory-scale research on the determination of the impact of ball mill parameters and the feed directed to grinding on its effectiveness and comparing it with the efficiency of grinding in a rod mill. The research was carried out for grinding copper ore processed in O/ZWR KGHM PM S.A. The aim of the research was to evaluate the efficiency of grinding in a drum mill dependably on the grinding media used (balls, rods) for various feed parameters - i.e. particle size distribution, mix of lithological type generating different susceptibility to grinding. Determination of the influence of the grinding media type on the efficiency of material grinding is of great practical importance in the designing of optimal grinding circuit. Furthermore, thanks to the acquired knowledge in this area it allows to improve the technological and economic indicators of the copper ore enrichment process.

The paper presents guidelines and recommendations for conducting the grinding process of mineral raw material (copper ore) enabling the achievement of optimal technological indicators. An important practical part of the recommendation is determination of the conditions for using a given type of grinding media, such as rods or balls.

1. Introduction

Grinding is a very important process of mineral processing of raw materials constituting an significant part of the technological system of comminution in which the process of preparation of the feed for further operations occurs. Dependably on many factors and requirements related to the enrichment process, various technical and technological solutions can be used. The important factors determining the choice of specific solution for enrichment process are such properties of the feed as mineral composition, particle size distribution, lithological type of material, moisture, content of substances hindering the screening and crushing process as well as processing costs. An example of changes in existing technological systems resulting in lower costs is the Mount Isa processing plant [1]. The Mount Isa processing plant started working with a grinding system based on two-stage crushing in jaw crushers and grinding in rod mills. The second regrinding stage consisted of two ball mills with a diameter of 5 m in a closed circuit with a 505 mm hydrocyclone. The expected capacity was 800 Mg/h, assuming that the 80-percent of particle size of the hydrocyclone overflow will be approximately of diameter 140 μm. In 1991, two self-grinding mills were installed to replace two crushing stages - rod mills and ball mills for the first regrinding. The original self-grinding mills were unable to achieve 800 Mg/h, so they were adapted to the SAG (semi autogenous) operation by adding a grinding medium. The energy consumption of the AG / SAG mill system was 30% higher than the old crushing and first grinding system. At a capacity of 800 Mg/h, 80% of the passing particle size as a product of the AG/SAG system was about 120 μm in size, i.e. 20 μm smaller comparing with the old system. Mineralogical studies of the comminuted products indicated that the current energy consumption of the second stage of size reduction was higher than required for proper liberation of chalcopyrite, therefore, one or two mills from this system could be removed without affecting the productivity of the plant. According to this the costs reduction by more than $ 2 million a year was expected. However, it required proper tests of the efficiency of the work of ball mills. Simulations of the grinding circuit were carried out for various system configurations. The analyzed mill parameters were the size of the balls, mill filling and density of the feed to the mill. Also the diameters of the hydrocyclones, the angle and the overflow nozzles were selected at the same time. The technological circuit based on a two-stage classification with regrinding of the underflow of the second hydrocyclone, despite the replacement of the original diameter of hydrocyclones from 505 mm up to 660 mm, did not meet expectations due to the unfavorable effect of coarse grains from overflows of hydrocyclones. To reduce this unfavorable phenomenon, changes were made to the system configuration and two underflows of hydrocylones were directed through the ball mill. Balls of size 65 mm were replaced with 80 mm balls to reduce the amount of coarser particles in circulation. Larger balls reduced the number of particles coarser than 1 mm by about 5%. In case the feed to ball mill was equal to 1200 Mg/h, it resulted in a reduction of 60 Mg/h of the weight of coarse particles in circulation. It also enabled the processing of a larger feed size without the risk of balls being thrown from the mill. No negative impact on the use of a larger size of balls to the production of particles smaller than 150 μm was noticed.

The circuit with one mill reached an average yield of 670 Mg/h over a 12 month period, with recovery of 93.2% Cu and a concentrate content of 27.2%. The results were better than expected and saved more than $ 2 million. It proved that the modification of the grinding circuit in 2nd stage to reduce energy consumption significantly lowers operating costs.

The authors Wei D., Craig I.K. [2] provided information on trends in the application of technical solutions in individual mines. A questionnaire asking for basic information related to the enrichment process for processing plants such as gold (27% - respondents), platinum (22% - respondents), silver (19% - respondents), copper (16% - respondents) and Cobalt, diamonds, lead, nickel, antimony, pyrite, magnetite, Fe, TiO_2, Cr_2O_3 and CrFe was sent. The work shows that the most commonly used mills in the world include ball mills, which are being used in around half of the processing plants in the world, SAG/AG mills are used less frequently (38%), while rod mills are used the least frequently. Half of the respondents indicated that the number of milling circuits varies from 1 to 3. There is often only one milling circuit (35%). Only a few plants have more than 5 circuits (10%). The most common configuration of the circuit is a single-stage and closed circuit (37%).

The second most commonly used configuration is a two-stage, first open and then closed grinding circuit (30%), in some plants multistage circuits exist. When asked what was applied to the classification if the circuit was closed, mainly cyclones with the division into branches (37%) were indicated, individual cyclones were used less frequently (27%). Spirals or incline classifiers, as well as combinations of sieves and cyclones, are less commonly used. In the major number of plants (92%) balls are used as grinding media, the ore itself is used less frequently (31%). The percentage profile of individual mill fillings depends directly on its type.

The latest technological solutions for grinding are based on high pressure grinding rolls (HPGR), which redefine the needs for determining particle size for the grinding process due to the additional effects of destruction inside particles and changes in their enrichment properties because of better opening of the mineral surface [3]. The use of HPGR in technological circuits creates the possibility of reducing the particle from the crushing process while ensuring high productivity and low costs. Interesting solutions in this field include two technological circuits for raw mineral materials processing presented in Fig. 1 (copper processing plant in Kazakhstan) and 2 (gold processing plant in

IOP Conf. Series: Materials Science and Engineering **427** (2018) 012031 doi:10.1088/1757-899X/427/1/012031

Irkutsk, Russia), which are in line with the global trend related to the use of HPGR. In the plant located in Kazakhstan, two high-pressure grinding rolls for milling and only one ball mill grinding stage were used. The maximum particle size for roll presses in this plant is equal to 38 mm and it is the result of the fragmentation of the material in a cone crusher. The product crushed in the first roll press is classified in two stages on a 5mm screen and hydrocyclones. The upper product of the screen is directed to the second roller press, including the underflow of hydrocyclone from 1st stage. The overflow of hydrocyclone goes to the 2nd stage classification, which overflow is a final product, and the underflow is directed to a ball mill co-working in a closed circuit with the second stage of classification in hydrocyclones. For the classification of the second stage conducted in hydrocyclones, the product is also directed after crushing in the second press after prior classification on the screen.

Kazakhstan: Roller Press RPS 13-170/140
Design and process data

Company/Location:	Kasachsmys/Kazakhstan
Roll diameter:	1.700 mm
Roll width:	1.400 mm
Feed material:	Copper ore
Feed moisture:	max. 3 %
Top feed size:	38 mm
Final grain size:	max. 3,0 %
Throughput rate:	945 Mg/h
Spec. energy consumption:	<2,0 kWh/Mg
Spec. press force:	5N/mm^2 (max)
Motor size:	2x1,150 kW
Lifetime of studded tyres:	8,500 operating hours

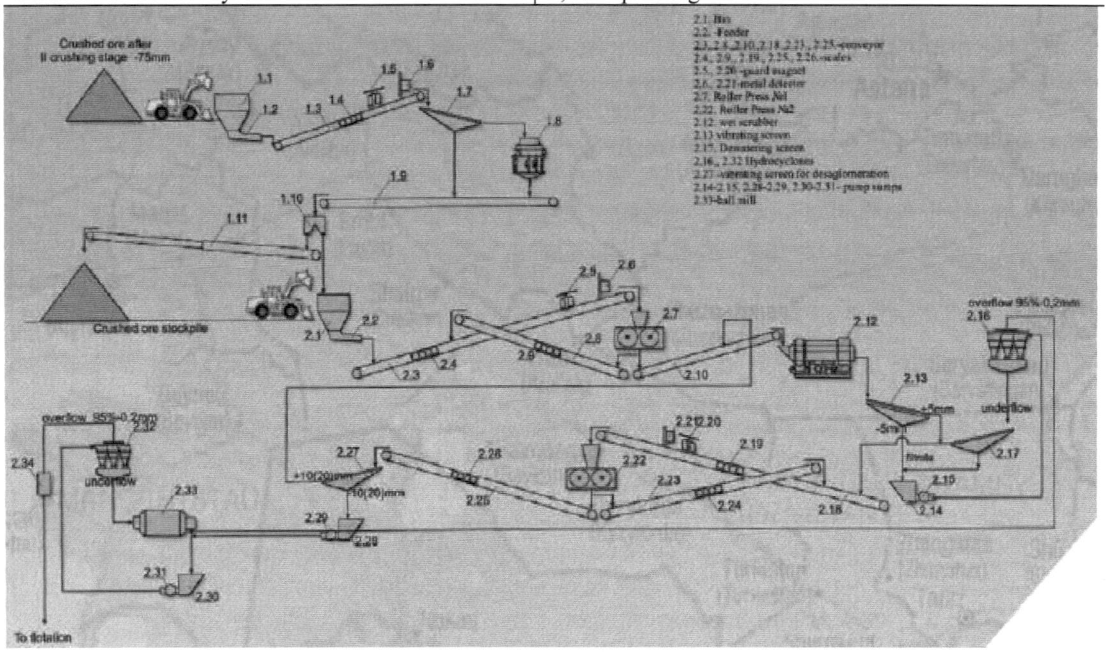

Figure 1. The technological scheme of the copper processing plant in Kazakhstan (advertising materials KHD Humboldt Wedag) [4].

The gold processing plant in Irkutsk, Russia, was designed to obtain a product below 1 mm in a three-stage crushing system (1st stage in jaw crusher, 2nd stage in impact crusher and 3rd stage in roll press). The roll press feed particle size is equal to 25 mm. This solution with double classification as

IOP Conf. Series: Materials Science and Engineering **427** (2018) 012031 doi:10.1088/1757-899X/427/1/012031

well with product circulation after the roller press on the first screen ensures its proper load and allows achieving the expected particle size while minimizing energy costs.

Russia: Roller Press 5-100/90 For grinding of gold ore
Design and process data

Company/Location:	Suchoj Log/Irkutsk Russia
Roll diameter:	1,000 mm
Roll width:	900 mm
Feed material:	gold ore
Feed moisture:	max. 6.5 %
Top feed size:	25 mm
Final grain size:	40 % < 1.0mm; 70 % < 5.0 mm
Throughput rate:	320 Mg/h
Spec. energy consumption:	<1.80 kWh/Mg
Spec. press force:	5 N/mm^2 (max)
Motor size:	2x400 kW
Lifetime of studded tyres:	8,800 operating hours

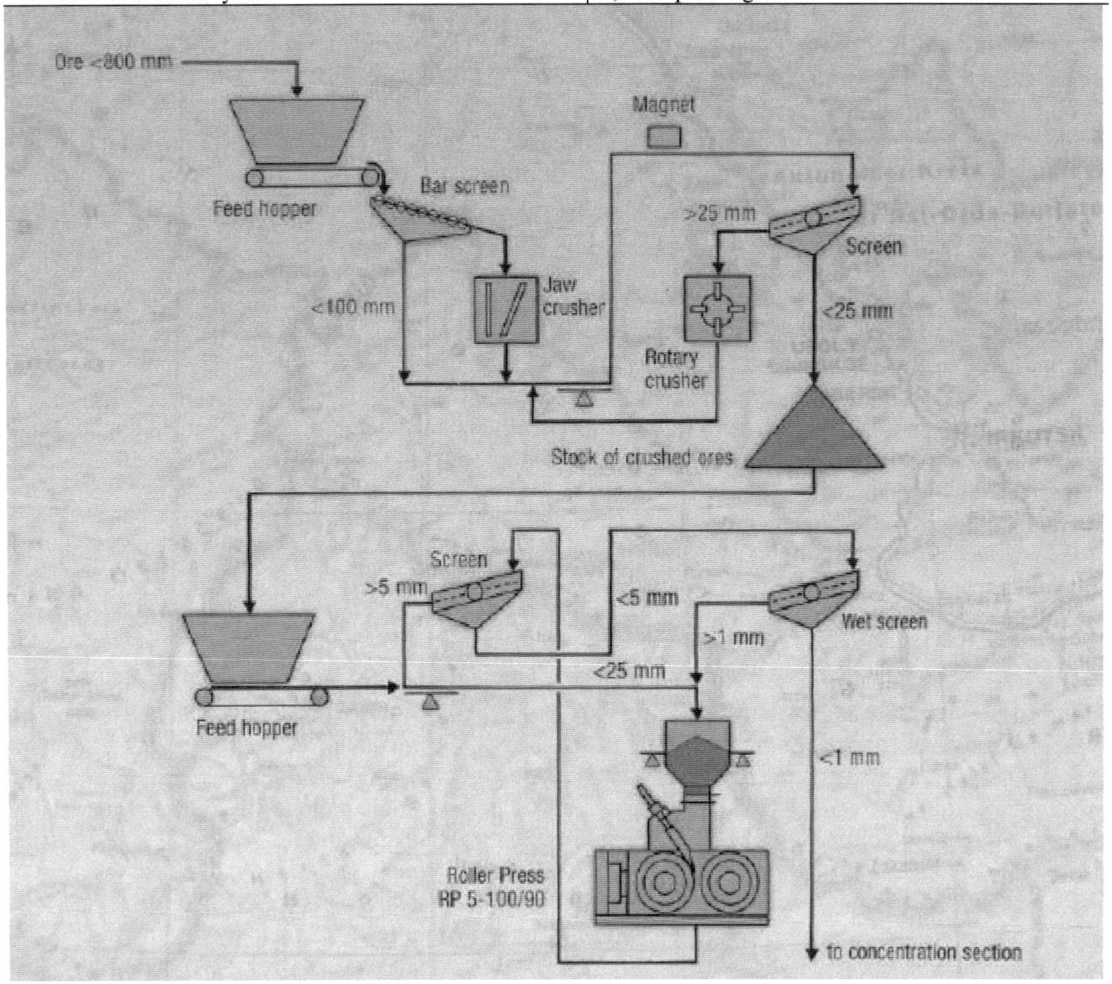

Figure 2. The technological scheme of the gold processing plant in Irkutsk, Russia (advertising materials KHD Humboldt Wedag) [4].

IOP Conf. Series: Materials Science and Engineering **427** (2018) 012031 doi:10.1088/1757-899X/427/1/012031

Summarizing, it should be emphasized that the mill's processing results are effect of its milling capacity, processing capacity of the technological circuit and susceptibility to grinding of raw material as well as the particle size distribution of the feed. Changing these parameters reduces or increases the capacity of the mill. Properly selected parameters, taking into account their mutual impact on the efficiency of work of grinding equipment are crucial in determining the working conditions and can lead to an improvement of the process of crushing (grinding) and lowering costs [5-9].

For each ore, there is an optimal size distribution of the feed to the grinding circuit. It is conditioned by the technological properties of the ore. The optimum particle size depends on the nature of the crushing and grinding costs dependably on the maximum particle size in the crushing product and it is often "fuzzy". Technological limitations (i.e. ore moisture) do not allow working in an area considered as optimal. Looking at the technological circuits operating in O/ZWR KGHM PM S.A. one should remember about the specific character of ore and the period when the plant was built, i.e. the 70's of the last century. Rod mills found application for relatively coarse grinding and were installed in the first grinding stage. The use of rod mills can be treated as the second degree of size reduction, the grinding product obtained contains about 30% of the size -0.075 mm. The rod mill discharge, as compared to the ball mill grinding product, is characterized by less concave (indented) character of the particle size of the product. The rods inside the drum during grinding retain a distance between each other thanks to pieces of ore and serve as screening grids. Small particles are transferred through the gaps between the rods and they quickly leave the mill. In contrast, coarser particles are retained between the rods and a relatively slower process of their grinding takes place. This avoids over-grinding of ore particles that are fragile in nature and are more susceptible to comminution. In rod mills, grinding as a result of the abrasion process is less intense than in ball mills. Therefore, a significant part of middle fractions is observed in the grinding product of rod mills. The content of coarse (not grinded) particles of the feed as well as the very fine particles is small.

Ball mills due to the different nature of the grinding process are used for fine grinding. Impact and abrasion cause the formation of a large amount of fine particles. The particle size distribution curve of the product can be obtained here. The grinding material stays in the mill for a relatively long time, which often causes the over-grinding of ore particles, which are usually characterized by greater brittleness than particles of gangue [10, 11]. Comparative analyzes for strength distribution for three lithological type of copper ore was investigated by authors [12-14]. The use of rod mills was justified by the need of crushing the coarse particles of dolomite and shale. At the same time, the concerns about possible over-grinding of ore minerals previously released from sandstone ore were eliminated.

The effectiveness of the mill's operation is determined by the parameters of its operation, i.e. rotational speed, load, filling with grinding media, size .distribution of grinding media as well as the parameters of the grinded feed. The most important feed parameters are the particle size distribution, including maximum particle size subjected to grinding and susceptibility to grinding.

The key technological issue related to grinding is the proper selection of both the type of mill/grinding media and their size enabling the proper efficiency of the entire grinding stage and classification circuit as well as the regrinding of the feed properties subjected to this process [3, 6]. For rod mills, the breaking of the feed particle occurs as a result of falling and rolling cylindrical rods as a grinding medium in which the feed particle can reach 50 mm size, the particle size of the product is <1 mm. The crushing occurs due to the large mass and surface of the grinding media. For ball mills, grinding media are balls of various diameters, the feed load is finely grained in relation to the feed of rod mills and its size is <15 mm. The product achieves a much finer particle size <0.2 mm. The process takes place by grinding and abrading the material with a large surface of grinding media and the number of contacts [10, 11].

IOP Conf. Series: Materials Science and Engineering **427** (2018) 012031 doi:10.1088/1757-899X/427/1/012031

The basic parameters determining the mechanics of the work of drum mills are:

- the relative frequency of rotation of the mill roll ψ , equal to the ratio of the actual frequency of rotation n to the conditional critical frequency of rotation n_{kr}:

$$\psi = \frac{n}{n_{kr}} \tag{1}$$

The conditional critical velocity is determined by the formula:

$$n_{kr} = \frac{30\sqrt{2g}}{\pi\sqrt{D}} \simeq \frac{42,3}{\sqrt{D}} \tag{2}$$

where: g – standard acceleration due to gravity, m/s2; D – the inside diameter of the drum; m.

- relative filling of the mill with grinding media ϕ , equal to the ratio of the grinding media V_m to the internal volume of the mill roll V:

$$\phi = \frac{V_m}{V} = \frac{4G_m}{\gamma_m \pi D^2 L} \tag{3}$$

where; G_m - mass of grinding media; Mg; γ_m - bulk density of grinding media depends on the composition.

The above parameters must be taken into account when selecting the operating conditions of the drum mill. It should also be marked that rod mills generally work at lower speeds than ball mills, ranging from 50 to 65% of critical rotation.

2. Materials and methods

The aim of the research was to determine the influence of the grinding media type on copper ore grinding efficiency, determine the conditions for the change of grinding media used in industrial conditions from rods to balls with the evaluation of the benefits resulting from them. Research may enable improvement of technological and economic indicators of the copper ore enrichment process.

Laboratory experiments were carried out for a mill filled with rods as basis and balls as the preferred solution to increase the efficiency of ore grinding. Varying feed parameters (mix of lithological type, grain size distribution) and grinding media sizes were used. In addition, for the mill filled with balls, its rotational speed has been changed.

The plan of experiments as well as the conditions and scope of their implementation are presented in Table 1. Three additional experiments were carried out for mix lithological type with O/ZWR KGHM a sample 1 in the fraction 0-8 mm with a lower proportion of sandstone (60P + 25D + 15L), with an increased speed of the mill (from 58 to 70 rpm) and three grinding times (10, 20 and 40 min). In total, 51 milling tests had been done.

Table 1. Plan and conditions of experience.

Feed particle size	Lithological composition	District	Grinding time t = 10 min Rods			t = 20 min Rods			t = 40 min Rods		
			Rods	Small balls	Large balls	Rods	Small balls	Large balls	Rods	Small balls	Large balls
0-8mm	mix (lithological type) (80%P+5%D+15%L)	O/ZWR KGHM sample 1	×	×		×	×		×	×	
	mix (lithological type) (60%P+25%D+15%L)	O/ZWR KGHM sample 1	×	×		×	×		×	×	
	natural mix (industry)	O/ZWR KGHM sample 1	×	×	×	×	×	×	×	×	×
		O/ZWR KGHM sample 2	×	×	×	×	×	×	×	×	×
0-16mm	natural mix (industry)	O/ZWR KGHM sample 1	×	×	×	×	×	×	×	×	×
		O/ZWR KGHM sample 2	×	×	×	×	×	×	×	×	×

where: P –sandstone ore, D – dolomite ore, L – shale ore

The milling tests were carried out on 6 mixes of ore types (four from O/ZWR KGHM O/ZWR KGHM a sample 1 and two from O / ZWR KGHM O/ZWR KGHM a sample 2) prepared according to the scheme (Figure 3). The lithological composition of the mix is presented in Table 2.

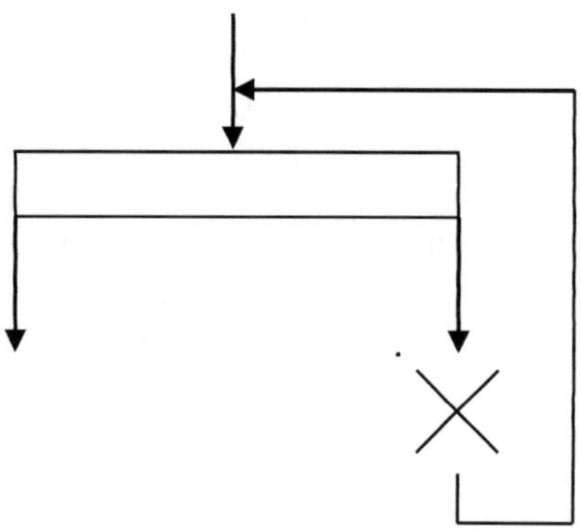

Figure 3. Scheme of sample preparation for tests.

Table 2. Percentages of individual lithological types in mix prepared for milling tests.

Place of sample		O/ZWR KGHM a sample 1	O/ZWR KGHM a sample 1	O/ZWR KGHM a sample 1	O/ZWR KGHM a sample 2
Mix of ore		80P+5D+15L mix	60P+25D+15L mix	natural mix (industry)	natural mix (industry)
Sandstone	%	80.0	60.0	64.5	62.9
Shale	%	15.0	15.0	12.0	8.6
Carbonates	%	5.0	25.0	23.6	28.5
Sum	%	100.0	100.0	100.0	100.0

The tests were carried out in a laboratory batch drum mill. The mill parameters were as follows:

- diameter of the mill $\phi = 305$ mm
- length of the mill $l = 440$ mm
- number of rotations n = 58 rpm, 70 rpm (respectively: 75.7% and 91.4% critical rotation)
- mass of grinding media – 14 kg

Three types of steel grinding media were used in the experiments: rods and balls of smaller and larger size. Detailed size distribution of grinding media is presented in Table 3.

Table 3. Size distribution of grinding media.

d mm	Rods	Small balls	Large balls	Rods	Small balls	Large balls
	Q [g]	Q [g]	Q [g]	γ [%]	γ [%]	γ [%]
15	5903	4800	1008	42.2	34.3	7.2
20	8090	1842	1984	57.8	13.2	14.2
30	–	7365	3900	0.0	52.6	27.8
35	–	–	3009	0.0	0.0	21.5
42	–	–	4110	0.0	0.0	29.3
Sum	13993	14007	14011	100	100	100

3. Analysis of results

The results of the 51 experiments were presented in the form of graphs of the curves of particle size distribution of the grinding product for particular conditions of the process and particular feeds. Exemplary curves for selected grinding conditions are shown in Figure 4.

IOP Conf. Series: Materials Science and Engineering **427** (2018) 012031 doi:10.1088/1757-899X/427/1/012031

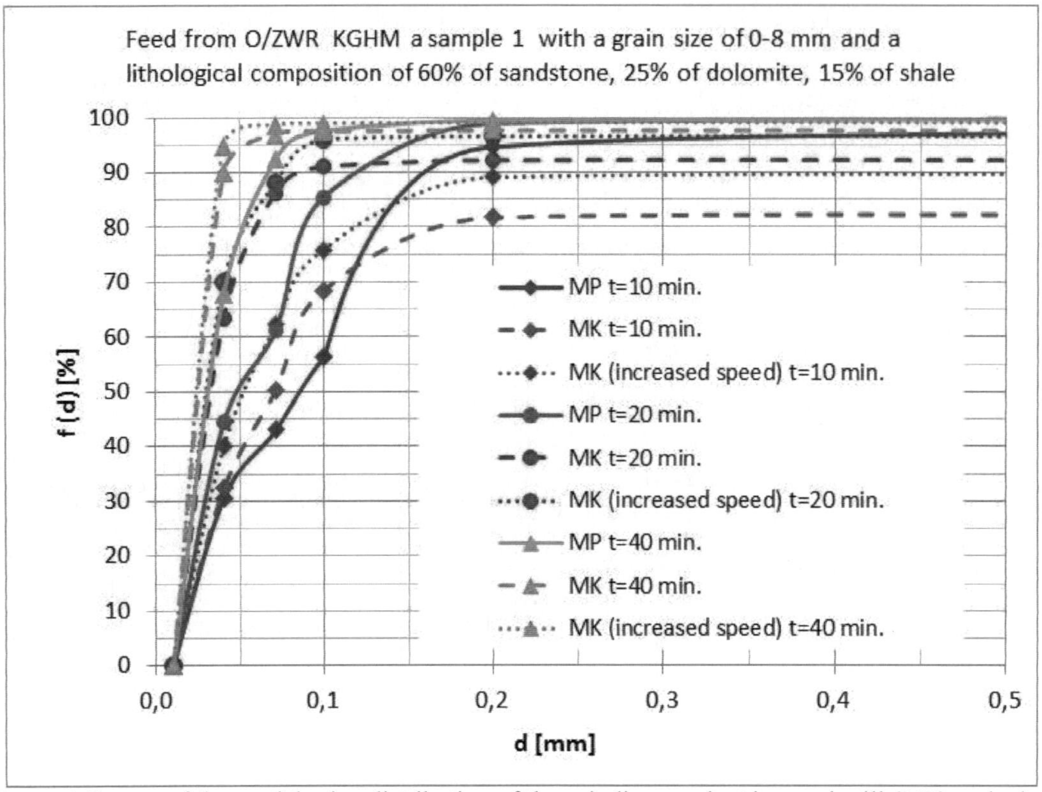

Figure 4 Curves of the particle size distribution of the grinding product in a rod mill (MP) and a ball mill (MK) for feed from O/ZWR KGHM a sample 1with a grain size of 0-8 mm and a lithological composition of 60% of sandstone, 25% of dolomite, 15% of shale.

The results in the form of particle size distribution curves enable a preliminary comparative analysis of the grinding efficiency depending on the grinding media used, as well as the rotational speed of the mill. The analysis shows that in the range of fine particle size <0.1 mm, the grinding efficiency with the use of balls is higher compared to the rod mill for the smaller feed size 0-8 mm. The increase in particle size of the feed to 0-16 mm resulted in a decrease in the milling efficiency of the ball mill, but still in the range of fine particles it is higher in relation to the rod mill. Increasing the size of the balls improves the ball mill's grinding efficiency, the grinding efficiency increases also for higher rotational speeds of the mill filled with balls.

The increments of particle size fraction such as: <0.04 mm; <0.071 mm; <0.1 mm; <0.2 mm in the grinding product after 10, 20 and 40 minutes were also analyzed. It indicates that the grinding efficiency measured by the particle size fraction growth depending on the grinding media used is strongly dependent on the particle size of the analyzed fraction and feed size, especially for fine particles below 0.2 mm and finely grained 0-8 mm feed. This efficiency is higher for a mill filled with balls. For example, the milling efficiency measured by the grain size increase <0.1 mm for feed with a particle size of 0-8 mm and a lithological composition of 80% sandstone, 5% dolomite and 15% slate depending on the grinding time is shown in table 4.

Table 4. Kinetic index of increasing of class grain size <0.1 mm in milling products.

Grinding time [minutes]	Rod mill	Ball mill
10 minutes	32.8%	46.4%
20 minutes	57.2%	73.5%
40 minutes	74.9%	79.0%

4. Conclusions

Based on the results of tests carried out in laboratory conditions, a higher efficiency of the ball mill comparing to the rod mill, in conditions of increment of small size fractions was found. The value of grinding efficiency in a ball mill depends on the particle size of the feed. For the small-grained feed, the grinding efficiency of the ball mill is increased by approx. 20-30%.

The effect of the grinding environment tested (ball and rod) on the grinding efficiency depends on the change in the lithological composition, i.e. sandstone and dolomite content indicates that the mill filled with balls after a longer grinding time better cope with the grinding of ore with increased sandstone content, in particular for increments fraction <0.071 mm and <0.1 mm. For shorter grinding time, as well as for the increment of the small size fractions, the effect of the various lithological composition on the change in efficiency is comparable.

The results of the tests and the analyzes carried out indicate, that for the small-grained feed, ball mills should be used. This is due to the higher efficiency of grinding ore measured by the increasing of small size fractions, and for longer milling time, the higher efficiency of grinding ore with an increased content of the sandstone fraction of the mill, filled with matched balls.

Applying grinding environment changes, and, in particular, the determination and optimization of parameters such as the grinding balls size, should be preceded by tests in industrial conditions confirming specific technological effects.

The results of the tests indicate higher milling efficiency in the ball mill in relation to the rod mill in conditions of increasing of small size fractions with comparable ranges of not grinded coarse particles for a 0-8 mm particle feed suggest changing of the rod grinding environment for ball mills in industrial conditions. The expected benefits, from changing the type of grinding media are: an increase of approx. 20-30% in relation to the current classifier overflow stream, and thus a smaller stream of grinding material in the second-stage ball mill, increased operating time of the ball mill, more convenient process of loading and unloading of grinding environment, no effect of rods jamming, lower efficiency of the mill filled with balls due to its lower sensitivity to impurities accumulated in the spaces between the rods, high grinding speed measured by the increasing of small size fractions over time, with relatively small disadvantages of this solution (balls application), i.e. smaller maximum permitted particles for grinding feed, higher rotational speeds, possibility of over-grinding coarsely grained minerals.

Conclusions from the analysis indicate, that the ball-filled mill is more resistant to changes in the lithological composition, which means that the particle size distribution of the feed is the key to changing the grinding environment. However, the condition of applying a ball mill is to meet specific process conditions, especially in terms of grain size and size of grinding environment. The maximum particle in the feed should not exceed 20 mm in the industrial conditions. In the field of industrial practice, it is recommended to evaluate the possibility of particle reduction, due to the positive effect of the grinding kinetics increase and obtaining much higher increment of small size particles.

The research results confirm the need to select optimal technical and technological solutions for the properties of the raw material in theoretical and experimental way in order to determine the process parameters and details of the technological circuit.

References

[1] Gao M W, Fountain C, Magee M and Way D 2006 Energy Efficiency Grinding Circuit Optimization *XXIII International Mineral Processing Congress Proceedings* **1** pp 283-9

[2] Wei D and Craig I K 2009 Grinding mill circuits – A survey of control and economic concerns *International Journal of Mineral Processing* **90** pp 56-66

[3] Saramak D, Mlynarczykowska A and Krawczykowska A 2014 Influence of a high-pressure comminution technology on concentrate yields in copper ore flotation processes *Archives of Metallurgy and Materials* **59** pp 951-5

[4] http://www.zkg.de/en/artikel/zkg_2010-10_KHD_Humboldt_Wedag_1022957.html

[5] Saramak D 2011 The influence of chosen ore properties on efficiency of HPGR-based grinding circuits *Gospodarka Surowcami Mineralnymi* **27** pp 33-44

[6] Gawenda T, Saramak D, Foszcz D, Krawczykowski D, Krawczykowska A and Lewandowski D 2018 An assessment of influence of selected characteristics of feed on the results of ball mill grinding process course for Polish copper ores; Ocena wpływu wybranych charakteFigtyk nadawy na wyniki procesu mielenia w młynie kulowym dla polskich rud miedzi *Archives of Mining Sciences = Archiwum Górnictwa* **63** pp 125–37

[7] Saramak D, Tumidajski T, Brożek B, Gawenda T and Naziemiec Z 2010 Aspects of comminution flowsheets design in processing of mineral raw materials (Aspekty projektowania układów rozdrabniania w przeróbce surowców mineralnych) *Gospodarka Surowcami Mineralnymi* **26** (Kraków: Wyd. IGSMiE PAN) pp 59–69

[8] Tumidajski T, Kasińska-Pilut E, Gawenda T, Naziemiec Z and Pilut R 2010 Badania energochłonności procesu mielenia oraz podatności na rozdrabnianie składników litologicznych polskich rud miedzi. Gospodarka Surowcami Mineralnymi **26** (Kraków: Polska Akademia Nauk. Komitet Zrównoważonej Gospodarki Surowcami Mineralnymi; Instytut Gospodarki Surowcami Mineralnymi i Energią) pp 61–72

[9] Foszcz D and Gawenda T 2012 Analysis of efficiency of grinding in ball and rod mills dependably on contents of fine particles in feed (Analiza efektywności procesu mielenia w młynach kulowych i prętowych w zależności od zawartości ziarn drobnych w nadawie) *AGH Journal of Mining and Geoengineering* **36** pp 17-30

[10] Sidor J and Klich A 2018 Współczesne maszyny do rozdrabniania – kruszarki i młyny (Gliwice: Instytut Techniki Górniczej KOMAG)

[11] Lowrison G C 1974 Crushing and grinding (London: Butterworths)

[12] Nad A, Brożek M and Naziemiec Z 2012 The tensile strength properties of lithological variety of Polish copper ores *AGH J. Min. Geoeng.* **36** pp 101–8

[13] Nad A and Saramak D 2018 Comparative analysis of the strength distribution for irregular particles of carbonates, shale and sandstone ore *Minerals* **8** pp 1–13

[14] Nad A and Brożek M 2017 Application of three-parameter distribution to approximate the particle size distribution function of comminution products of dolomitic type of copper ore *Archives of Mining Sciences* **62** pp 411–22

IOP Conf. Series: Materials Science and Engineering **427** (2018) 012032 doi:10.1088/1757-899X/427/1/012032

Increasing efficiency of technological process by limiting impact of corrosive environment on operation of spiral classifiers

M Czekajło[1] and K Żakowski[2]

[1] KGHM Polska Miedź S.A. – Division of Concentrators

[2] Gdańsk University of Technology – Department of Electrochemistry, Corrosion and Materials Engineering

marcin.czekajlo@kghm.com

Abstract. Most of the technological operations related to the preparation of the output to be enriched and to the production of the final copper concentrate take place with the use of water environment. Water management, besides using innovative technical and technological solutions, is a significant factor in the whole copper ore enrichment process. Mine water resources and surface water of the tailing pond named "Żelazny Most" are the two sources of technological water. Its physico-chemical composition is not insignificant for both the flotation process and the machinery and equipment maintenance. The concept of electrochemical protection of the spiral classifier presented in this article is the supplement and alternative for anti-corrosion protection of the machinery and equipment used in KGHM Polska Miedź S.A. Ore Enrichment Plant. The adequate appreciation of the spiral classifier characterisation and the working conditions will allow for the optimal adjustment of cathodic protection installations with the anode polarization system. Research into the development of an effective protection method will combine two complementary types of anti-corrosion protection systems, that is the passive (a protective coating) and the active (a cathodic protection). The effect of the planned research will be the evaluation of the potential possibilities of the corrosive and erosive impact limitation of the spiral classifier working condition. And, moreover, it will lead to an extension of working hours and the reduction in the costs of the technological circuit.

1. Introduction

The Ore Concentration Plant (OCP/ZWR) is a single organizational unit of KGHM Polska Miedź S.A. and consist of three production areas: Lubin, Polkowice and Rudna which fulfils a special role in the technological lines of copper ore processing. The basic task of the plants is to maximize metal recovery and produce the concentrates with the quality parameters required for effective smelting at the lowest possible cost.

The copper ore deposits mined by KGHM in Poland comprise three lithological types: shale, dolomite, sandstone, which contain, on average, 1.50 % of the actual copper used. The copper ore is subject to process whose main operation is the flotation enrichment, which, in turn, results in the obtaining of the concentrate with the copper content enabling its metallurgical processing while smelting. The remaining part of the output, with a small metal content, constitutes the floatation waste, which is further transported to the tailing pond named "Żelazny Most". The technics and technology that are used in concentration plants ensure a very effective recovery of copper, silver and other elements, even up to 90 % of copper with high copper content in the concentrate up to 26 %. The Concentration Plants operating at each of KGHM's three mines are able to process an average of 33 million tonnes of ore per year.

The basic production processes in OCP/ZWR include: sieving, crushing, grinding and classification, flotation, concentration, filtration and thermal drying of the concentrate. The implementation of the above tasks requires that enrichment processes should be carried out in a manner that is safe for employees, as smooth as possible and effective. That is why, it is vital to keep machines and devices in good technical condition. It should be understood as carrying out all the planned maintenance and repairs as well as applying modern and effective methods of their protection.

This could be achieved by using the proper anti-corrosion protection system, that is the passive (a protective coating) and the active (a cathodic protection) or by combining both methods.

2. The conditions of the technological process

The grinding and classifying processes are the stages which prepare ore to flotation enrichment. Starting with this stage, all subsequent operations of the production stages are carried out using water.

The main source of water used in the technological process is the surface water of the tailing pond and the mine water. The amount of water pumped to all the ore enrichement plants is variable and results directly from ore throughput. This relationship is determined by water usage per one tonne of ore dried mass, m^3/Mg (figure 1, figure 2).

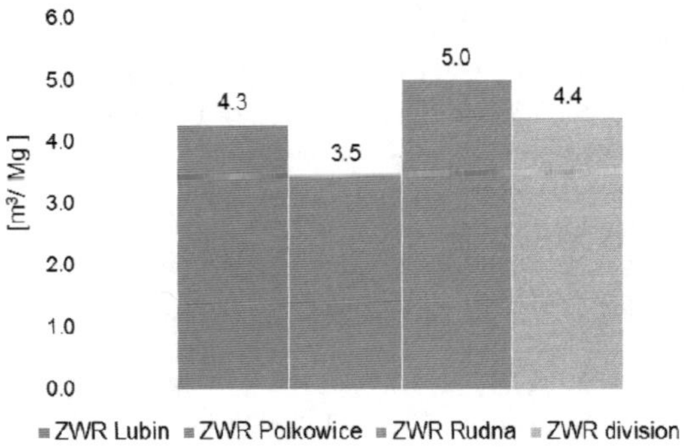

Figure 1. The process water usage indicator of one tonne of ore dried mass on average in 2017.

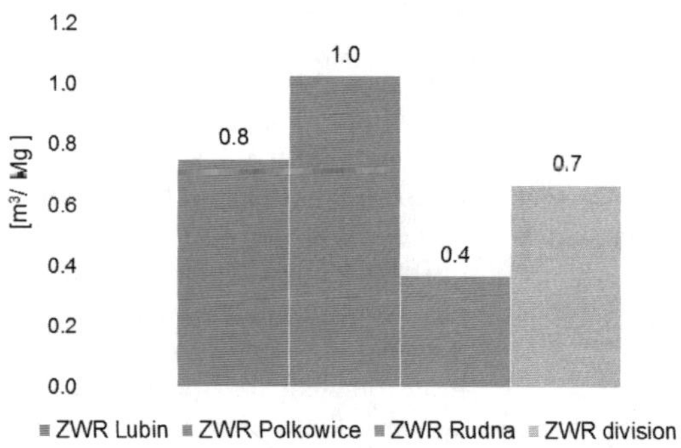

Figure 2. The mine water usage indicator of one tonne of ore dried mass on average in 2017.

Currently the largest participation of the water used in the technological process is the surface water from the tailing pond (86 %). The mine water amounts to about 14 %. In addition to monitoring the amount of water directed to the process, another important aspect is water quality. In the last few years dissolved substances in the processed water has been monitored especially for sodium chloride (NaCl). It is estimated that mine water salinity over the next few years will be slightly, but gradually, increasing [1].

Current data from the monitoring system of chlorides and sulphates created for the needs of KGHM Polska Miedź S.A. has shown the presence of chloride ions (Cl^-) at around 2.0 g/dm^3, which along with mine water are directed to OCP/ZWR Lubin, 7.8 g/dm^3 to OCP/ZWR Polkowice and 100 g/dm^3 to OCP/ZWR Rudna. In the case of tailing pond's surface water the concentration of chlorides nears 25 g/dm^3.

Another component of water-soluble salts is sulphate ions (SO_4^{2-}). Their current concentration in the mine water directed to OCP/ZWR Lubin and Polkowice is 1.3 g/dm^3 and 2.3 g/dm^3 to OCP/ZWR Rudna, whereas the tailing pond's surface water contains about 3.1 g/dm^3. For many years the sulphate concentration in water directed to OCP/ZWR has been maintained at a similar level and occasionally exceeded 3.0 g/dm^3.

3. The influence of industrial water on the technological process

Industrial water quality is not without significance for the flotation process, where the hydrophobicity of the mineral surface is essential. The flotation reagents usage regime can affect not only the hydrophobicity of the grains, but also the foam formation and stability. The effect of stable foam is obtained mainly by adding foaming agents, which also improve gas dispersion and accelerate the flotation process [2]. When using saline technological water, a similar effect is observed. The water affects the quality of the air bubbles produced, as well as the mineral particles surface wettability and the interaction between the air bubble and the grain [3]. Smaller bubbles are generated themselves better in high salinity solutions, which may lead to a reduction in reagent consumption [4].

The research on commission of KGHM's Ore Enrichment Plants carried out by the University of Technology in Wrocław has shown that the addition of sodium chloride usually does not change the selectivity of copper and organic carbon flotation, but slightly affects the selectivity of silver and lead flotation. In turn, conducted research as part of a doctoral thesis entitled "The enrichment of output in KGHM Polska Miedź S.A." showed an increase in the efficiency of copper and organic carbon enrichment as the concentration of salt in the water added to the flotation process increases [5]. Other attempts to determine the influence of salinity of technological water on the flotation process were made in the past by Żmudziński, Lekki [6] and Karpisz [7]. They showed an adverse effect of salinity increase on flotation of sandstone ore in the concentration range from 6.0 to 14 g/dm^3 in relation to flotation in tap water. In the case of shale ore, the most favorable enrichment was obtained with salinity of water with concentrations of 6 and 40 g/dm^3, while worse for tap water. On the other hand, in the case of carbonate ore a linear improvement in flotation results is observed along with the increase in sodium chloride content in water.

The influence of salt on the efficiency of the copper flotation process has not been yet clearly defined and described in the literature. Therefore, it is necessary to acquire knowledge and carry out further research in this area.

Taking into account all the above, it can be stated, with some simplification, that the observed changes in the concentration of chloride ions do not involve technological problems in achieving the assumed quality parameters of the final copper concentrates. Their presence, within certain concentration limits, can have a positive effect on the flotation process course. The control and carrying out of the flotation enrichment process are the easier the smaller variability of process water quality is observed.

The situation, however, is different in the case of the impact that salt has on machinery and equipment, as well as the technical and technological infrastructure. The research carried out by Gdańsk University of Technology [8] has confirmed the high corrosivity of the atmosphere as defined according to PN-EN ISO 12944-2 as C5-I (industrial) and C5-M (marine). Its aggressive character is observed both in the lower and upper parts of the production sites. Difficult climatic conditions, in particular, occur at the floor level, where the relative humidity nears or exceeds 80 %. In addition to the corrosion conditions mentioned above, the corrosion and erosive character of the ore enrichment process should also be listed. Its negative impact is especially visible at the stage of classification in spiral classifiers. The significant amounts of the aqueous medium transporting the crushed and milled ore produce a clear mechanical effect, mainly through friction, cavitation and impact attack.

The inspection of the construction, installations and equipment operating in the plant showed the need for effective methods of corrosion protection.

Taking into consideration all the above, it is important to use the sources of technological water skilfully, supporting the flotation process on the one hand, and on the other limiting its negative impact on the maintenance of machinery and equipment in motion.

4. Initial tests for the possibility of using cathodic protection

The aim of the conducted research was to assess the degree of corrosivity of the feed - an aqueous slurry, which is the natural working environment of spiral classifiers.

The classifiers that classify output, pre-crushed and milled grains up to a grain size of less than 5 mm copper ores work under conditions that until recently have not been examined for corrosivity.

The presence of a large amount of both chloride and sulphate ions has also been confirmed in the conductivity (the inverse of resistivity) studies of classifiers feed. The knowledge of electrolyte conductivity, in this case an aqueous medium, is of great importance to the calculations of the amount and the size of anodes, as well as the location of the sources of protective current for the cathodic protection system.

The measurements of the feed conductivity (comminuted, ground ore in the form of an aqueous slurry) were carried out using a digital resistance meter, to which a "soil-box" vessel was connected (figure 3, figure 4).

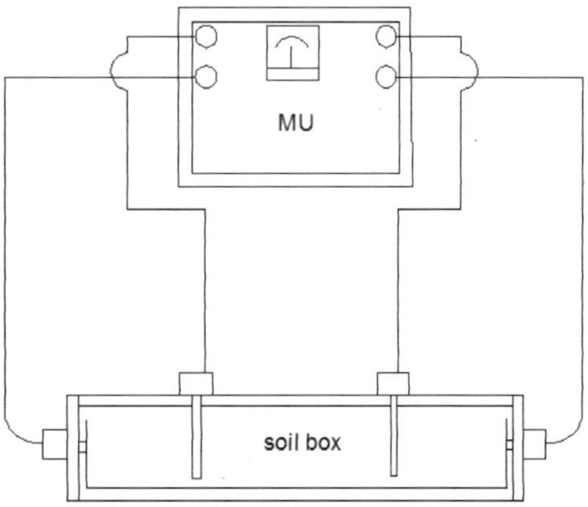

Figure 3. Diagram of the apparatus used to measure the conductivity of the feed.

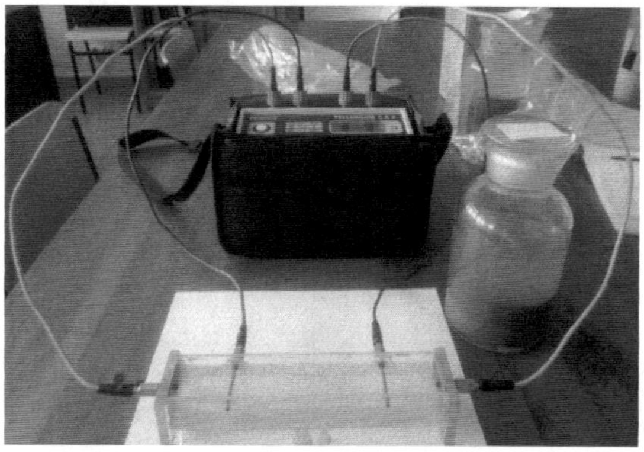

Figure 4. A measuring set used to determine the conductivity of the feed. Digital earth resistance meter, type TELLUROHM C.A 2 and „soil-box" vessel.

The conductivity of samples of the feed taken from three regions: Lubin, Polkowice and Rudna were measured separately for decanted water and mixed feed containing solids.

The above results (table 1) indicate that the feed coming out of OCP/ZWR Rudna region is characterized by the lowest resistivity and thus the highest conductivity. Therefore, the use of cathodic protection in the case of OCP/ZWR Rudna classifiers will require providing the largest protection current. It confirms the presence of the largest amount of soluble mineral salts (mainly chlorides) and their corrosive nature as well.

Lower values of conductivity were achieved in the case of mixed feed, which is related to the influence of the output's suspended grains.

Table 1. The conductivity of the spiral classifier's feed.

Ore Enrichement Plant area	Decanted water conductivity (mS/cm)	Mixed feed conductivity (water + solids) (mS/cm)
Lubin	42.0	37.3
Polkowice	45.0	44.4
Rudna	70.4	61.0

Transferring the conductivity results into the amount of dissolved salts, the following figure 5 showing the dependence of seawater resistivity on salinity and temperature can be used. Reading the values from the graph, it can be concluded that the obtained resistivity results (~ 0,22 Ωm, measurement temperature 25 °C) correlate with salt water salinity at 3.0 %. It means that in 1 dm^3 of sea water, 30 g of mineral salts is dissolved. Similar values are achieved for the tested samples, which confirms the high contents of soluble salts in technological waters. The result is a high corrosivity of the work environment of spiral classifiers, which determines the use of specific corrosion protection measures.

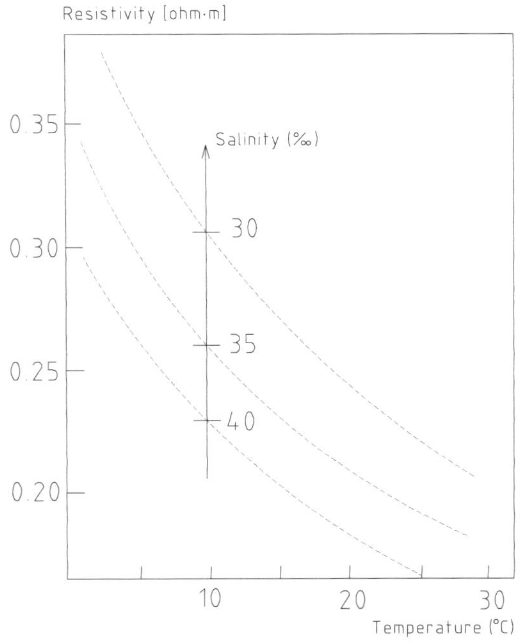

Figure 5. Resistance of sea water as a function of temperature for salinity in the range of 30 - 40 ‰ [9].

5. The concept of implementation of cathodic protection of spiral classifiers

Coatings which are currently used for classifiers protection are inadequate, which results in the necessity to carry out periodic repairs or replacements of complete helixes.

Further research into the possibility of protecting these devices, planned for the future will be a combination of passive (a coatings protection) and active method of anti-corrosion protection (a cathodic protection). The selection of the appropriate abrasion resistant coatings will be carried out by the corrosion coupon (metal specimens) testing in the classifier's natural working environment. At the same time, impedance tests in the laboratory test stand will be conducted. The selection of

IOP Conf. Series: Materials Science and Engineering **427** (2018) 012032 doi:10.1088/1757-899X/427/1/012032

effective wear-resistant coatings will protect metal elements from the erosive impact of the feed and will significantly reduce the demand for protective current in the cathodic protection system.

Cathodic protection research will be based on the surface polarization of the protected element. The first method will involve the creation of a galvanic link, that is a combination of a structure with a metal of a lower potential, the so-called galvanic anode (protector), the other will entail the flow of protective direct current from an external source, the so-called cathodic protection station (SOK). The ultimate method of protection of the classifier's spiral will be determined on the basis of a detailed analysis of the device operating conditions and initial polarization measurements. With some probability, however, one can point out to the choice of cathodic protection technology using an external power source. This is due to the high current demand for the helix protection and the classifier bath. Presumably, galvanic anodes would not provide enough electricity, because each of them would have a relatively low current. Therefore, it would be necessary to install a large number of anodes, which would be difficult due to the limited space of their assembly.

Cathodic protection is probably the most effective method of preventing corrosion of underwater and underground constructions. At the appropriate assumptions, corrosion processes occurring on the metal surface can be practically stopped, for example up to 0.01 mm per year. At the current corrosion rate of the classifier spiral, which nears 0.3 mm per year [8], the search for effective protection methods seems to be indisputable.

When designing the cathodic protection, it will be necessary to collect data on the protected structure, its surface and the type of metal from which it has been made, as well as the surrounding of electrolytic environment and its corrosion potential. It will be equally important to properly assess the economic aspect of the device's life expectancy. It is significant to research the right selection of the anodes supplying cathodic protection current, as well as conducting a series of studies on the selection of wear-resistant protective coatings cooperating with electrochemical protection.

The scheme of the planned connection of cathodic protection with the use of an external source of protection current is shown below (figure 6). The system will be built from a cathodic protection station (SOK), a cathode and potential connection, an anode system and a reference electrode.

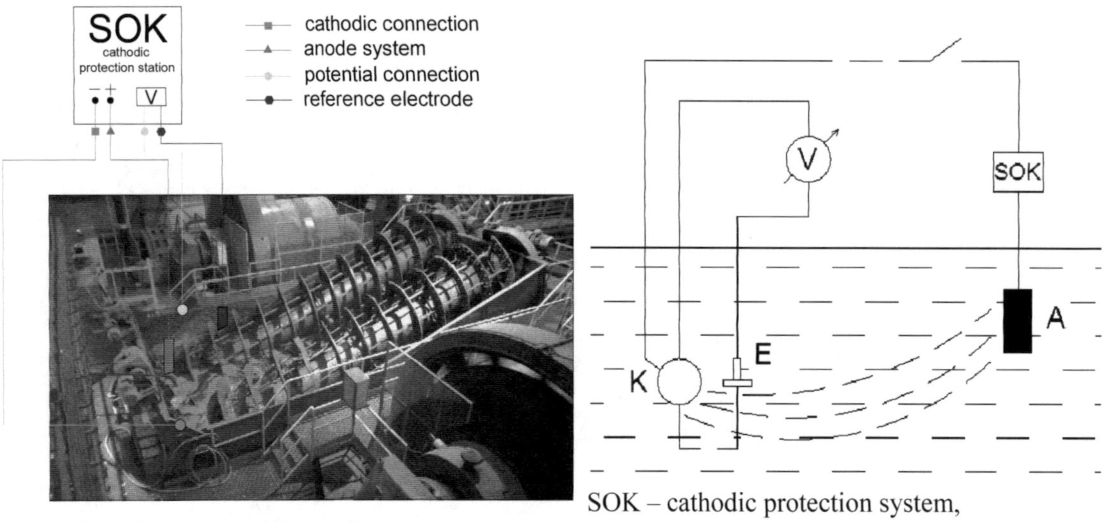

SOK – cathodic protection system,
K – protected construction, A – anode [10].

(1) (2)

Figure 6. An overview of the planned cathodic protection of the classifier using an external current source – a cathodic protection station (SOK) - real view (1), a schematic diagram (2).

Any type of corrosion protection applied should offer a possibility of assessing its effectiveness in a simple, relatively cheap manner, not requiring long-term research. The first criterion for the effectiveness of electrochemical protection will be the potential measured in relation to the reference electrode, which is placed close to the protected structure. Using the potential measurement, the rate of reduction of the corrosion rate of metal can be predicted [11]. The value of the protective potential should be achieved for the entire surface of the construction, which in the case of spiral classifiers may be difficult due to the fact that the rotating rotor periodically emerges from the electrolyte. Preliminary studies [8] indicate that despite the periodic ascent of the spiral from the feed, the cathodic protection can reduce the corrosion rate of the classifier up to 5 times. Research in this area is due to continue.

6. Summary

The limitation of corrosion processes should be considered in the economic aspect, security issues as well as direct and indirect losses. The effective protection of structures and devices exposed to corrosion will reduce the costs associated with the replacement of the corroded elements, carry out renovation works and, above all, extend the machine and equipment cycles. The elimination of these factors will contribute to the reduction of material loss, improve the safety and comfort of work and allow for the rational management of natural resources.

Therefore, in ore enrichement plants a number of detailed tests should be carried out in order to determine the possibility of effective corrosion reduction through the use of cathodic protection. The choice of the method of protection implementation should be individually adapted to the existing corrosion risk and working conditions of the protected metal surfaces. In order to confirm the effectiveness of the application of the innovative method of machinery protection and processing equipment it is required to develop a detailed protection technology, as well as the implementation and launching of the pilot installation.

An extremely important issue in the application of cathodic protection is the ability to assess the correctness and effectiveness of the actions taken. In addition to periodic check-ups of the installation's operation parameters, it could be advisable to perform additional tests by using corrosive coupons, simulating electrodes, or even resistive corrosion resistance, which determines corrosion loss based on the determined increments of electrical resistance [12]. An excellent complement to control activities should be a comprehensive system of managing the corrosion protection process, in which one should define, among others, the areas of responsibility of cathodic protection personnel [13].

In addition to classifiers, potential devices that can be secured using the methods described in this article include Dorr thickeners and flotation machines. Both devices work in an electrically conductive environment.

References

[1] Łuszczkiewicz A, Drzymala J, Henc T, Konopacka Ż and Duchnowska M 2011 *Określenie wpływu związków chemicznych zawartych w wodach przemysłowych na process wzbogacania w O/ZWR* Sprawozdanie z badań: raport **i-11**/ (Wrocław) p 42

[2] Drzymała J 2001 *Podstawy Mineralurgii* (Wrocław: Oficyna Wydawnicza Politechniki Wrocławskiej)

[3] Wang B and Peng Y 2014 The effect of saline water on mineral flotation–a critical review *Miner. Eng.* **66-68** pp 13-24

[4] Quinn J J, Kracht W, Gomez C O, Gagnon C and Finch J A 2007 Comparing the effect of salts and frother (MIBC) on gas dispersion and froth properties *Miner. Eng.* **20** pp 1296–1302

[5] Księżniak K 2017 *Wzbogacalność urobku w KGHM Polska Miedź S.A. – rozprawa doktorska* (Gdańsk)

[6] Żmudziński K and Lekki J 1967 Badania nad wpływem składu wód kopalnianych w legnicko-głogowskim okręgu miedziowym na flotację rud miedzi *Sprawozdanie* **1216/67**(Gliwice: Zakład Przeróbki Rud – Instytut Metali Nieżelaznych)

[7] Karpisz A 2010 Wpływ parametrów fizykochemicznych wody technologicznej na flotowalność

rud miedzi *Praca dyplomowa* (Wrocław: Wydział Geoinżynierii, Górnictwa i Geologii Politechniki Wrocławskiej)

[8] Darowicki K 2015 *Określenie możliwości zastosowania nowoczesnych metod ochrony antykorozyjnej obiektów technicznych i technologicznych w O/ZWR* (Gdańsk)

[9] Cathodic protection design. Recommended practice RP-B401 2017 (DET NORSKE VERITAS DNVGL)

[10] Czermiński J (red.) 1986 *Ochrona przed korozją. Poradnik* (Warszawa: Wydawnictwo Komunikacji i Łączności)

[11] Sokólski W *Nierozwiązane problemy ochrony katodowej* (Gdańsk: SPZP CORRPOL, Polski Komitet Elektrochemicznej Ochrony przed Korozją SEP)

[12] Sokólski W 2003 Norma EN 12954 – nowe kryterium ochrony katodowej? Ochrona przed korozją **8/2003**

[13] Peabody A W 2007 *Peabody's Control of Pipeline Corrosion* (Houston, Texas: NACE)

IOP Conf. Series: Materials Science and Engineering **427** (2018) 012033 doi:10.1088/1757-899X/427/1/012033

Analysis of the working chamber size influence on the effectiveness of grinding in electromagnetic mill

D Krawczykowski[1], D Foszcz[1], S Ogonowski[2], T Gawenda[1], M Wołosiewicz-Głąb[1]

[1] AGH University of Science and Technology, Faculty of Mining and Geoengineering, Department of Environmental Engineering and Mineral Processing, Mickiewicza 30 Av., 30-059 Krakow, Poland

[2] Silesian University of Technology, Institute of Automatic Control, Akademicka 16, 44-100 Gliwice, Poland

foszcz@agh.edu.pl

Abstract. The paper presents study on the working chamber size influence on the technological and economic effectiveness of mineral material grinding in electromagnetic mill. Performance at desired throughput and technological effectiveness is the key issue for the processing apparatus application in the specific technological circuits. The problem of process effectiveness assessment for the specific mill's construction is important for the grinding circuit design and the process operating points' selection. It also allows for the proper control system design and the process optimization according to the specific criteria function. The research with electromagnetic mill was performed in order to assess the influence of the physical parameters of the mill's working chamber on the grinding process effectiveness. Quality factor was based on the chosen grain size fraction increase in the final product. Experiments were performed on electromagnetic mills with the working chamber diameter 200 mm and 320 mm respectively.

1. Introduction

Technological operations of mineral processing determine the efficiency of mining companies' production lines. The processing operations play an important role in shaping the properties of the processed material which finally becomes a commercial product. Mineral materials, depending on their final technological use, must be characterized by properties and parameters that meet the expectations of the recipient. In that regard, the most important material properties are: grain size, shape of grains, surface features, impurities content, oversize and undersize share, etc. From the final industrial application point of view, the geometrical properties of grains like shape, volume, surface area are of key importance [1]. These properties are shaped for a given raw material as a result of appropriate processing operations within the particular technological system.

The growing demand for raw materials from primary and secondary deposits, in particular those classified as strategic or critical, requires the appropriate selection of technological processes in terms of increasing their efficiency and effectiveness. Innovative solutions in the field of construction and processing technologies lead to development of modern solutions allowing to produce the products of desired quality taking into account a number of aspects, including economical or environmental ones. These solutions are particularly important due to the post-production waste management.

The scattering and depletion of natural resources in terms of the content of raw materials desirable for the economy, with the rapidly growing demand forced by the development of technology and consumer demand, forces the search for modern solutions in the technology and technology of processing raw materials and waste. Solutions in this area focus on adapting process engineering operations to the complex characteristics of mineral raw materials, ie their physical, physicochemical and chemical properties. The raw material situation and technological needs as well as economic considerations inspire the continuous development of machinery and technological equipment construction. This development has a strong economic basis due to the energy consumption of processing processes. Directions of development, in addition to increasing the efficiency of machines, are based also on appropriate adaptation of technologies and devices to the properties of raw materials, which enables reduction of costs in relation to the volume of its processing. For the effective conduct of the process of transformation, it is necessary to know the characteristics and factors affecting the course of a given process. Full identification of processes along with the geometallurgical model of the deposit enables appropriate control and optimization of the technological process at every stage from extraction to metallurgical processes for metal ores enabling the introduction of a production management system in industrial conditions. Such activities are necessary for the full exploitation of exploited deposits by mining enterprises, as they will help to minimize the unavoidable losses of a useful component in waste.

One of the most energy-intensive processes is the process of grinding in particular when it concerns the needs in terms of obtaining fine graining of the product by grinding. Depending on the raw material, the purpose of the grinding operation is different. In the case of ores and raw materials requiring enrichment processes, grinding is aimed at the liberation of ore minerals. The course of the enrichment process depends on the degree of release of useful minerals. They are carried out so as to achieve an appropriate degree of liberation, and the granulation is determined by the dimensions of the mineral grains, the more precisely the useful minerals are liberated, the easier it is to enrich the given feed. The process of mineral liberation in grinding operations depends, among other things, on the properties of rocks and the minerals forming those rocks, and on the nature of the grinding process. Preparation of the feed for enrichment is a complex system of comminution (crushing, grinding) and classification with dizziness. The basic operation is grinding which, depending on the conditions and the adopted technological assumptions, can take place in drum mills (e.g. rod, ball or cylpebs), semi-autogenous (SAG) and autogenous (AG) [2-4]. At the subsequent stages of milling, vertical, vibrating and other rotor mills are used. However, bringing the material (feed, ore, rock raw material) to the grain size below 0.1 mm is very energy intensive. The amount of energy required for proper fragmentation depends mainly on the mechanical properties of the mineral raw material, but also on the type of mills used and the technological rules. It is assumed that, for example, drum gravity mills absorb up to 50% of the energy costs of the processing plant. An electromagnetic mill, due to its capabilities in the field of grinding raw materials is a device with a high potential for effective grinding of raw materials at low energy costs. An important element of the introduction of an electromagnetic mill to the technological system is to determine the parameters of the feed determining the efficiency of its operation. This means that it is necessary to determine the efficiency of the work of mills on the first stage working in the systems of preparing the raw material for enrichment or for further technological operations [3,4].

The enrichment of non-ferrous ores in the flotation process is not an easy matter. It requires certain conditions that guarantee obtaining high-quality flotation concentrates. One of the basic factors that are of key importance for the effectiveness of this process is the grain characteristics of the feed. The appropriate graining of the feedstock is achieved thanks to the use of mills in classification and fragmentation nodes. However, conventional grinding equipment (ball mills, rod mills, cylpebs mills) are characterized by high energy consumption. The reduction of process costs can be achieved by introducing into the technological systems electromagnetic mills, characterized by the reduction of energy consumption to a large extent, which is their undeniable advantage. The key element, however, is the surface condition of the mineral grain and its release from the gangue. Verification of this state takes place during the beneficiation process, e.g. flotation and evaluation of the ore beneficiation characteristics.

2. Features and advantages of electromagnetic mill

The results of laboratory tests on Polish copper ore showed that the electromagnetic mill allows obtaining the expected grain size of the feed directed to the flotation enrichment nodes in processing plants. An important problem associated with the main process of separation of minerals from gangue is the size of the ore grains. This factor forces the feed to the appropriate grain size, most often the grain size is below 0.1 mm, although in some cases this limit decreases even to 10 μm. Tests of grinding media dedicated to electromagnetic mills showed that the best results were achieved for 10x1 mm stainless steel grinding media. Obtaining copper ore with a granulation of 0-45 μm was possible after almost 25 seconds of the grinding process. During the extension of the grinding time of the material to 30 seconds in an electromagnetic mill, it was possible to obtain a relative increase in the 0-0.02 mm class at a level of approximately 80%. It can be concluded that the use of smaller size grinding media provided faster growth of the studied grain grades in milling products [5,6,7-9].

Obtaining the right graining determines the technological efficiency of a given system and devices included in its composition. This makes it necessary to select the right machines and their system in order to obtain the appropriate technological effect while maintaining the cost minimization criterion. The key is to ensure high efficiency for high cost operations, i.e. grinding. The selection of the mill and its technical parameters determines the degree of fragmentation to be obtained and the efficiency is strictly dependent on the type of feed. When considering a particular material, the final grinding efficiency is also affected by the moisture content of the feed and its initial particle size distribution. For this reason, direct comparison of different types of mills is difficult and their performance is difficult to rescale. It is worth to pay attention to such parameters as the efficiency and energy consumption of different types of grinding installations, since knowledge of those parameters is necessary to pre-assess the usefulness of a mill for specific applications.

The ball mill belongs to the group of gravitational mills, in which balls of various diameters are used as the grinding media. Grinding media are made of high density materials and low abrasiveness, the most commonly used grinding media are steel and ceramic grinding media. Ball mills are currently the basic devices used in technological processing systems of ores, hence they can be a reference element in the case of assessment of new technical solutions in the field of grinding. The main disadvantages of the mill include its low energy efficiency, which results in ongoing work on the development of new solutions [5,10]. One of the examples of such works may be an electromagnetic mill developed as part of the PBS project financed by NCBiR - SYSMEL [6,11]. Quality of the mineral processing products depends highly also on the quality of the process control [12]. For this reason, research on measurement and control systems for electromagnetic mill were also important part of the SYSMEL project. Models for the soft-sensing measurements and control purposes were identified [13,14] and dedicated control structures and algorithms were proposed [15,16].

The electromagnetic mill is a device in which the mill's ferromagnetic grinding media are moved by the purposely generated rotating electromagnetic field as a carrier of energy. The mill is made up of the exciter of the rotating magnetic field and the tube placed in its axis, constituting the working chamber [6,11]. The work chamber is filled in about 20% with ferromagnetic grinding media. The effectiveness of the grinding process depends on the proper selection of the parameters of both the work coil and the working chamber [17,18,19]. The solution allowing for significant intensification of many technological processes is leading them in the works space in which the small ferromagnetic rods (grinding media with a properly selected ratio of length to diameter) move due to the action of the rotational electromagnetic field [18]. Devices with electromagnetic rotation of a very large number of grinding media, allow to intensify a number of technological processes due to the simultaneous and comprehensive impact on the load treated in them, many physical variables of force fields, including:

electric, magnetic, acoustic (also in the field of ultrasounds), thermal, high pressure and friction [20]. The electromagnetic mill has been a little-known grinding device used in laboratories rather than in industry. This limitation of applications resulted from the very low efficiency and low efficiency of this type of mills. The low technical and operational parameters were influenced by improper design assumptions leading to very large losses in the rotating magnetic field coil. The change in the design assumptions allowed for the construction of a prototype electromagnetic mill at the Częstochowa University of Technology [18]. The research on the milling process in this mill has shown its high efficiency as manifested by short several-second grinding time for one portion of material and low particle size distribution including colloidal one. The industrial use of this type of equipment will be determined by the possibility of building a high efficiency mill (several dozen Mg per hour) [18,19]. The possibilities of such a highly efficient mill depend primarily on the properly designed magnetic exciter of the rotating field. The mill works efficiently in terms of both high and low humidity of the feed. Material with humidity in the range of 15% -30%, especially if it has a high viscosity, reduces the grinding efficiency by about 30%.

The basic condition for the use of an industrial electromagnetic mill is the possibility of its continuous operation. Grinding media should match grain size of granular material. In the case of a large variation in grain size, a mix of grinding media of various sizes is used [21]. The possible size changes of the grinding media match the physical properties of the ground substance and the final graining of the milling product. The electromagnetic mill technology is particularly useful in the case of ultra fine milling, with grain sizes beyond the reach of conventional solutions, or when obtaining such a small particle size distribution would require disproportionate energy expenditure [7]. The competition may be the IsaMill mill developed by Xstrata Technology. It is a mill used for very fine grinding of ores for the graining of products with a size of several or several micrometers. The mill is characterized by a relatively low energy consumption compared to ball mills, even 10 times smaller when crushed to a size below 10 micrometers (figure 1).

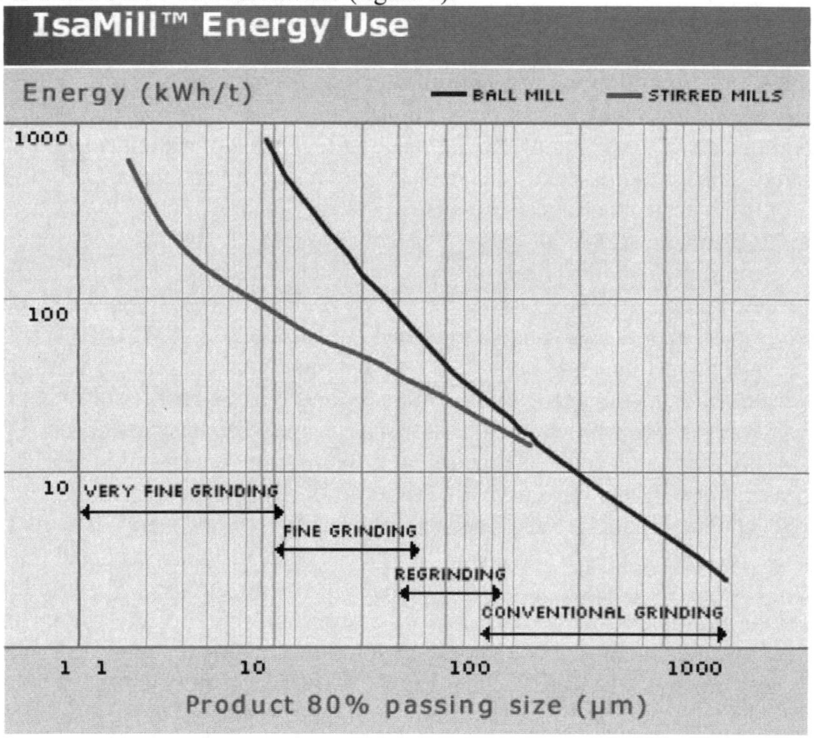

Figure 1. Energy consumption for ball mill and IsaMill [22].

Consideration of the energy consumption of the grinding process in various mills and technological systems allows to determine the general demand for grinding energy depending on the grain size of the feed as well as the desired products. According to figures 1, 2 and 3, the finer the product we want to achieve, the more energy should be put into its fragmentation, the optimal grinding can be obtained by increasing the grinding time, selecting proper grinding media physical parameters, reducing particle size (introducing the previous crushing or grinding device) or introducing additional milling stage or changing the type of mill (figure 3).

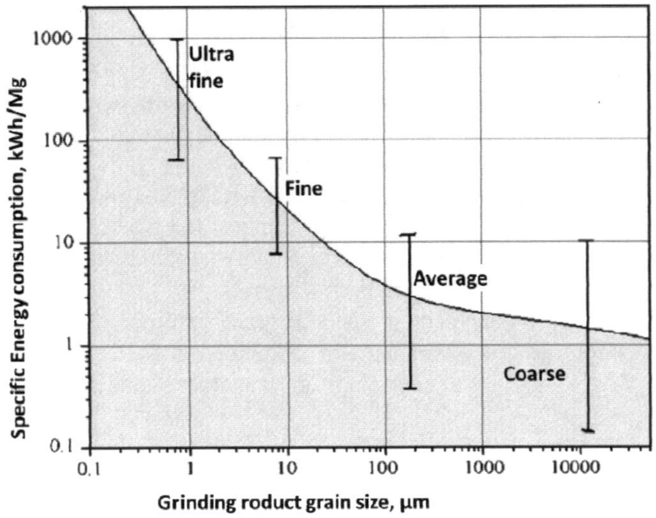

Figure 2. Relation between energy consumption and product grains size [23].

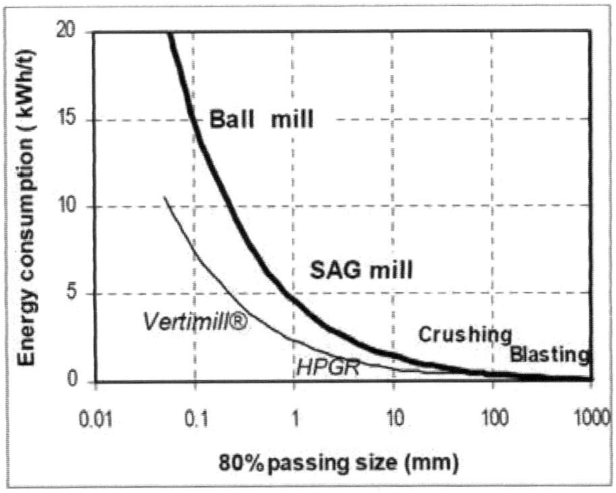

Figure 3. Energy consumption comparison for different grinding devices [24].

Comparative laboratory tests show that in the case of an electromagnetic mill, its efficiency is about 28% higher in relation to a ball mill, but in the range of a fine feed of less than 1 or 2 mm. This means that under industrial conditions an electromagnetic mill could bring tangible benefits in the case of raw material grading at the final stages of comminution. The advantage of this mill is also the short grinding time of the material associated with the short residence time of the material in the mill, which, using a recycle classification system, will selectively separate the product capable of flotation, protecting the metal components from being milled [8, 9].

An important element is to determine the efficiency of grinding depending on the load of the electromagnetic mill with ground raw material and the possibilities of increasing the capacity of the working chamber of the electromagnetic mill.

3. Analysis of grinding process in electromagnetic mills with different working chambers

Research in the evaluation of grinding efficiency depending on the mill load was carried out in an electromagnetic mill with a diameter of 200 mm (D200). In terms of increasing the working chamber, grinding tests were carried out in a mill with a diameter of 320 mm (D320) along with reference tests in a mill with a diameter of 200 mm. The material for testing was copper ore from KGHM PM S.A.

3.1. Influence of the material load on D200 mill effectiveness

In order to determine the possibilities of loading the mill with ground material, tests were carried out on a mill with a diameter of 200 mm with different loading of the quantity of ground material - 1 kg, 2 kg, 2.5 kg and 3 kg with solid filling with grinding media weighing 1.5 kg and size used grinding media with a diameter of 1 mm and a length of 10 mm.

The obtained test results were analyzed for specific grain size increments in the grinding product after 15, 25 and 35 seconds for grinding 1 kg and 2 kg feed and 15 and 25 seconds for grinding material weighing 2.5 and 3 kg. Increments of grain class were analyzed: 0.045 mm; 0.071 mm; 0.1 mm; 0.2 mm.

Analysis of the grain class growths after time made it possible to determine the milling speed of the electromagnetic mill, which allowed to determine the time and load of the mill with the material after which for the given grain class there is an end to the possibility of further growth of this class due to the decline of the mill's efficiency. In figure 4, the increments of analyzed particle size distributions for the mill are summarized for each of the samples. The analysis was based on relative increments as the value of the difference in the content of a given grinding product class in relation to the potential maximum value of this class increase resulting from the difference: 100% - the contents of this class expressed as a percentage in the feed:

$$class\ relative\ increase = 100 * \frac{\phi_{class\ in\ product} - \phi_{class\ in\ feed}}{100 - \phi_{class\ in\ feed}}\ \% \tag{1}$$

The relative increase analysis makes it possible to assess the end of the mill's potential in terms of further growth of the analysed particle size fraction. Comparison of the work of a mill filled with specific grinding media through relative growth makes it easier to evaluate the efficiency of grinding - achieving 100% of the increment of the analysed grain class means high efficiency of the mill in the grinding of a given grain class. An important element of the analysis is also the time in which the high increment values appear and the state of saturation (lack of increment) occurs, which in case of increments with values lower than 100%, sets the limit of the mill capacity in terms of further material fragmentation.

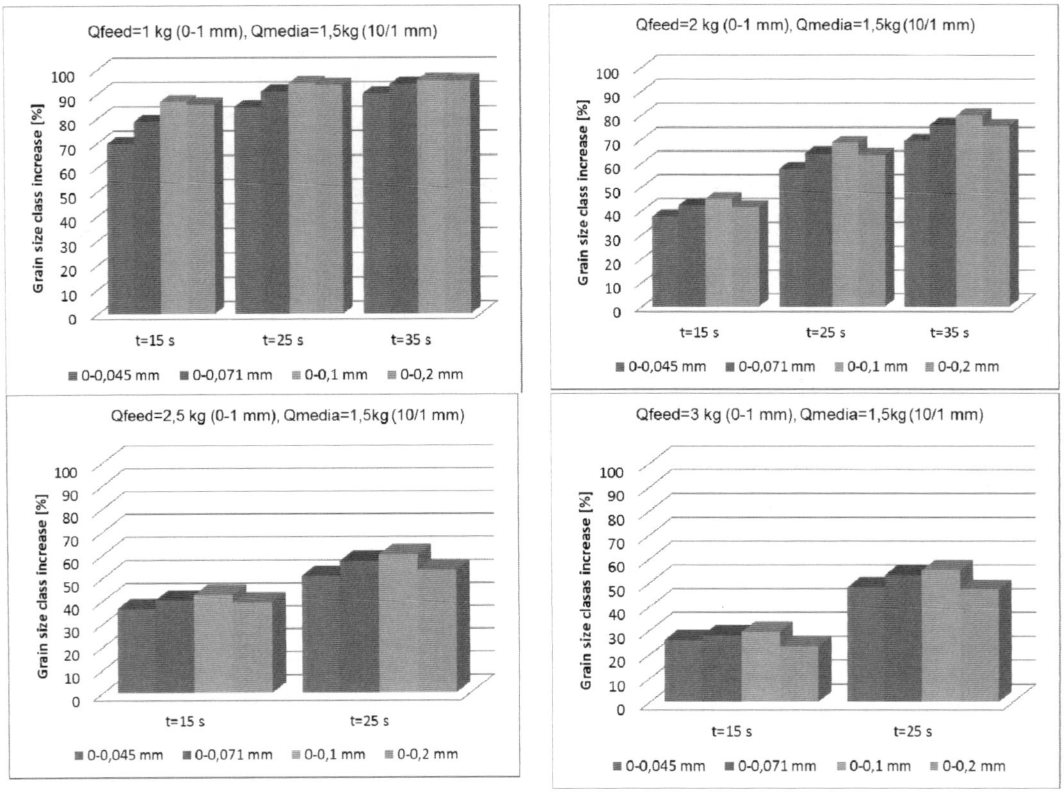

Figure 4. Grain size class increase for different processing time and material load in EM.

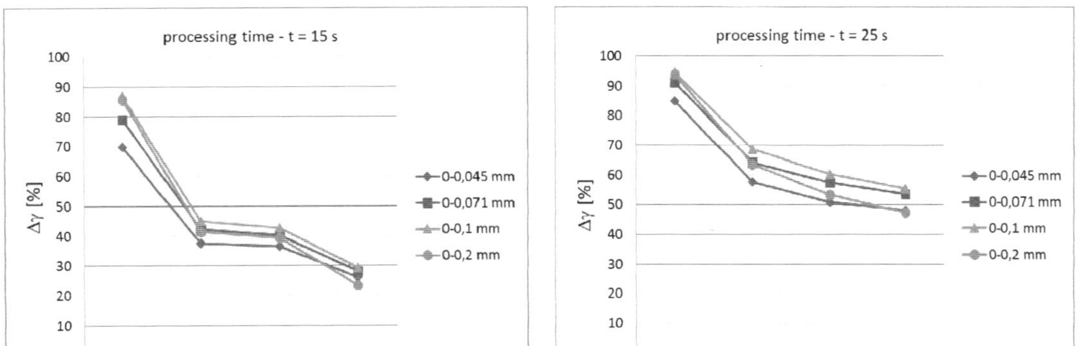

a) b)

Figure 5. Grinding effectiveness in function of mill load for different processing time: a – 15 seconds, b – 25 seconds.

3.2. Grinding effectiveness in mills D200 and D320

In order to assess the efficiency of the electromagnetic mill with increasing dimensions of its working chamber, a sample of copper ore with particle size 0-1 mm was tested, with a higher content of particles above 0.5 mm (feed 1), and with a lower content in the mill feed grade above 0.5 mm (feed 2).

For this purpose, copper ore was prepared in the grinding and classification process and grinding in a rod mill and then the raw material prepared in this way was divided into a sample with 1 kilogram weight for grinding in a mill D200 and 6.5 kg for grinding in a mill D320. The mass of the samples resulted from the calculation of the load on the mill D200 and D320 with such masses of material that gave the same level of filling the chamber with the material. Grinding tests were carried out at 10, 15, 20 seconds - the choice of time resulted from previous work optimization experiments carried out on the D200 mill. The results of sieve analysis of the feed and the grinding product is presented on figures 6-9. For easier analysis the results were grouped according to the feed type and the mill size. Feed 1 and 2 differs by a larger and smaller share of the class above 0.5 mm.

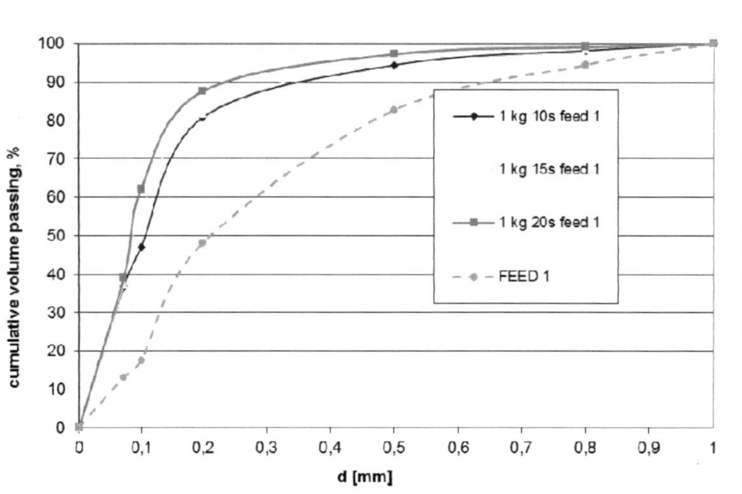

Figure 6. Cumulated graining curves for feed 1 grinding process in D200 mill.

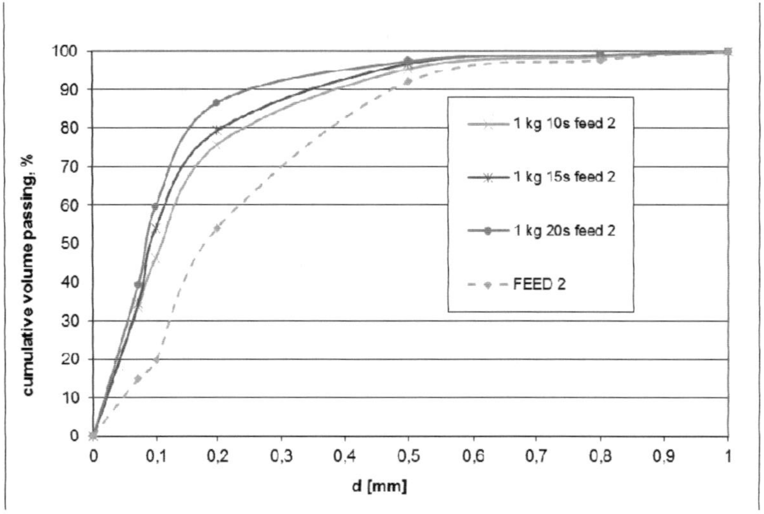

Figure 7. Cumulated graining curves for feed 2 grinding process in D200 mill.

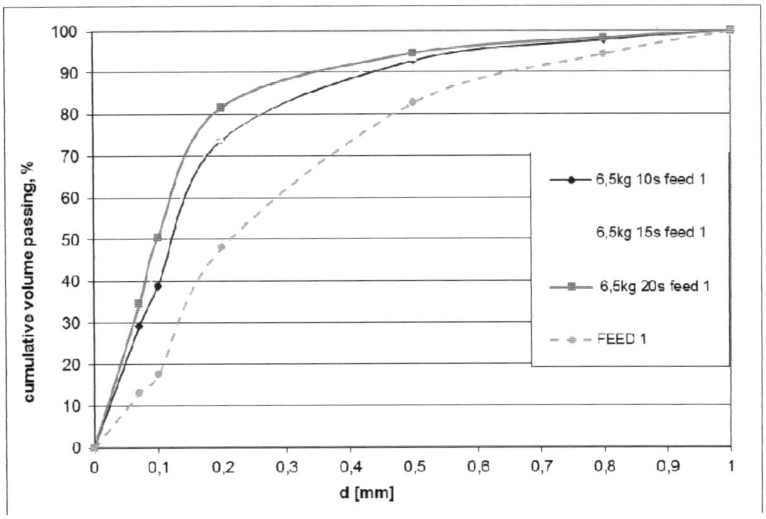

Figure 8. Cumulated graining curves for feed 1 grinding process in D320 mill.

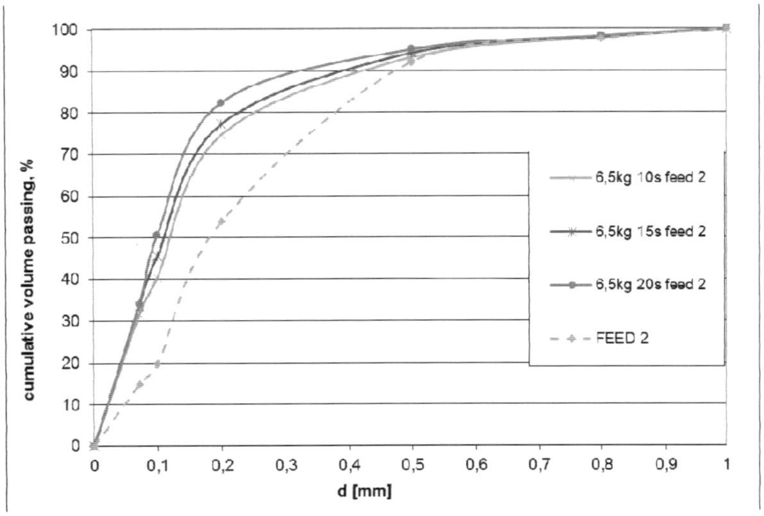

Figure 9. Cumulated graining curves for feed 2 grinding process in D320 mill.

The grinding efficiency was measured by the relative increase of selected grain classes. The analyzed grain classes are 0-0.071 mm, 0-0.1 mm and 0-0.2 mm. The results of the growth along with the grinding time of the analyzed grain classes for grinding ore in mill D320 are shown in figures 10 and 11 for the mill D200 in figures 12 and 13 and collectively for both mills for the 0-0.2 mm class in figure 14.

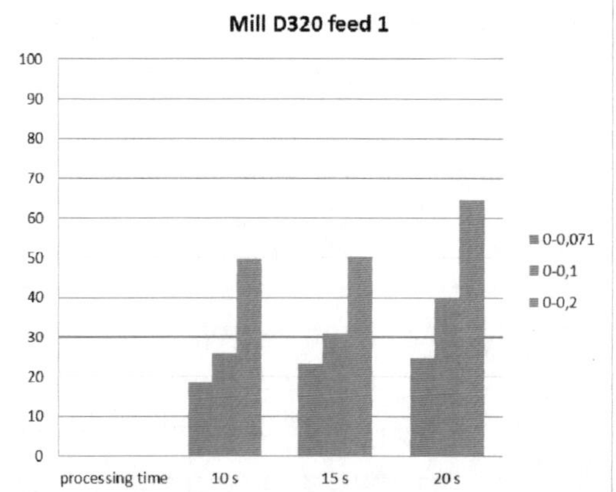

Figure 10. Kinetics of the analyzed grain size class increase in grinding product for mill D320 and feed 1.

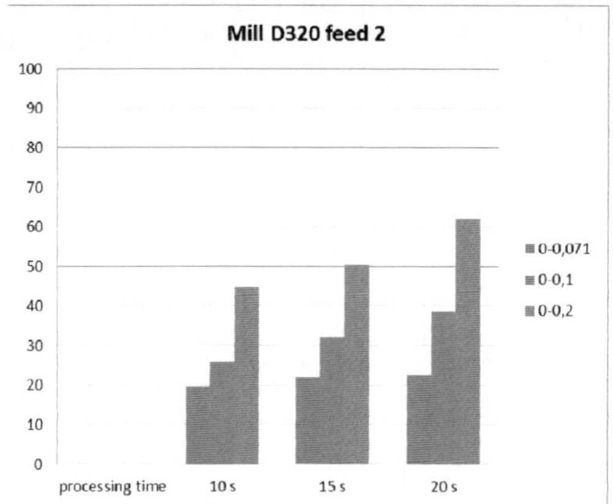

Figure 11. Kinetics of the analyzed grain size class increase in grinding product for mill D320 and feed 2.

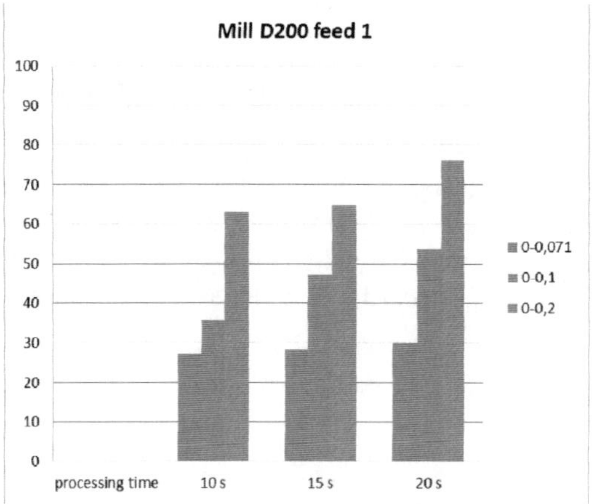

Figure 12. Kinetics of the analyzed grain size class increase in grinding product for mill D200 and feed 1.

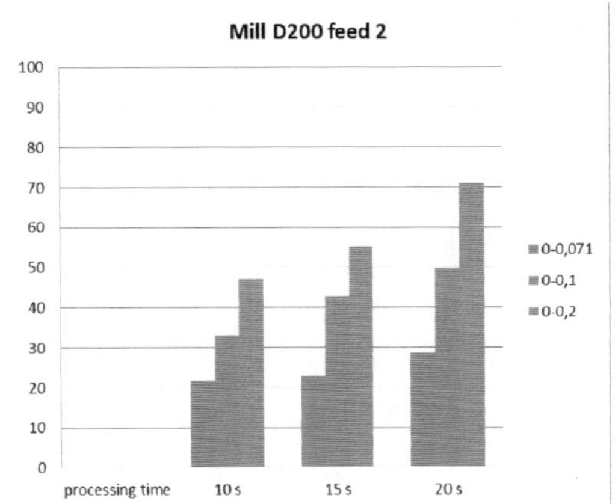

Figure 13. Kinetics of the analyzed grain size class increase in grinding product for mill D200 and feed 2.

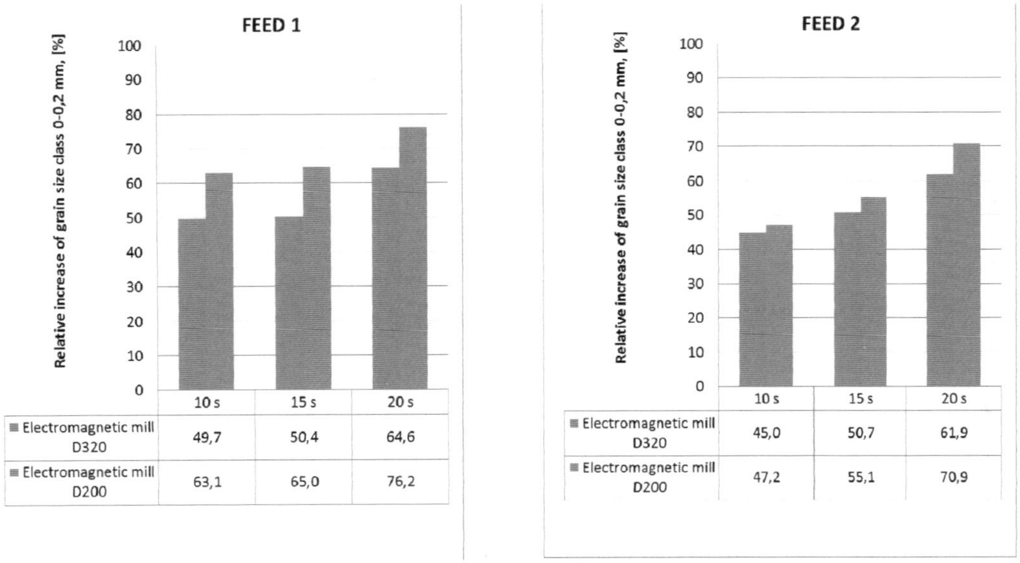

Figure 14. Comparison of kinetics of grain size class 0-0.2 mm increase in product of mills D320 and D200, for feed 1 and 2

The results clearly indicate that the mill with a smaller diameter D200 in the tested mass load range in a volume unit (the same for both mills) worked more efficiently. The relative increase of the analyzed grain classes in relation to the larger D320 mill was several percent (about 15%) higher for a coarse feed and a few percent for a feed with a smaller class content of more than 0.5 mm. This may indicate that the larger mill, due to the obtained kinetic energy, requires finer feed for grinding.

Analyzing the unit energy consumption with respect to the possibility of individual loading of both mills, we obtain values representing the material grinding time, respectively (table 1). The mill with a larger diameter has a lower energy consumption per ton of raw material by about 8% in relation to the smaller mill.

However, if one looks closer to particular grain size classes, the difference is opposite in some cases. The same energy index calculated for the production of 1 tone of the finest 0,071 mm grain size class is higher for the bigger mill (table 2). Authors must stress that such results were obtained in the batch experiments. Such approach, on the one hand, allows for the precise repetition of the experiments' conditions. On the other hand, when such short processing times are considered, different start-up dynamics of D200 and D320 mills influence the results. One can notice that effect, while compering results for feed 1 – the difference is about 35% for 10 seconds of processing time and only about 10% for 15 and 20 seconds. Above results requires further analysis of mill scaling effect in continuous processing experimentation, that will neglect start-up effect.

Table 1. Comparison of unit energy consumption for D200 and D320 in relations to the grinding time.

Processing time (s)	D200 load (Mg/h)	D320 load (Mg/h)	D200 unit power consumption (kW/h)	D320 unit power consumption (kW/h)	D200 unit energy consumption (kWh/Mg)	D320 unit energy consumption (kWh/Mg)
10	0.36	2.34	22	132	61.1	56.4
15	0.24	1.56	22	132	91.7	84.6
20	0.18	1.17	22	132	122.2	112.8

Table 2. Comparison of unit energy consumption for D200 and D320 for 0,071 mm grain size class.

			Feed 1		Feed 2	
Processing time (s)	D200 load (Mg/h)	D320 load (Mg/h)	D200 unit energy consumption (kWh/Mg)	D320 unit energy consumption (kWh/Mg)	D200 unit energy consumption (kWh/Mg)	D320 unit energy consumption (kWh/Mg)
10	0.36	2.34	258.9	352.4	326.2	339.3
15	0.24	1.56	376.5	413.9	470.0	449.2
20	0.18	1.17	469.3	527.8	500.5	590.7

4. Conclusions

Studies on the efficiency of grinding process in electromagnetic mills with different diameters of 200 and 320 mm were carried out in the Eltraf laboratory in Lubliniec. Analysis of the impact of mill load with mass of ground material indicates that the mill with a diameter of 200 mm works most efficiently with a load of 1 kg of material (relative growth of the studied grain classes in the grinding product after 15 seconds was from 70 to 85% and after 35 seconds from about 85 up to 95% - fig. 5). Increasing the mass of ground material to 2 kg (or two times) causes a rapid drop by nearly half of the relative growth value of a given grain class for the grinding time of 15 seconds and about 30% for a grinding time of 35 seconds. Further increase of the material mass already causes a slight decrease in efficiency, which means that it is very beneficial to introduce the work of this mill in a closed system with the dizziness of unmilled material. Such operation may result in an increase in the efficiency of the mill, however, it requires appropriate mill control systems to ensure adequate efficiency of the entire grinding system.

Analysis of the obtained results allows to conclude that increasing the size of the mill has a positive effect on the mill's efficiency increase of about 6.5 times and a reduction of energy consumption by about 8% with a slight decrease in grinding efficiency measured by the increase of fine grain sizes in the milling product. Further research should focus on the verification of the maximum feed grain directed to the D320 mill due to the kinetic energy of grinding media and thus the ability to grind larger size granules.

Acknowledgment

The research reported in this paper was co-financed by statutory activities AGH no. 11.11.100.276 and grant of Institute of Automatic Control, Silesian University of Technology no. 02/010/BK17/0060. The research was performed on the installations co-financed by the National Centre for Research and Development, Poland, under Applied Research Program, project no. PBS3/B3/28/2015.

References

[1] Gawenda T 2011 Produkcja Surowców Mineralnych O Wąskim Zakresie Uziarnienia w Dwustadialnym Zamkniętym Układzie Rozdrabniania i Klasyfikacji *Górnictwo i Geologia* **6 (2)** pp 26-9

[2] Saramak D, Tumidajski T, Brożek B, Gawenda T and Naziemiec Z 2010 Aspekty projektowania układów rozdrabniania w przeróbce surowców mineralnych *Gospodarka Surowcami Mineralnymi* **26 (4)** pp 59–69

[3] Tumidajski T, Kasińska-Pilut E, Gawenda T, Naziemiec Z and Pilut R 2010 Badania energochłonności procesu mielenia oraz podatności na rozdrabnianie składników litologicznych polskich rud miedz *Gospodarka Surowcami Mineralnymi*, **26 (1)** pp 61–72

[4] Fuerstenau M C and Han K N 2003 Principle of Mineral Processing *Society for Mining, Metallurgy and Exploration*, Inc.: Englewood, CO, USA, pp. 92–171

[5] Wolosiewicz-Glab M, Foszcz D and Gawenda T 2016 Analysis of possibilities of obtaining the fine particle size in mills of various designs *Journal of Polish Mineral Engineering Society* **17 (1)** pp 223-31

[6] Wolosiewicz-Glab M, Foszcz D, Gawenda T and Ogonowski S 2016 Role of classification in grinding using the electromagnetic mill: a case study *E3S Web of Conferences* 8(01065) pp 1-8

[7] Ogonowski S, Ogonowski Z, and Swierzy, M 2017. Power optimizing control of grinding process in electromagnetic mill *Proceedings of the 2017 21st International Conference on Process Control, PC 2017*. Institute of Electrical and Electronics Engineers Inc., pp. 370–5

[8] Wolosiewicz-Glab M, Ogonowski S, Foszcz D, and Gawenda T 2018 Assessment of classification with variable air flow for inertial classifier in dry grinding circuit with electromagnetic mill using partition curves *Physicochemical Problems of Mineral Processing* **54 (2)** pp 440–7

[9] Wolosiewicz-Glab M, Foszcz D, Saramak D, Gawenda T and Krawczykowski D 2017 Analysis of a grinding efficiency in the electromagnetic mill for variable process and feed parameters *E3S Web of Conferences*, **18** p 01012 pp 1–6

[10] Naziemiec Z and Saramak D 2012 Analiza energochłonności procesów rozdrabniania kruszyw mineralnych *Prace Naukowe Instytutu Górnictwa Politechniki Wrocławskiej* **134 (41)** pp 209-220

[11] Wolosiewicz-Glab M, Ogonowski S, and Foszcz D 2016 Construction of the electromagnetic mill with the grinding system, classification of crushed minerals and the control system. *IFAC-PapersOnLine* **49** pp. 67–71

[12] Tatjewski P 2007 Advanced Control of Industrial Processes: Structures and Algorithms, Springer-Verlag, London, doi: 10.1007/978-1-84628-635-3.

[13] Ogonowski Z and Ogonowski S 2017 Estimation problems of pneumatic transport system for electromagnetic grinding, *22nd International Conference on Methods and Models in Automation and Robotics*, doi:10.1109/MMAR.2017.8046818

[14] Budzan S 2017 Automated grain extraction and classification by combining improved region growing segmentation and shape descriptors in electromagnetic mill classification system, Proc. SPIE 10696, Tenth International Conference on Machine Vision (ICMV 2017)

[15] Ogonowski S, Ogonowski Z, Swierzy M, and Pawelczyk M 2017 Control System of Electromagnetic Mill Load *25th International Conference on Systems Engineering* pp 69–76.

[16] Ogonowski S, Ogonowski Z, and Pawełczyk M 2018 Multi-Objective and Multi-Rate Control

IOP Conf. Series: Materials Science and Engineering **427** (2018) 012033 doi:10.1088/1757-899X/427/1/012033

of the Grinding and Classification Circuit with Electromagnetic Mill. *Applied Sciences* **8 (4),**

[17] Sławiński K, Knas K, Gandor M, Balt B, and Nowak W 2014 Młyn elektromagnetyczny i jego zastosowanie do mielenia i suszenia węgli *Energetyka. Piece przemysłowe i kotły I-II/2014*

[18] Sosiński R, Nowak W, Rolicz P, and Szczypiorowski A 2007 *Młyn* Elektromagnetyczny. Zgłoszenie patentowe w Urzędzie RP z dnia 08.06.2007r., Nr P.382610, Warszawa 2007.

[19] Wolosiewicz-Glab M, Foszcz D, Gawenda T, and Ogonowski S 2016 Design of an electromagnetic mill. Its technological and control system structures for dry milling *E3S Web of Conferences*, **8**

[20] http://www.eltraf.eu (30.05.2015)

[21] Ogonowski S, Wolosiewicz-Glab M, Ogonowski Z, Foszcz D, Pawełeczyk M, 2018 Comparison of Wet and Dry Grinding in Electromagnetic Mill *Minerals* **8 (4)**, 138,

[22] http:// www.isamill.com

[23] Wang Y and Forssberg E 2007 Enhancement of energy efficiency for mechanical production of fine and ultra-fine particles in comminution, *China Particuology* **5** pp 193-201

[24] Jankovic A, Valery W, Maloney K, and Markovic S 2006 Improving Overall Concentrator Performance with Stirred Milling, *IMPC 2006 in Istanbul*, Turkey

IOP Conf. Series: Materials Science and Engineering **427** (2018) 012034 doi:10.1088/1757-899X/427/1/012034

New technology used to prepare ore for flotation based on impact comminution technique

A Wieniewski, B Skorupska and M Osoba

Institute of Non-ferrous Metals, Sowińskiego 5, 44-100 Gliwice, Poland

bozenas@imn.gliwice.pl

Abstract. The paper presents a new design of a high-energy mill and describes a pilot installation equipped with a high-energy mill with an output of 3-7(8) Mg/h. The mill design was developed on the basis of a beater wheel mill used to pulverize coal. The potential of reaching good grinding results has been confirmed for the new mill. Achieved degrees of grinding for copper ore initially crushed in hammer and cone crushers range from over 80 to more than 150. The concept of a procedure based on a single-stage method of ore preparation for flotation process is discussed herein. The main elements of the technology are: preliminary flotation as an element of effective preparation of ore for the roughing flotation process, intensive roughing flotation, conventional cleaning circuit, separate processing of middlings with a grinding circuit involving the new type of mills and flotation with separation of tailings as the second component of the final waste.

1. Introduction

Ore enrichment using the flotation method requires proper comminution, which is usually performed in multi-stage processes of crushing followed by grinding. Grinding is dominated by tumbling mills filled with grinding media or autogenous and semi-autogenous mills. Mills with grinding media are widespread in mineral raw materials processing. At KGHM, grinding is based on mills of this type, as well. Comminution is effected in them by grinding media movement forced by a rotating mill shell. Abrasion is the prevailing grinding method, available in each of possible systems of tumbling mill operation.

Currently, at KGHM research is conducting on applying the new method to prepare ore for flotation using a high-energy mill. This type of mill belongs to impact mills, which use kinetic energy of a beater wheel in the grinding process. It is a new structure, developed on the basis of a beater wheel mill used to pulverize coal.

The examined option using a high-energy mill is planned to replace the system used until now, which involves two-stage grinding in tumbling mills in a wet process, with a single-stage grinding executed by a beater wheel mill in a dry process.

The paper presents the results of grinding tests in a beater wheel mill and a proposed technological solution for the process involving preparation of ore for flotation and enrichment based on the new equipment.

2. Current method used to prepare ore for flotation in the KGHM enrichment plants

The technology used to prepare ore for enrichment constitutes an important element of mineral raw materials processing. It determines the effectiveness of usable minerals recovery resulting from the

IOP Conf. Series: Materials Science and Engineering **427** (2018) 012034 doi:10.1088/1757-899X/427/1/012034

adequate degree of liberation of minerals being enriched from waste rock, and generally affects concentrate production costs.

Currently, the processes involving preparation of ore for enrichment (flotation) include crushing, screening, grinding in rod and ball mills, as well as hydraulic classification in spiral classifiers and hydrocyclones. Power consumption for comminution and classification processes in the case of KGHM PM SA is ca. 15 kWh/Mg of ore. Therefore, crushing, grinding and classification costs range from 40 to 50% of concentrate production costs.

The product of the system used to prepare ore for enrichment is the flotation feed – its particle size distribution is adapted to the size of copper sulphide grains and specific floatability characteristics [1].

Crushing and screening processes are dry, while grinding with classification is carried out in water environment.

In all three plants, preparation of ore for flotation includes crushing operation in hammer crushers to reach grain size below 25-40 mm, and then screening [3]. Only in Lubin Area, preparation of ore for flotation includes the operation of the 2nd stage of crushing in a cone crusher to which the oversize of screen (class over 20 mm) classifying the crushing product of a hammer crusher is subjected.

Therefore, in terms of preliminary comminution, ore preparation includes:

- ore screening in high-yield vibration screens (sieve mesh 25 mm),
- crushing of coarse screening product in hammer crushers.

Worldwide, graining of ore constituting the feed material for mills usually remains under 15-20 mm, sometimes even 5 mm. In Polish plants, the volume of class exceeding 15 mm, which is sent to the mills, is high and amounts to:

- for Lubin Area: 13-15%,
- for Rudna Area: 22-25%,
- for Polkowice Area: 15-20%.

The new technology used to prepare ore for enrichment is completely different from the solutions used so far in the processing of non-ferrous metal ores, and involves using a high-energy mill in the dry grinding process. The mill is characterized *inter alia* by high kinetic energy of beater wheel working tools [2]. The comminution process is a result of repeated hitting of grains by these tools and ground grains bouncing off the mill housing. The linear speed of the main working tool (beater) exceeds 100 m/s, which guarantees high grinding efficiency. An example solution of a high-speed mill is shown in figure 1.

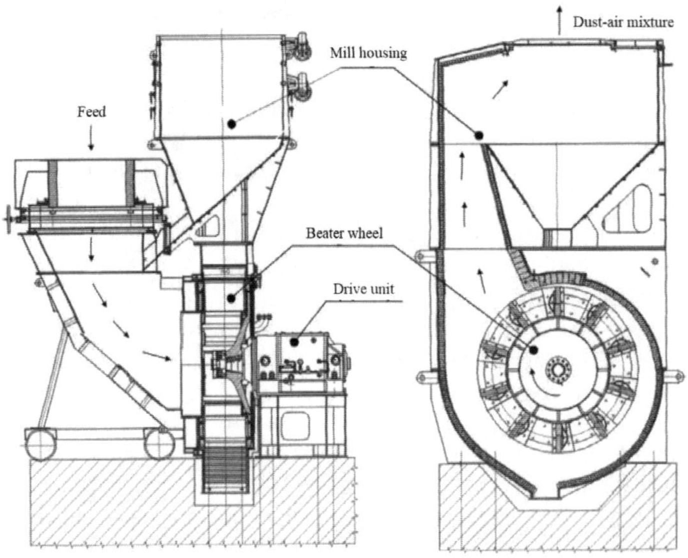

Figure 1. Diagram of a high-energy mill.

The basic components of a high-energy mill with a horizontal beater wheel axis are:

- a two-stage beater wheel, outrigger type, with outside working diameter of ca. 1 m, strengthened for the purpose of grinding non-ferrous metal ores,
- strengthened working chamber walls,
- covering the mill working elements with special materials characterised by high tribological resistance,
- gearless drive using an asynchronous (induction) motor with a thyristor system executing max. speed of the outside beater rim: 150-180 m/s,
- choking the flow through the inlet to the working chamber with a flap,
- aerodynamic emptying of the mill working chamber.

3. Pilot installation equipped with a beater wheel mill

A pilot installation equipped with a high-energy mill has been built in the grinding plant in the Lubin Area (figure 3). It should be stressed that this type of mill for dry impact ore grinding is a prototype design intended for testing purposes.

The installation includes technological circuit from crushed ore container (grain size below 25 (45) mm) to the conditioner preparing pulp for the enrichment process in flotation machines. Schematic diagram of a pilot installation preparing ore for flotation using a high-energy beater wheel mill is shown in figure 2, while in figure 4 the view of the installation in the technology hall is presented.

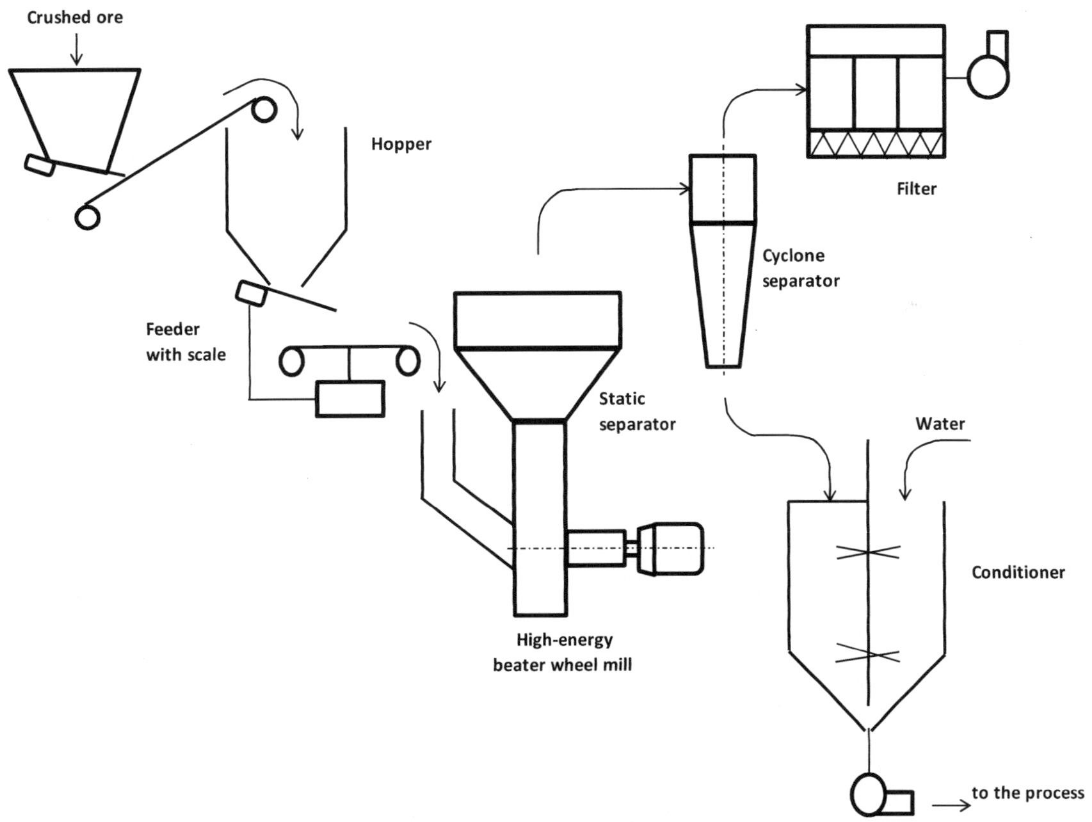

Figure 2. Schematic diagram of the pilot installation.

Copper ore is delivered to the container, and then, through a belt conveyor equipped with a weighing system, it is transported to the feeding opening in the beater wheel mill. The weighing system with a feeder allows for the adjustment of the delivered ore volume between 3 Mg/h and

IOP Conf. Series: Materials Science and Engineering **427** (2018) 012034 doi:10.1088/1757-899X/427/1/012034

10 Mg/h. Ore ground to grain size below 1(0.5) mm is sent via a cyclone separator to the conditioner, which prepares the suspension of ground ore and water for flotation. The weight of 1 dm^3 of the suspension is ca. 1200 g (max. 1400 g).

In order to guarantee the proper air purity, the coarse product of the cyclone separator is sent to the ejector through a filter.

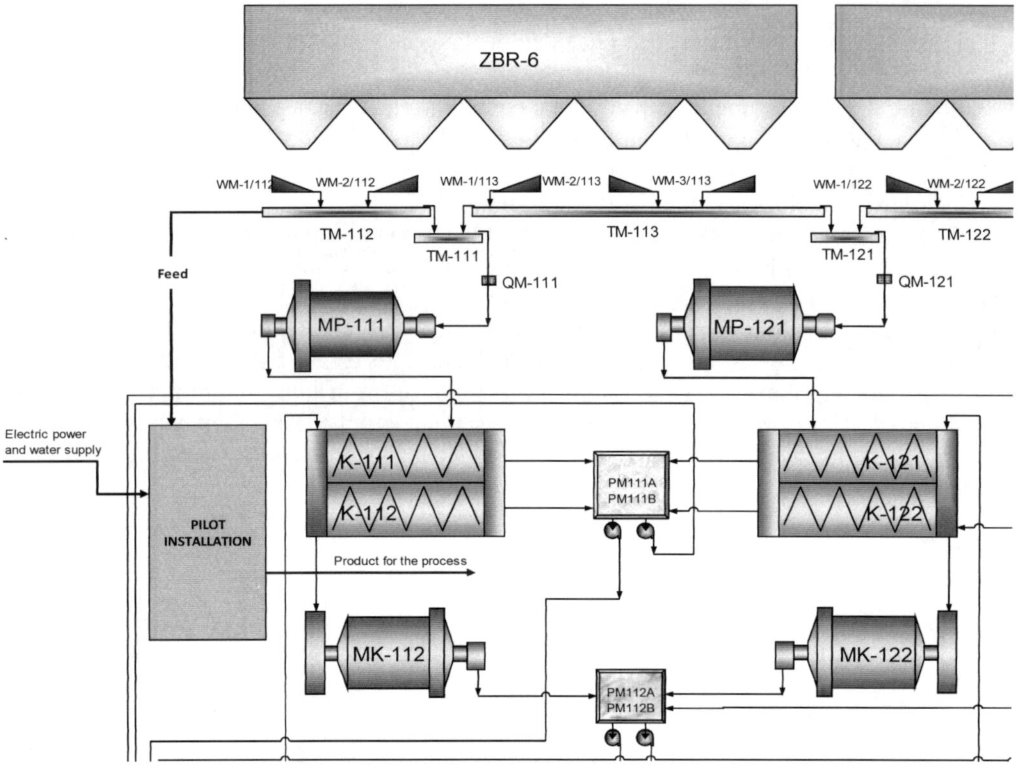

Figure 3. Location of the pilot installation for ore grinding including the beater wheel mill.

Figure 4. View of the pilot installation preparing ore for flotation using the impact comminution technique in Lubin Ore Enrichment Plant.

4. Technological testing of the wheel beater mill

Currently, technological tests of a new high-energy mill in Lubin Area are in progress. Among other things, the following parameters are being determined:

- mill output and energy consumption for grinding,
- grinding product size analysis,
- strength parameters of the mill and material wear of the mill working components.

The mill operating parameters during technological tests:

- motor speed –1630-2600 rpm,
- motor current – 240-280 A,
- motor voltage – 200-400 V,
- motor power – 90-154 kW,
- motor frequency – 33 Hz,
- mill output – 3-7 (max. 8) Mg/h,
- position of classifier flaps 30-60°.

Samples of feed and grinding product are taken during technological tests. Mill feed consists of the product of hammer crushers or a cone crusher.

The continuity of feed delivery to the mill is checked because of gradual adherence of the material to the container walls and resultant breaks in the material stream on the belt conveyor supplying the ore to the mill. A significant impact of the flap in the drawer-type feeder on the installation output has been observed – the opening adjustment is carried out by screwing down a bolt in the proper place on the arm, which is shown in figure 5.

Figure 5. Drawer-type feeder with visible flap opening adjustment.

The maximum installation output has been determined during preliminary tests. The impact of fed material volume on the mill speed has been checked. The output ranging from 4 to 6 Mg/h has been reached for motor speed up to 2600 rpm. Temporary drops in mill speed down to ca. 1650 rpm have been observed, although most of the time at constant loading with feed the speed was between 2180 and 2200 rpm. Increasing preset output above 7 Mg/h caused a drop of the mill beater wheel speed from 1800 to 1600 rpm, which could have resulted in its covering with feed material. At flap maximum opening, the drawer-type feeder was unable to operate in the automatic mode and it was not possible to set the required feed volume. This caused output leaps reaching 10 Mg/h, and instantaneously even 12 Mg/h, which resulted in the mill "choking" and the speed drop to 700-800 rpm. In this case, it was necessary to shut down the feeder and the belt conveyor until the mill was able to process the feed material and return to its high speed.

To sum up, it can be stated that the installation has been working in the most stable way at the output of up to 6 Mg/h in the automatic mode. Above this value, it was necessary to carry out

IOP Conf. Series: Materials Science and Engineering **427** (2018) 012034 doi:10.1088/1757-899X/427/1/012034

operation in the manual mode, continuously controlling the mill speed so as to prevent its complete stop.

The results of grain size analyses for the feed and grinding product are shown in table 1 and table 2, and their graphical presentation is contained in figure 6.

Table 1. Grain size analysis of the feed – hammer crusher product and high-energy mill product.

Grain class [mm]	Feed of beater wheel mill – product of hammer crusher		Product of grinding: Test 1 Capacity 3.5 Mg/h Rotations 2000 min^{-1}		Product of grinding: Test 2 Capacity 3.5 Mg/h Rotations 1630-2200 min^{-1}	
	γ, %	Σγ, %	γ, %	Σγ, %	γ, %	Σγ, %
+30	21.54	21.54				
30-25	6.92	28.46				
25-15	16.04	44.50				
15-10	8.49	52.99				
10-5	9.91	62.90				
5-3	3.46	66.36				
3-2	2.36	68.72	3.11	3.11	2.92	2.92
2-1	2.36	71.08	2.60	5.71	2.46	5.38
1-0.5	1.73	72.81	3.67	9.38	4.59	9.97
0.5-0.3	0.94	73.75	3.11	12.49	3.07	13.04
0.3-0.2	4.40	78.15	6.33	18.82	5.07	18.11
0.2-0.10	14.94	93.09	30.61	49.43	25.54	43.64
0.10-0.075	0.94	94.03	8.04	57.47	5.91	49.55
0.075-0.045	1.57	95.60	11.02	68.49	13.15	62.70
0.045-0.036	0.31	95.91	2.82	71.31	1.50	64.20
0.036-0	4.09	100.00	28.69	100.00	35.70	100.00
Σ	100.00		100.00		100.00	
d_{80} mm	30.71		0.196		0.193	

Table 2. Grain size analysis of the feed – cone crusher product and high-energy mill product.

Grain class [mm]	Feed of beater wheel mill – product of hammer crusher		Product of grinding: Test 3 Capacity 3-4 Mg/h Rotations 2000^{-1}		Product of grinding: Test 4 Capacity 4-6 Mg/h Rotations 2400^{-1}	
	γ, %	Σγ, %	γ, %	Σγ, %	γ, %	Σγ, %
+30	2.96	2.96				
30-25	2.53	5.49				
25-15	16.03	21.52				
15-10	10.55	32.07				
10-5	14.56	46.63				
5-3	5.70	52.33			0.84	0.84
3-2	5.06	57.39			5.88	6.72
2-1	4.22	61.61	1.88	1.88	7.27	13.99
1-0.5	4.64	66.25	2.62	4.50	7.68	21.67
0.5-0.3	3.59	69.84	2.43	6.93	4.35	26.02
0.3-0.2	12.03	81.87	11.56	18.49	17.37	43.39
0.2-0.10	2.74	84.61	36.06	54.55	36.57	79.96
0.10-0.075	3.16	87.77	5.73	60.28	4.40	84.36
0.075-0.045	2.74	90.51	7.87	68.15	2.92	87.28
0.045-0.036	0.63	91.14	1.23	69.38	0.28	87.56
0.036-0	8.86	100.00	30.62	100.00	12.44	100.00
Σ	100.00		100.00		100.00	
d_{80} mm		15.95	0.196		0.602	

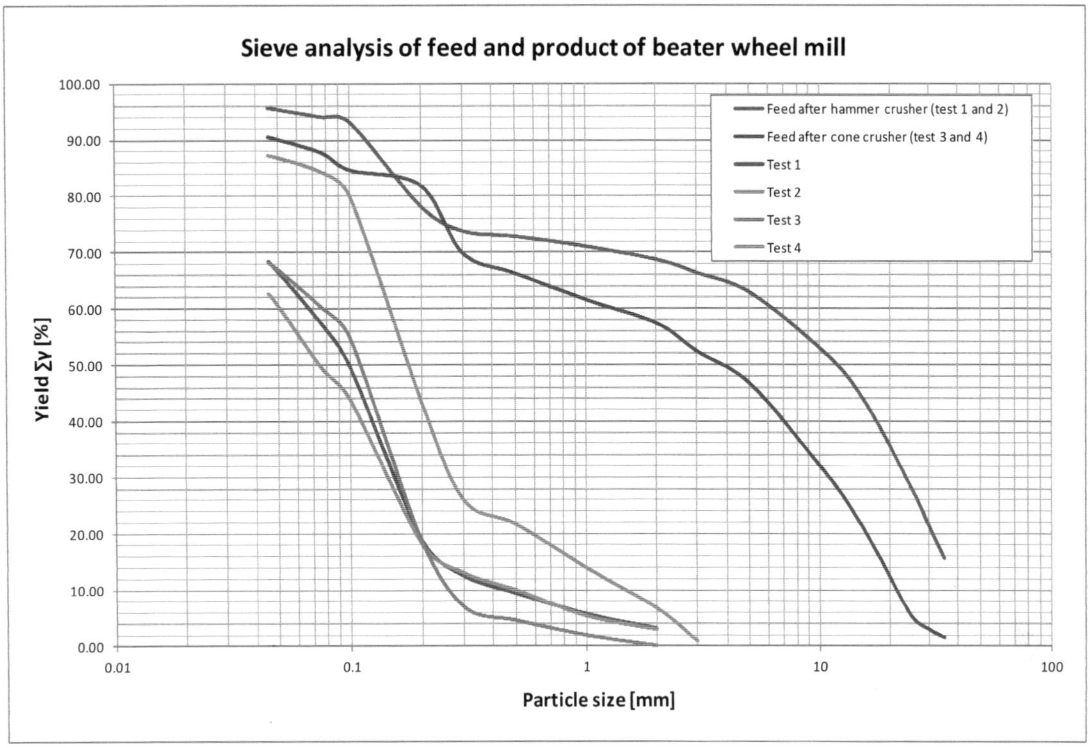

Figure 6. Size analysis of the feed and the product of the beater wheel mill.

The grain size analysis of the grinding product allowed to observe good grinding efficiency of the new mill – the content of the class prepared for flotation range from ca. 40 to 55%. The achieved grinding degrees are very high and range from above 80 to more than 150.

The durability of coating on structural components of the mill, i.e. the beater wheel and the housing reinforcement, is examined simultaneously with conducted technological tests. This gives an opportunity to develop a technology to coat working elements of the mill with modern materials resistant to tribological wear.

5. Summary

As a result of unavoidable deterioration of the quality of mined ores, reduction of enrichment indexes is a natural trend. Therefore, a necessary element of the KGHM activity is to analyse possible actions preventing these disadvantageous trends. Apart from modernisation involving the introduction of new equipment, it is also necessary to make efforts aimed at adjusting flowsheets to changing ore properties.

An important element of this is the technology used to prepare ore for` flotation. The problems of ore crushing and grinding are of particular interest at KGHM. Solutions are being explored which would allow to reduce the number of unit operations connected with multi-stage grinding and classification operations. The research project on grinding in a high-energy mill is an effort aimed at finding these solutions.

Preliminary tests of size reduction in the new mill are positive, although there are still many issues to be clarified with regard to the technology using this type of mill in the conditions of continuous enrichment plant operation.

Nevertheless, the single-stage method of preparation of ore for the flotation process is an advantageous solution, which can be applied on the basis of conventional high-yield systems.

The concept of this technology is shown in figure 7. It includes the following major components:

- the new system to prepare ore for flotation using a high-energy mill,

- preliminary flotation of raw ore - it is an essential element that prepares the feed for roughing flotation (feed decopperisation) and stabilises the production of the final concentrate, *inter alia* by maintaining reduced organic carbon content in the final concentrate,
- change in middlings processing through grinding in new type mills, e.g. Vertimill®, IsaMill and others, and then separate flotation in an open cycle, including removal of tailings – the second component of final waste,
- implementation of a cleaning circuit in the conventional cleaning circuit,
- changing the method of the final concentrate dehydration by way of conducting the process at a higher temperature, which allows to reduce the water content to ca. 8-9% and to eliminate the drying process.

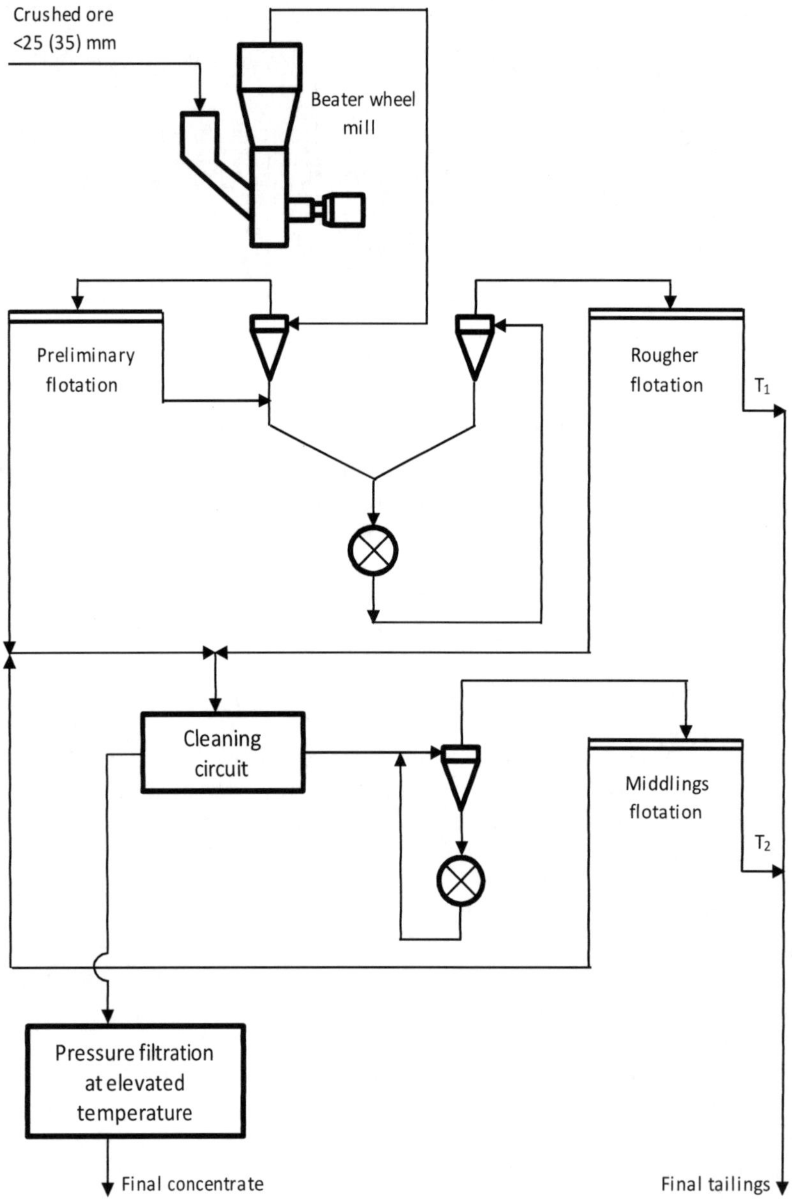

Figure 7. Flowsheet showing the new enrichment technology with a single-stage ore preparation for flotation.

Modernisation of the system used to prepare ore for flotation, which involves effective grinding, would allow to simplify the enrichment flowsheet, and thus ensure lower ore processing costs. At the same time, the proposed flowsheet is characterised by good technological efficiency in case of low-grade ores and ores which are difficult to enrich.

Research in the scope of grinding in the high-energy mill are continued. This also applies to the study of the flotation process of grinding products in the new mill. Results of flotations will be published.

References

[1] Wieniewski A and Skorupska B 2016 Technology of Polish copper ore beneficiation – the future from the past *E3S Web of Conferences* **8**,01064

[2] Szolc T, Konowrocki R and Michajłow M 2015 On research&development – aspects of the highly-energetic concept for the copper ore comminution *ICNOP Trzebieszowice, Conf. Proc.* pp 47-59

[3] *Monografia KGHM PM Lubin* 2007, p 474

IOP Conf. Series: Materials Science and Engineering **427** (2018) 012035 doi:10.1088/1757-899X/427/1/012035

Assessing the benefits of automatic grinding control using PST technology for true on-line Particle Size Measurement

R Maron[1], C O'Keefe[1] and J Sepulveda[2]

[1] CiDRA Minerals Processing, USA

[2] J-Consultants, Ltd. Chile

rdopieralla@cidra.com, rmaron@cidra.com

Abstract. Despite the recognized significance of ground product fineness on both line capacity and downstream process performance (typically flotation), less than 10% of mineral ore concentrators today use real-time particle size measurements in automatic control applications for the final product particle size. Although "near-line" particle size measurement instrumentation has been in place for several decades and is commonly installed in most modern concentrators, its availability and low measurement frequency have typically been inadequate for reliable use in automatic control. Their low availability is often associated with problems with slurry sample collection and handling systems required to present the samples to the size measurement instrument. Recently, an innovative technology for real-time, on-line particle size measurement has been developed by CiDRA Minerals Processing, under its commercial name Particle Size Tracking (PST), already proven in several commercial installations, demonstrating near 100% availability with minimal maintenance, thus overcoming the limitations of previous technologies. Investment decision makers require convincing, reliable estimates of the expected economic value that automatic grind control projects will deliver. This paper presents a methodology for estimating such value from the installation of the new PST particle size measurement technology, capable of tracking particle size on each individual hydrocyclone overflow stream, thus adding significant new options for improved process stability and performance. Based on actual plant operating records, accumulated over long periods of time, the evaluation approach herein described assesses and highlights the significant potential contribution to be expected from this unique PST technology.

1. Introduction

Optimizing the metallurgical performance of an existing industrial-scale mineral processing plant as a function of grind product size is an important but often complex task. Changing the grind product size in a full-scale plant and observing the effects on flotation is a direct way, but the real effects can be masked by the multiple and unexpected disturbances common to grinding and flotation circuit operations [1]. Simulators for both grinding and flotation have been developed and are very useful, but the integration of both can be complex, and they are still only models of the full-scale industrial plant and thus have inherent limitations [2]. Methods have been developed using a series of laboratory tests conducted on industrial circuit samples that have also proven useful and gained acceptance [3,4].

These three methods, and combinations of the above are certainly useful but involve considerable time and expense. In addition, they introduce uncertainty in the results due to the limitations of simulations, differences between plant and laboratory tests, and extrapolations from data taken from small sample sets and over small time frames. But often it is desirable to obtain an assessment of performance of an existing plant as a function of grind product size with less time and expense, and with more confidence that the results are representative of the industrial scale operation.

The methodology presented herein uses typically available historical plant data over a long time period to assess the current performance, and predict the performance improvement that may be possible if automatic control of the grind product size is implemented by the addition of reliable on-line particle size measurement technology.

2. Impact-based real-time hydrocyclone particle size measurement

2.1. Principle of Operation

Acoustic impact-based particle size tracking is a unique method for measuring and controlling a reference product mesh size in cyclone overflow streams. The implementation of this technology is centered upon a sensor probe that is inserted into the overflow slurry stream via a two-inch (50 mm) hole in the overflow pipe (Figure 1). Particles within the slurry stream impact the surface of the probe generating travelling stress waves within the probe. A sensor converts these travelling stress waves into an electrical signal and proprietary signal processing techniques convert these signals into a particle size measurement that is output every four seconds. The sensor is constantly in contact with many particles in the slurry stream, thus obtaining information from orders of magnitude more particles than traditional sample-based technologies. Also, because of the location of the sensor downstream of the hydrocyclone and the presence of an air core at that point, the sensor produces no change in the back pressure seen by the hydrocyclone and thus does not affect hydrocyclone performance. The probe has a useful life of approximately 18 months due to the abrasive wear caused by the direct slurry impact. The probe life is related to the particle hardness and size which is obviously finer in the overflow stream compared to the feed stream, thus enabling an acceptable probe life.

Figure 1. CYCLONEtrac PST particle sizing sensor mounted on hydrocyclone overflow pipe.

2.2. Plant Implementation of CYCLONEtrac PST Technology

The plant implementation of the PST particle sizing system includes a measurement probe with its associated electronics on each cyclone overflow pipe, a junction box that consolidates the processed

IOP Conf. Series: Materials Science and Engineering **427** (2018) 012035 doi:10.1088/1757-899X/427/1/012035

signals from the individual sensors, and sends them to a computer that is typically located in the control room. Final processing takes place in that computer and the resulting particle size information is transferred to the plant system via an integrated OPC server. Currently, the software provides a single reference mesh size that is to be incorporated into the overall process control strategy. This system is outlined in Figure 2 [5].

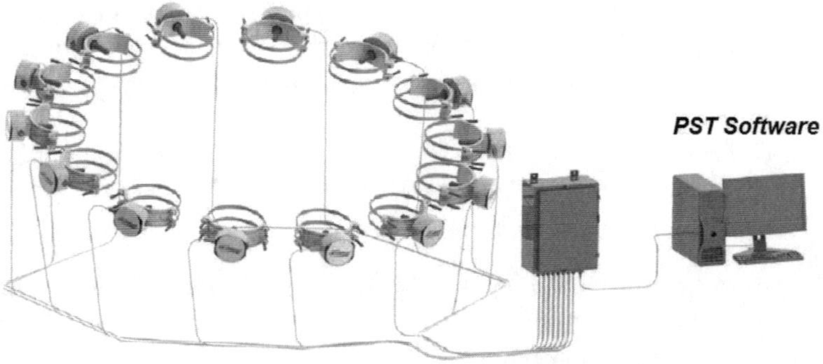

Figure 2. PST plant scale installation diagram.

On-line impact-based PST or alternative "near-line" cyclone overflow particle size measurement methods, whether laser diffraction, ultrasonic or caliper, all require empirical calibration by correlating their signals against reference particles or actual slurry samples analysed with standard laboratory procedures [6]. The impact-based CYCLONEtrac PST also requires calibration to compensate for influences from cyclone type and sensor installation location. To ensure a good composite calibration that can be applied across all the cyclones in a cluster, calibration samples must be taken from the overflow of each cyclone in a cluster. Once such calibration takes place, it does not have to be performed again when the probe is replaced due to wear, as long as the same model of replacement probe is used. In addition, samples must be taken beyond the expected operating range of the cyclones to ensure accurate measurements when the cyclone is operating outside its normal operating range. Calibration ranges must cover above and below the usual operational conditions including but not limited to roping events, startups, shutdowns, and grind outs. This avoids the measurement uncertainty that occurs when calibration models are used to extrapolate measurements beyond their calibrated range. For rapid processing of samples, a single sieve size is used with a custom wet sieving procedure to generate a calibrated number such as percentage of material passing or retained by a reference sieve size, e. g. X% +100#.

Experience shows that the resulting calibrated signals exhibit a standard deviation that is less than 4.5 percentage points from the corresponding true values. An undetermined but significant portion of this standard deviation can be attributed to the sample collection procedure. Cyclone overflows typically have limited access for sampling, so a full cross-stream sample is normally difficult. Instead plunge cuts or partial cross-stream cuts are performed. As a way of example, the results from the commissioning at a phosphate concentrator is shown in Figure 3 [7].

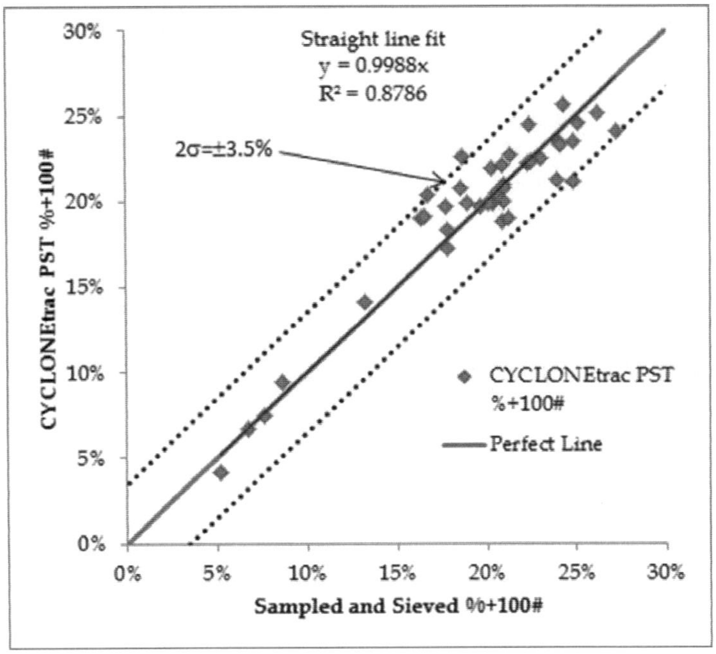

Figure 3. Example of PST results from single cyclone compared to sampled and sieved measurements

3. Value assessment methodology

3.1. Objective

The objective of this work is to quantify the value of incorporating the PST on-line particle size measurement into the automatic process control system for a grinding/classification circuit using normally available historical plant data. The methodology uses daily data over a long enough time period (ideally over 1 year) to obtain a high-level assessment of the general plant operating performance, mainly focusing on throughput (T), recovery (R) and net production of the valuable metal (NMP) to be the final objective function for process optimization purposes, as determined by the simple expression:

$$NMP = h\ T\ R \quad (1)$$

where: h represents the head grade of valuable metal being recovered.

Inherent in this methodology is the usual observation that throughput (T), recovery (R) and NMP are strongly correlated to ground product size, in such a way that there would exist in every application an optimal grind size that maximizes NMP; that is, the net value to be achieved from the operation. Therefore, better measurement and control of grind product size is here highlighted as a necessity for effective process control. Correlations of throughput and recovery to other variables such as feed grade and ore hardness must also be examined in every case, whenever the proper information is available. Finally, operational plant constraints (like, maximum tonnage or flowrates, product coarseness limitations for slurry transport, etc.) should also be taken into consideration as they may limit the maximum NMP values to be consistently obtained. This methodology is significantly simpler than others that rely on labóratory tests and sampling from an industrial circuit [3,4].

3.2. Input Data Set

Daily operating records are to be obtained for a minimum of 6 months with one year or more preferred. The quality of all data is important, but particularly the product size data. The frequency and method of obtaining the slurry samples should be well understood, and ideally should use a modern well-designed automated sampler that obtains samples every two to 4 hours, which are then combined to produce a daily composite sample. The composite samples should be processed in a metallurgical laboratory using proper quality control to obtain the grind product size. This is typically defined by a single number, the P80, which may be readily obtained from a full particle size distribution measurement. However, for convenience, the percent passing a single mesh size is typically used, and the mesh size chosen is typically the approximate P80 size. An example of the Input Data Set and the calculated NMP is shown in Figure 4, constructed on the basis of equation 1 above.

INPUT DATA					CALCULATED
Date	Size (%+Mesh#)	Throughput (tph)	Feed Grade (%Cu)	Recovery Total (%Cu)	NMP (tph)
2-Jan	16.8	2,775	0.69	82.1	15.7
3-Jan	17.3	2,381	0.60	77.5	11.0
4-Jan	20.7	2,864	0.57	79.0	12.8

Figure 4. Example, input data set and calculated output.

3.3. Data Filtering, Analysis and Data Binning

Days with obviously faulty data are first eliminated. The primary grinding capacity indicator is the average dry tons of ore ground per actual operating hour (tph) for each day. Special attention must be given to days of low % plant utilization, since on these days the tph averages may easily be outside normal operating ranges. The objective of this data filtering method is to be left with maximum number of reliable data points.

Because most plants run at a nominal performance point, the data points for each day tend to be clustered around a central range as can be clearly seen when the data is plotted. However, this centrally weighted cloud of data makes it difficult to detect an underlying trend. Data binning is a technique for dealing with this problem because it places equal weight on each bin, rather than equal weight on each data point [8]. The equally weighted bins can more accurately reveal the underlying trend in the data, providing there are sufficient data points per bin across the entire range of interest. As an example of how non-binned data can give misleading results, if most of data points are concentrated in a central region where the underlying trend is mostly flat compared to the non-central regions, then the trend fitted to the non-binned data will be flat because the data points outside the central flat region will have relatively low influence and be unable to reveal the underlying trend. In this case, where we are searching for trends in throughput, recovery and NMP related to product size, the data was binned to the product size as shown in Figure 5. It was found that equal widths of 2% +Target Mesh Size (TMS) produced the most statistically valid results.

IOP Conf. Series: Materials Science and Engineering **427** (2018) 012035 doi:10.1088/1757-899X/427/1/012035

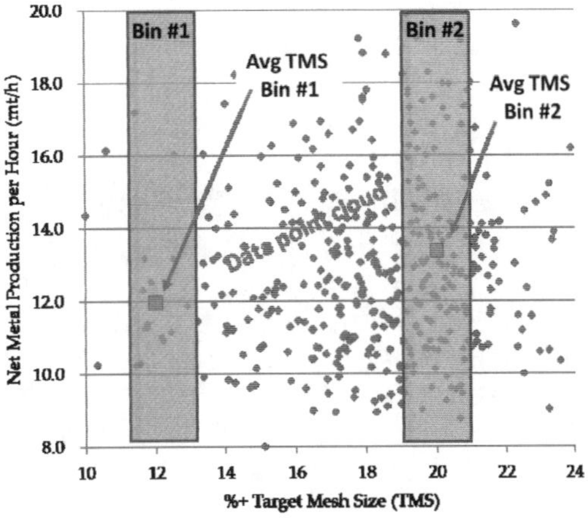

Figure 5. Example of data binning; non-binned points and average from binning.

3.4. Throughput vs Product Size

The first basic correlation to examine is the relationship of ore throughput vs. product size as shown in Figure 6. Typically, this relationship will show that a coarser grind size (higher % + TMS) corresponds to a higher throughput. This graph shows the grinding circuit is typically operating between 15% and 25% +TMS with a corresponding throughput increase at coarser product size.

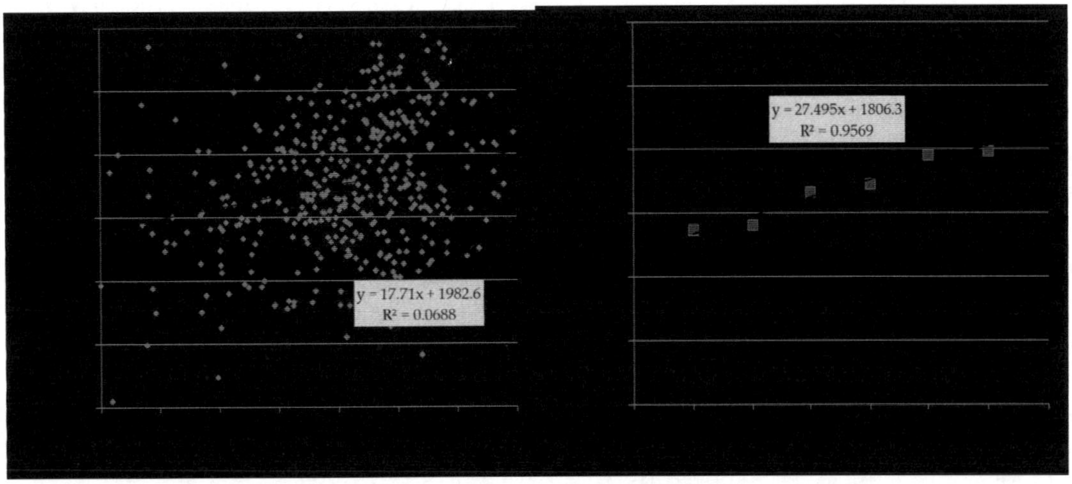

Figure 6. Throughput vs Product Size, with and without data binning.

As indicated, the bin averages shown on the right plot reveal an acceptable (high R^2) linear trend for the throughput vs. TMS correlation, which could not be as clearly observed in the left plot showing all data points with their natural day-to-day process variability. As described previously, the high uncertainty of fitting a regression line to the cloud of points without using data binning means that the slope of the regression line in the left graph is very unreliable as indicated by the very low R^2 value. However, when data binning is used in the right graph the slope can be considered reliable as indicated by the high R^2 value.

3.5. Recovery vs Product Size

The next basic correlation to examine is the relationship of recovery (of the target metal) to product size as shown in Figure 7. Typically, the relationship will show a maximum recovery at a particular product size, with lower recovery in the finer and coarser size ranges due to losses in flotation recovery.

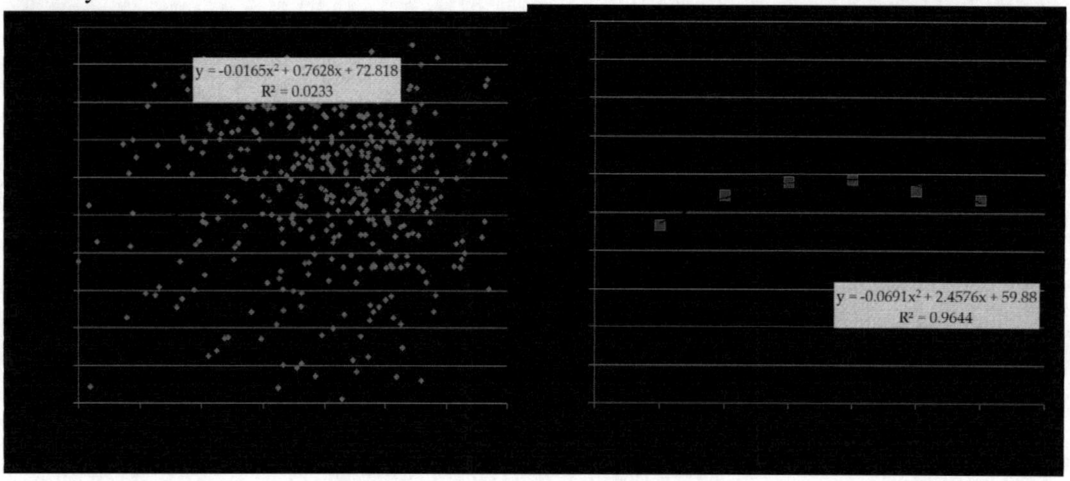

Figure 7. Recovery vs product size.

Again, bin averages show a clear correlation trend which is hard to identify directly from the raw data.

3.6. Net Metal Production (NMP)

Having examined the relationships of throughput and recovery vs product size, the NMP can be calculated using the data binning technique previously described, as illustrated in Figure 8.

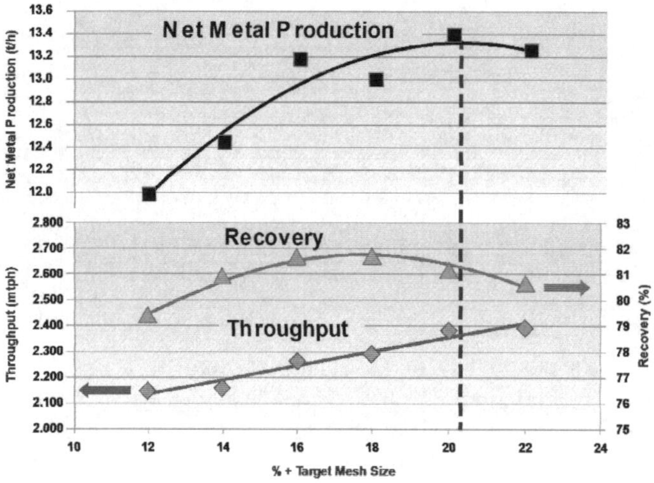

Figure 8. Net Metal Production (NMP) objective function (top) per equation 1.

3.7. Net Metal Production Gain from Improved Product Size Control

Average NMP for the example operation is 13.05 tph using the current product size measurement and control strategy. If the plant were to operate at the peak of 20.3% +/- 1% of TMS, the NMP would increase by 2% to 13.3 tph as illustrated in Figure 9. Furthermore, it is well recognized that process stability relates to improved overall circuit performance [9]. In other words, there is more gain in NMP to be achieved by reducing process variability through better product size control than has been shown in the previous analysis.

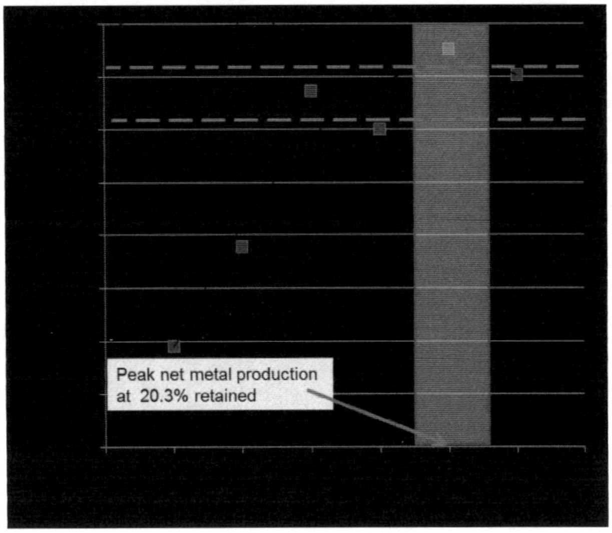

Figure 9. Net Metal Production optimized by operating at +/-1% of 20.3% +TMS.

4. Summary and conclusions

Availability of reliable product size measurements is key for the effective operational control of grinding/classification systems, particularly in cases where the following downstream stage is selective flotation concentration. As discussed in the present publication, CiDRA's PST technology is being shown to be most valuable in providing true on-line particle size information which, properly incorporated into the respective overall system control strategy, allows for the continuous maximization of the attainable process benefits.

Recognizing that product size is strongly correlated to both line throughput and valuable metal recovery, it is then possible to assess the target product size set point that will maximize at all times the objective function defined as the Net Metal Production (NMP) rate, to be expressed in tons of valuable metal produced per operating hour.

In the value assessment methodology here proposed, "data binning" techniques were found to be very useful and illustrative in the analysis of actual operational records, normally affected by fairly large normal process variations and unplanned instabilities which tend to obscure the underlying interactions amongst the most relevant process conditions. An important advantage of the here described methodology is that it relies only on existing plant-scale data and does not involve the execution of cumbersome laboratory work.

The 2% increase in Net Metal Production predicted by this analysis, plus whatever benefits can be achieved from reducing process variability by incorporating PST technology in the control strategy, is consistent with increases previously reported from existing PST installations.

References

[1] McKay J, Ynchausti R and Gritton K 2014 Benefit Analysis of Implemented Supervisory Control Systems Knowledge Scape (Salt Lake City UT USA) https://kscapedotcom.files.wordpress.com/2014/05/benefits.pdf

[2] Schwarz S and Richardson J 2013 *Modeling and Simulation of Mineral Processing Circuits Using JKSIMMET and JKSIMFLOAT* SME Annual Meeting (Denver CO)

[3] Bazin C, Grant R, Cooper M and Tessier R 1994 Prediction of Metallurgical Performances as a Function of Fineness of Grind Canadian *Mineral Processors Annual Conference CMP Proceedings 1964-1998*
[4] Edwards R, Vien A and Desbiens A 1999 *Application of a model based size-recovery methodology* (Canada: Canadian Inst of Mining, Metallurgy and Petroleum)
[5] Cirulis D 2015 Process Optimization Using Real-Time Tracking of Coarse Material in Individual Cyclone Overflow Streams *METPLANT 2015 Conference (Perth, Western Australia, Sept 7-8, 2015*) Australian Institute of Mining and Metallurgy Perth
[6] Wills B and Finch J 2016 *Mineral Processing Technology* 8th Edition (Elsevier, Ltd UK) Section 4.5 p 105
[7] O'Keefe C, Maron R, Cirulis, D and Medina J 2016 Survey of Real-Time Individual Cyclone Particle Size Tracking (PST) in Multiple Concentrators *Proceedings of 5th International Congress on Automation in Mining, Autominig 2016* Nov 30-Dec 2 2016 (Antofagasta Chile) https://www.cidra.com/resource-center/survey-of-real-time-individual-cyclone-particle-size-tracking-PST-In-multiple-concentrators
[8] Wikipedia 2017 *Data Binning* https://en.wikipedia.org/wiki/Data_binning.
[9] Wysowl J 2017 *Optimising Continuous Operations* Newsletter No. 33 (Wysowl Pty Ltd Queensland Australia)

Substitution of magnetite in dense medium separation by Zinc-Lead waste

D Gajda[1], M Lutyński[1] and M Kujawska[2]

[1]Silesian University of Technology, Faculty of Mining and Geology, Department of Mining, Akademicka 2, 44-100 Gliwice, Poland

[2]Silesian University of Technology, Akademicka 2, 44-100 Gliwice, Poland

dawid.gajda@polsl.pl

Abstract. Dense Medium (or Heavy Medium) Separation (DMS) is one of the most efficient and commonly used method of separation in different mining branches. Dense Medium for coal processing is based on the suspension of fine and dense material, usually magnetite or ferrosilicon, in water. Magnetite for that purpose is the most commonly used mineral in Polish mines, but has to be entirely imported from abroad. Ferrosilicon, recovered from Zinc-Lead waste could be an alternate to magnetite for use in DMS coal processing. In this study an attempt of recovering that material from "Miasteczko Śląskie" Zinc Plant wastes was made. By simple separation and comminution processes, large amount of magnetic fraction, which could be successfully used in dense medium as the substitution for the magnetite was obtained. Almost 70% of collected waste from lead refining process is proper for the use in the DMS process. Obtained fraction is relatively brittle and easy to grind, which should generate minor costs of production. Almost entire fraction is also highly reactive for magnetic field, which will be helpful in the recovery from dense medium solution, using existing magnetic separators in coal processing plants. Chemical analysis show that concentration of a few metals is significant, which may cause environmental obstacles. But that can also be solved, using simple method of recovery of metal ions from process water.

1. Introduction

Reuse and recovery of waste is the most desired way of waste management, following the prevention of waste formation. At the same time, storage and disposal are the less desired and most harmful for the environment [1]. That principle is essential not only in the industry but in general consideration of waste management. Searching for the ways of reusing wastes, instead of disposal, can improve the waste management in the companies and help saving natural resources.

There are a lot of known, common and also unconventional technologies of recovering different metals from industrial solid [2,3] and liquid wastes [4,5]. Metallurgical waste can also be reused as an aggregate or concrete component, which can even upgrade the concrete properties [6]. Some research was also performed in the field of recovering magnetite dust from carbon combustion by-products [7].

Outstanding way of utilize certain sort of zinc industry waste is recovering the ferromagnetic fraction, which can then be used as the substitute for magnetite in Dense (or Heavy) Medium Separation (DMS/ HMS). Average use of magnetite in single processing plant at the Polish coal mine is approx 700-800 grams per Mg of processed coal. What's more magnetite used in DMS in Polish coal industry, has to be entirely imported from abroad.

In this study an attempt was made to verify if properties of Zinc-Lead slags disposed in Miasteczko Śląskie site in Poland could be of use as an alternative material to magnetite in DMS process.

2. Production of Zinc-Lead in Miasteczko Śląskie Plant

Production of Zinc-Lead in the "Miasteczko Śląskie" Plant is based on the Imperial Smelting Process (Blast Furnace Process). The Blast Furnace Process technology consists of Sinter Plant, Furnace Plant and Refining Plant. Process in the Furnace Plant is charged with solid sinter from Sinter Plant, and coke. The products are vapour zinc and slag with lead, collected at the bottom of the furnace. Next step of production takes place in the Refining Plant where selected metals like Copper, Silver, Tin, Arsenic, Antimony, Bismuth are recovered. Refining process generates slag, which is deposited in Hazardous Waste Landfill [8].

According to Polish regulations [9] the code of generated waste is 10 05 01. Black to gray slag are block-shaped and contain ceramic and metallic inclusions. Plant generates about 700 Mg of refining slags per year [8].

Hazardous waste landfill occupies the area of approx. 50 000 square meters It contains about 40 000 Mg of refining slag (estimated value). Significant content of Fe in the landfill at "Miasteczko Śląskie" Zinc Plant was mentioned in the estimation of heavy metals quantity research. Content of Fe in samples oscillated between 7% and 19%, depending on the layer [8]. Other research show the content of Iron (FeO) at the level of 33,95% [10].

3. Sampling and research methods

Samples of refining slag waste were taken in April 2018 from the Hazardous Waste Landfill of "Miasteczko Śląskie" Zinc Plant. Approximately 70 kilograms of sinters and blocks were collected. Slags were collected from recently disposed part of wastes which have not yet been completely buried with the aggregate. Collection of the waste samples is presented in figure 1.

Figure 1. Collection of waste samples from "Miasteczko Śląskie" Zinc Plant Hazardous Waste Landfill.

Preliminary preparation was done by mechanical crushing of the samples, with additional use of 25 mm sieve, to separate the friable fraction from sinters. Then, a friable fraction was averaged by quartering method (4 times) to obtain the representative sample from the whole amount of collected waste. At the end of the quartering process, 2 samples of 2700 grams each were obtained. One of the

IOP Conf. Series: Materials Science and Engineering **427** (2018) 012036 doi:10.1088/1757-899X/427/1/012036

mentioned samples was then ground, using vibration mill. Duration of grinding was 15 minutes, and single cell contained about 200-240 grams of feed. Milling was used for the magnetic separation and chemical composition tests. Consequently, the material was dried at 100 Celsius degrees.

Magnetic separation of ground friable fraction was done, using Magnetic Disc Separator , type LOG1-4, manufactured by Boxmag – Rapid Limited, Birmingham. Separator was equipped with vibration feeder from The Sinex Engineering Co. Ltd. Chemical composition tests were performed at the JY 2000 Inductively Coupled Plasma Atomic Emission Spectrometer (ICP-AES). Both of these tests were done at the Faculty of Mining and Geology of Silesian University of Technology. Additionally, the bulk density of samples was measured using the pycnometer method.

4. Results and discussion

4.1. Separation tests

On the basis of preliminary separation process of the waste, we could divide two characteristic fractions. First fraction is the friable and easy-crushing black material, with numerous bright, shiny grains. Second, named "sinters", consists mostly of hard, crushing-resistant sinters and malleable, metallic plates. Both fractions are presented in the figure 2 and figure 3. Preliminary separation results are presented in the table 1.

Table 1. Pre – separation results.

Fraction	Mass [g]	Percentage [%]
Feed	67 857	100.00
Friable	46 160	68.03
Sinters	21 697	31.97

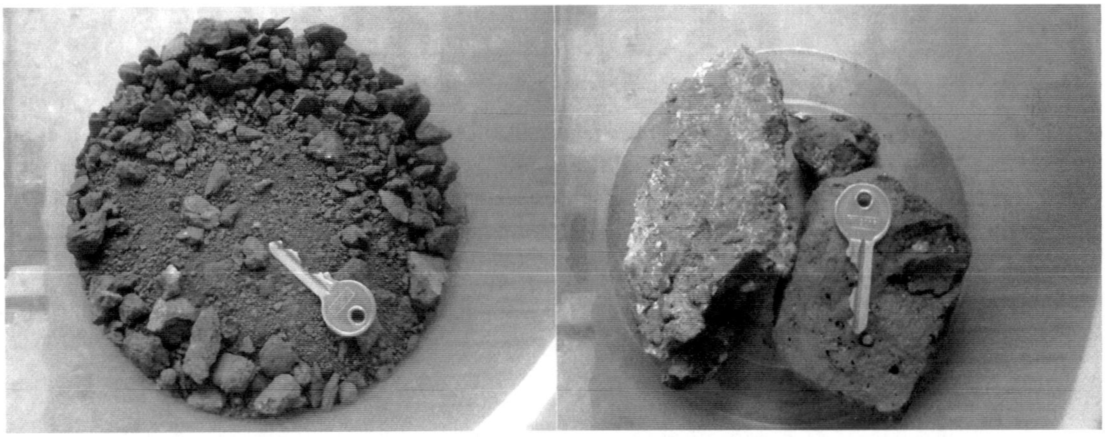

Figure 2. Friable fraction (left) and sinters (right) obtained from pre-separation.

Figure 3. Sinters with malleable metallic plates.

Friable fraction was then quartered, milled and dried, as described before. Sample was directed to the magnetic separator. Results of separation are presented in the table 2.

Table 2. Magnetic separation results.

Fraction	Mass [g]	Percentage [%]
Ferromagnetic	256.9	91.30
Semi-ferromagnetic	24.1	8.56
Non-magnetic	0.4	0.14
Total	281.4	100.00

According to [8], amount of refining slag, deposited in "Miasteczko Śląskie" waste landfill in 2008 was approx. 32 500 Mg. Estimated amount of iron in the landfill was 22 400 Mg, which gives 69% of Iron in refining slag. Obtained amount of magnetic fraction shown in table 3 is similar to the amount of Iron presented in the mentioned article [8].

Table 3. Overall separation results.

<table>
<tr><th colspan="2">Fraction</th><th colspan="2">Percentage [%]</th></tr>
<tr><td colspan="2">Sinter</td><td colspan="2">31.97</td></tr>
<tr><td rowspan="2">Magnetic</td><td>Ferromagnetic</td><td rowspan="2">67.93</td><td>62.11</td></tr>
<tr><td>Semi-ferromagnetic</td><td>5.82</td></tr>
<tr><td colspan="2">Non-magnetic</td><td colspan="2">0.10</td></tr>
<tr><td colspan="2">Total</td><td colspan="2">100.00</td></tr>
</table>

Table 4 shows, that concentration of iron in friable fraction is approx. 35-40%. That value refers to the amount of Iron, presented in other mentioned research [10]. Despite that, almost entire friable fraction is separated in magnetic field and can be easily recovered. Only negligible amount of sample was non-magnetic (less than 0.2%). That suggests, the considered material can consists of ferrosilicon (FeSi). It lowers the concentration of Iron, but the whole compound is still highly reactive for magnetic field. Gray colour of the magnetic fraction also indicates that compound. Ferrosilicon is the sort of ferroalloy, used as the supplement for metallurgical processes.

Sinters may also contain iron, but due to their high crushing resistance, that fraction is not considered as the potential material for recovery, because of high costs of crushing and grinding.

Amount of Alumina, Titanium, Chrome and Manganese was at the comparable level with magnetite composition. Content of Iron is slightly lower in the magnetic fraction of the waste. Concentration of other investigated elements, which are Nickel, Copper, Zinc and Lead, is significantly higher, comparing to the magnetite. Higher concentration of these metals are obvious in Zinc-Lead ores, from which the investigated waste comes from.

Table 4. Chemical composition of magnetic fraction, compared with Swedish magnetite used in polish coal preparation plants.

Selected elements	Percentage [%]		
	Friable fraction (feed)	Ferromagnetic	Swedish magnetite
Al	1.51	1.50	1.44
Ti	0.05	0.14	0.10
Cr	0.07	0.08	0.01
Mn	0.14	0.18	0.11
Fe	35.2	39.6	59.1
Ni	0.23	0.18	0.01
Cu	4.18	3.65	0.02
Zn	2.11	2.14	0.01
Pb	5.97	4.5	0.02
As	0.02	0.02	5.41 ppm
Sb	0.03	0.02	1.23 ppm

High concentration of Nickel, Copper, Zinc and Lead is undesirable, because it might cause the environmental contamination. It refers to the process water, where these metals circulate after removing it from the processing coal. Proper coal processing in the dense medium should eliminate the residual dense medium on the coal. The concentration of metal ions will appear in the process water, where metals from the solid fraction may be leached. Installation of simple device for recovering metal ions from water, should solve this problem. Iron reactor, described in [4,5] is cheap and non-mechanical, which makes it failure-free and simple in maintaining. It also has a high effectiveness for removal of mentioned metals [4,5]. These qualities will not increase the costs of coal production, and can even increase the quality of process water in the processing plant circuits.

IOP Conf. Series: Materials Science and Engineering **427** (2018) 012036 doi:10.1088/1757-899X/427/1/012036

Concentration of Copper in the magnetic reactive fraction is approx. 4%. Average concentration of Copper in the feed of Polish Copper Processing Plants is usually below 2%, when concentration in flotation slag is approx. 0.15-0.30% [11]. Recovery of Copper should also be considered.

4.2. Density, grain size and other properties
The density of obtained magnetic fraction is 5.16 g/cm^3. Density of magnetite, commonly used in the Dense Medium Separation is approx. 4.9-5.2 g/cm^3. Experience in coal processing in DMS shows, that density of magnetite or other material in heavy media should be at least twice of the required medium density. Since majority of dense media in coal processing is 1.5 g/cm^3 and 1.8 g/cm^3 [12,13], density of obtained magnetic fraction is proper for preparation of dense medium for coal enrichment. Grain size of investigated fraction has not been tested. Despite that, grain size depends on the time of exposition in the mill. That parameter is easy to adjust, if necessary.

Simple visual test of dense medium sedimentation was done. Solution of 100 g of material per 1000 ml of water was prepared and left for sedimentation. Sedimentation stages after preparing of sample, and after 5, 15 and 60 minutes are shown in the figure 4. That simple test has shown, that grain size of investigated sample is uneven. There are grains, which sediment fast and small enough to suspend in the solution even after 60 minutes. For better evaluation of material for use in dense medium, precise grinding and grain size tests should be done.

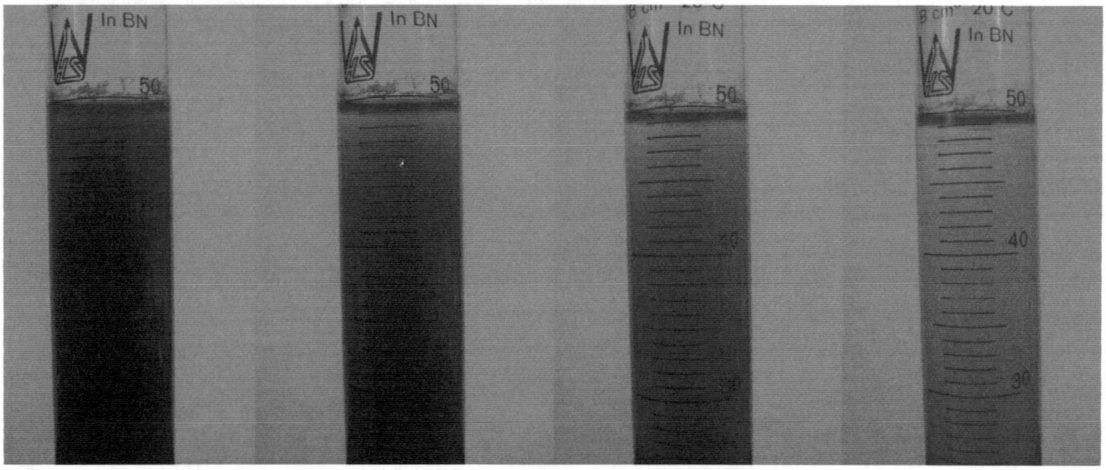

Figure 4. Sedimentation of ferrosilicon dense medium after preparation (left) and after 5 minutes, 15 minutes and 60 minutes of sedimentation.

Block-shaped ferrosilicon is susceptible to self-decay. That process can be accelerated by high concentration of Alumina (higher than 1%) [14]. Self-decay has also been observed in collected waste. That property is desirable, because the comminution process is essential for recovering investigated material for DMS.

5. Conclusions

Ferrosilicon, recovered from Zinc-Lead waste, seem to be a good substitution for magnetite for the Dense Medium Separation purposes. Physical properties are suitable for adaptation of the waste and ferrosilicon recovery. What's more, waste is susceptible for comminution. Density of material is also proper for use in the dense media. The whole material is highly reactive for magnetic field, so effective recovery from the process water is also possible, using the existing infrastructure at the coal processing plants. The chemical composition seem to be the biggest problem, because of the water contamination possibility. This problem can be solved, by using additional, simple and efficient Iron Reactor for recovering the metal ions from water.

Recovering of the ferrosilicon for dense media can be conducted directly at the Zinc-Lead Plant. Pre – separation process of the waste could separate the friable fraction, which could be used as the intermediate for the dense media purposes. Sinters could be deposited in the landfill, as it takes place now. That solution would reduce the amount of waste directed to the landfill. Separated material could be sold as the intermediate or even ready-to-use material, after proper adaptations (grinding to proper grain size). That material contains significant amounts of Copper, Zinc and Lead, so it could also be directed to the secondary enrichment processing for recovering these metals, if that process would be economically justified.

Aknowledgements
Authors would like to thank dr inż. doc. Stanisław Błaszczyński, emeritius professor of Silesian University of Technology, for help in preparing that paper.

6. References

[1] Act of waste (Polish law) 2012 (Ustawa z dn. 14.12.2012 r. o odpadach (Dz.U.2013 poz. 21))
[2] Ostrowska-Popielska P and Sorek A 2012 *Prace Inst. Met. Żel.* **4** pp 39-46
[3] Miller V R and Paulson D L 1982 *Resourc. And Conserv.* **9** pp 95-104
[4] Suponik T and Gajda D 2014 *Geochem. i Geol. Środ. Ter. Uprzemysł.- Monografia* ed. M Pozzi (Gliwice: PA Nova) pp 53-67
[5] Suponik T and Gajda D 2014 *Proc. Of the Min. Eng. Conf. MEC2014* ed. M Lutyński and T Suponik (Gliwice: SITG) pp 204-211
[6] Zieliński K and Rzeszowski M 2007 *Hutnik – Wiad. Hutn.* **74** (11) pp 610-5
[7] Hycnar J, Kochański B and Tora B 2012 *J. of Pol. Min. Eng. Soc.* **13** (2) pp 1-10
[8] Pozzi M and Nowińska K 2010 *Górn. i Geol..* **5** (4) pp 221-9
[9] Ordinance of waste catalogue (Polish law) 2014 (Rozp. Min. Środ. z dn. 9.12.2014 r. w sprawie katalogu odpadów (Dz.U.2014 poz. 1923))
[10] Adamczyk Z and Nowińska K 2013 *Civ. and Environ. Eng. Rep.* **11** pp 14-21
[11] Stadnicki K and Duchnowska M 2016 *Łupek Miedzionośny II* ed. P B Kowalczuk and J Drzymała pp 216-21
[12] Blaschke W 2009 *Przeróbka węgla kamiennego – wzbogacanie grawitacyjne* (Kraków: Wyd. IGSMiE-PAN) pp 117-8
[13] Laskowski T, Błaszczyński S and Ślusarek M 1979 *Wzbogacanie kopalin w cieczach ciężkich* (Katowice: Wyd. Śląsk) p 85
[14] Bulkowski L, Pogorzalek J, Wyrobek A and Goraždža J 1995 *Solid. of Metals and Alloys* **24** (1) pp 11-8

IOP Conf. Series: Materials Science and Engineering **427** (2018) 012037 doi:10.1088/1757-899X/427/1/012037

The behavior of oxidized copper during the sulphide Cu ore flotation in KGHM Polska Miedź S.A.

A Grotowski

KGHM Cuprum sp. z o.o. CBR, Gen. Wł. Sikorskiego 2-8, 53-659 Wrocław, Poland

Abstract. The work included a synthetic review of the occurrence of oxidized Cu minerals in the concentrators belonging to KGHM Polska Miedź S.A. and problems occurring in the flotation of oxidized Cu ores. The course of flotation of oxidized copper in concentrators A, B and C was examined on the basis of archival results and specialized sampling campaigns. It was found that the content of oxidized copper in the flotation tailings ranges from 0.015-0.05%, and in the concentrate from 0.23 to 0.72%. The share of oxidized copper is highest in the case of flotation tailings and ranges from 10.7 to 25.7%. Cu_{ox}'s share is only a few percent in concentrates. Analyzing the average annual balance samples, the Cu_{oxt} contents were clearly higher for all concentrators than for the sampling campaign samples, which, as indicated, is caused by the long storage time of the analytical samples. Evaluations of revenues from Cu_{ox} recovery were made and a further directions of research on the recovery of oxidized copper from flotation tailings were proposed.

1. Introduction

Flotation, and then smelting the concentrate obtained in this way, has been the most popular method of producing basic metals for more than 100 years, in particular Cu. The standard in the copper industry is currently the quality of flotation concentrates at the level of approx. 25% Cu, with a recovery at the level of 90%. Such a high level of indexes can be achieved, however, only when the predominant form of occurrence of copper are sulphides. In practice, copper sulphides are often accompanied by oxidized copper minerals, and sometimes they can be the only form of Cu occurrence. The results of flotation of ores with an elevated content of oxidized Cu minerals, even after applying special flotation conditions, are usually worse, or much worse, than sulphide ores and are frequently characterized by high variability.

The primary reason for the presence of oxidized copper minerals is the genesis of the deposit, but in certain circumstances, secondary oxidation of mineral surfaces may occur, especially in the finest classes, which generally causes deterioration of concentration indexes. The reasons for the secondary oxidation of sulphide Cu minerals are most often: long-term storage of ore, too fine grinding and/or long flotation times.

The method of processing of oxidized Cu deposits depends each time on the type and manner of occurrence of oxide forms, gangue composition, variability of oxidized minerals, content levels of individual Cu forms, etc. In the case of clearly defined, oxidized parts of the deposit, the most appropriate method of their processing is usually the selective exploitation and separate processing of both types of ores by methods adapted to their properties: e.g. by separate flotation of sulphides and oxides or hydrometallurgy. Depending on specific conditions, other solutions are also possible.

The matter of flotation of oxidized Cu minerals contained in the deposits of KGHM Polska Miedź S.A., due to their minimal content and expected, quite limited economic effects is a difficult issue and requires the inclusion of a number of restrictions. For these reasons, it has so far been recognized only fragmentarily, while conducting other research.

This paper presents the results of research on the occurrence of total and oxidized copper in the balance products of all concentrators belonging to KGHM Polska Miedź S.A., the analysis of its behavior in the process of industrial flotation and the estimation of feasibility Cu_{ox} recovery improvement.

2. Occurrence of oxidized Cu minerals in copper ores converted in KGHM

More than 120 oxidized copper minerals are known, but as practice indicates, commercial meaning is primarily malachite and azurite, to a lesser extent cuprite, tenorite and chrysocolla. The most important properties of selected oxide copper minerals are given in table 1.

Table 1. The most important properties of selected copper oxide minerals found in the KGHM Polska Miedź S.A. [1].

Mineral	Chemical formula	Cu content, (%)	Mosh hardness	Specific weight, (Mg/m^3)	Color
Cuprite	Cu_2O	88.8	3.5-4.0	5.9	Brick-red
Tenorite	CuO	80.0	3.5	6.5	Black
Azurite	$Cu_3(OH)_2(CO_3)_2$	55.3	3.5-4.0	3.7	Blue
Malachite	$Cu_2(OH)CO_3$	57.4	3.5-4.0	3.9	Green
Chrysocolla	$(Cu;Al)_2H_2Si_2O_5(OH)_4 \cdot nH_2O$	10-36	2.0-4.0	2-2.4	Blue

In the geological documentation of the copper deposit KGHM Polska Miedź S.A., made in 1958-1960 by the Geological Enterprise and the Polish Geological Institute from Warsaw, only cuprite and tenorite were identified from the oxidized copper minerals. It was only in later years that azurite, malachite and chrysocolla were discovered [2]. Due to the minimal content of individual Cu oxide minerals and the variability of oxide mineralization, the identification of oxide forms - even using the latest mineralogical techniques (e.g. MLA) - is difficult and usually requires confirmation by other methods.

Researching the influence of dry grinding on the results of Cu copper ore flotation, identification of the mineral composition and the degree of release of Cu minerals in the feed for Polkowice concentrator were carried out using the MLA method. Of the oxidized Cu minerals, only cuprite (0.01% by weight) was identified [3].

In mineralogical studies of a larger number of tailing and concentrate samples from laboratory flotation of Rudna ore, the presence of cuprite at the level of 0.06% by weight - concentrate, and in tailings of 0.001 - 0.0001% by weight, was shown [4]. The presence of atacamite was also found: in tailings at 0.005-0.010% by weight, and in the concentrates of 0.035-0.017% by weight. So far, none of the mines of KGHM Polska Miedź S.A. it was not possible to discover the oxidized parts of deposit that could be subject to exploitation or even the place for sampling for laboratory tests. This indirectly indicates the low level of oxidation of the deposit and the possibility of occurrence of only minimal content of oxidized Cu minerals, both in the feed and in the concentration products. The phenomenon of oxidation of sulphide Cu minerals is well known both in the mines of KGHM Polska Miedź S.A. - as so-called "green walls" as well as on rework - as so-called "weathered ore". Flotation results of the ore exposed to air are the worse the longer the ore storage time [5]. However, the studies carried out have shown that this process is quite slow and in the initial phase minimally affects the results of flotation. Such dependencies were found for both Lubin and Polkowice ores [6, 3]. In addition, it should be noted that the degree of deterioration of the flotation results depends to a large extent on the adopted research methodology and/or flotation technology.

3. Global practice of processing oxidized copper ores

Poor floatability of oxidized copper ores has been widely known for a number of years (e.g. [6-10]. Generally, for the processing of oxide copper ores, two methods are used: 1) hydrometallurgical, and; 2) flotation, but for a long time the hydrometallurgical method (SX-EW) has dominated the technique of processing oxidized and mixed copper ores [11-13]. The hydrometallurgical method is by far the most appropriate in the form of heap or dump leaching [11]. In the case of mixed ores, leaching is usually supported by the activity of microorganisms [13, 14]. The flotability of individual Cu oxide minerals is very diverse [7-10, 15, 16]. Of the oxidized copper minerals, chrysocolla (Cu; Al) $_2H_2Si_2O_5$ (OH) $_4$ • nH_2O), was the most common object of flotation research, but for Cu production the most important mineral from this group is malachite ($Cu_2(OH)CO_3$). Its relatively good - against other oxide minerals Cu - flotability has been confirmed by many studies. Azurite ($Cu_3(OH)_2$ $(CO_3)_2$) also floats quite well, similar to malachite, which is reflected in the data from a number of enrichment plants. The Kuprite (Cu_2O) requires higher doses of the sulphidising agent than other minerals, and the flotation behavior of the tenorite (CuO) is close to cuprite. As for other Cu oxidized minerals, such as atacamite ($Cu_2Cl(OH)_3$, or brochantite ($Cu_4(OH)_6SO_4$), their flotability was not investigated.

Although there are many examples of good results of oxidized Cu flotation in the laboratory scale, the practice is quite poor, due to a number of difficulties occurring in industrial conditions. The most important of them include natural variability of the deposit and brittleness of oxide minerals. Due to the brittleness of many oxide minerals, copper tends to crumble in the process of processing and going to the fine classes, which further complicates their flotation. The individual flotation reagents have their own specific limitations, which can also contribute to the deterioration of the flotation results.

Generally, three basic methods of flotation of oxidized Cu ores are used:

1. Flotation after previous sulphidisation of the surface of Cu oxide minerals.
2. Direct flotation of Cu oxidized minerals using fatty acids and their soaps
3. Direct flotation using a new type of reagents.

The most popular method of flotation of Cu oxide minerals is the prior sulphidisation of oxide mineral surfaces using different types of sulphides (NaHS, Na_2S and $(NH_4)_2S$), followed by flotation using thiol collectors. This method has been used in a relatively large number of concentrators, e.g. in China, Africa and other countries [17-20].

The practical implementation of flotation with sulphidisation is difficult because too little addition of the sulphidising reagent will cause only some oxide particles to be flotated, and too much - their depression [21, 22]. In both cases the flotation results will be worse than they could be. In laboratory conditions, the controlled redox method (CPS - Controlled Potential Sulphidisation) is very helpful, but it fails in industrial conditions. In addition, the optimal dose of the sulphidising reagent depends primarily on the mineral composition of the ore to be treated, on the time and manner of the ore conditioning, the method of mixing and other variables [8]. These are factors that are largely out of control in industrial conditions, which results in an unacceptably high fluctuation in the oxide ore flotation results. In the largest part, this is due to the natural variability of the deposit, which seriously hinders, or even practically prevents, applying the appropriate dose of the sulphidising agent.

For flotation of sulphidized Cu oxide minerals, the most commonly used are xanthates or other reagents designed for sulphide flotation - usually with a longer hydrophobic chain. In relation to sulphide ores, flotation of oxidized ores is characterized by reduced selectivity and reduced yields. For example, in the Cu-Co ore processing plant using NaHS, in Katanga, Democratic Republic of the Congo, about 77% Cu and 75% cobalt were recovered, with an concentration ratio of just 3.0 [23]. If the NaHS and $(NH_4)_2S$ mixture was used as the sulphidising agent in the ratio 1:1, the concentration results were slightly better: the recovery was about 80% for Cu and Co, and the enrichment ratio was 3.5.

IOP Conf. Series: Materials Science and Engineering **427** (2018) 012037 doi:10.1088/1757-899X/427/1/012037

The second, less popular method of flotation of oxidized Cu ores, without the need for prior sulphidisation, is the use of fatty acids and their soaps. Such reagents are best suited for quartz or silicate gangue, because they cannot be used if gangue contains Ca^{+2} ions in its composition, e.g. limestone, dolomites, etc.

Due to specific flotation disadvantages when both sulphidisation and fatty acids are used, long-lasting attempts to find other reagents for direct flotation of oxide copper minerals have been sustained. On the laboratory scale, satisfactory results were achieved with a number of reagents, such as complexing reagents, fatty amines, petroleum sulphates, but in an industrial scale, the number of applications of these new compounds was very limited [24]. In this group, one of the most interesting reagents is hydroxamic acid and its various alkyl hydroxamates (R-CHONHO) [20]. Its biggest advantage of hydroxamates is their higher selectivity than the reagents currently used for flotation of oxidized ores [23].

These difficulties in controlling the flotation process, fairly low concentration ratios, high costs of using special flotation reagents and the established position of hydrometallurgical methods mean that currently relatively few Cu oxide ores are processed using the flotation method [21].

4. Identification and analysis of the behavior of oxidized copper in the industrial enrichment process

4.1. Oxidized copper behavior in the flotation process - balance results

Work on identifying the behavior of oxidized copper in flotation circuits began with the collection and analysis of annual technological results for all concentrators of KGHM Polska Miedź S.A. Table 2 gives the average results of the content of oxidized copper in balance products i.e. Feed (F), Tailings (T) and Concentrate (C) for the years 2000-2015 and the calculated on this basis the average concentration ratios (CR) and recoveries (R). The same table also gives the fundamental statistics calculated for these results, such as minimum and maximum value, standard deviation (SD) and relative standard deviation (RSD.

Table 2. Average annual Cu_{ox} content in the balance products of concentrators, the most important concentration indexes and selected statistics for the concentrators A, B and C in the years 2000-2015.

Statistics	Concentrator A					Concentrator B					Concentrator C				
	F, %	T, %	C, %	CR	R, %	F, %	T, %	C, %	CR	R, %	F, %	T, %	C, %	CR	R, %
Average	0.15	0.5	0.79	5.6	68.1	0.23	0.09	1.07	4.7	64.0	0.29	0.09	1.21	4.23	73.7
Population	16	16	16	16	16	11	11	11	11	11	6	6	6	6	6
Max	0.22	0.08	0.98	7.7	76.8	0.26	0.10	1.23	5.3	72.9	0.30	0.10	1.44	4.86	76.43
Min	0.09	0.04	0.58	3.7	48.7	0.20	0.08	0.85	3.3	55.3	0.25	0.08	1.07	3.57	71.47
SD	0.04	0.01	0.12	1.3	7.59	0.03	0.01	0.11	0.65	8.05	0.02	0.01	0.13	0.48	1.73
RSD	24.4	20.2	14.6	23.4	11.1	12.2	8.2	10.0	13.7	12.6	6.0	7.5	10.5	11.4	2.30

The content of oxidized copper in tailings from the concentrator B and C amounts to 0.23 and 0.29% respectively, while the smallest content is observed for plant A, where it is 0.15%. The relatively highest - though still small - oxidized copper content is observed in concentrates: in the case of concentrate from plant C, Cu_{ox} content reaches on average to 1.21%, and in the case of B concentrate it is slightly lower and amounts to 1.07%. The fact of occurrence of Cu_{ox}'s largest contents in concentrate, and the smallest - in wastes is quite surprising.

The very good recoveries of oxidized copper, exceeding 60%, are also surprising, with the case of plant C equal to 73.7%.

Due to the mechanism of behavior of oxidized copper in the flotation process one would expect both lower Cu_{ox} contents in concentrates and significantly lower recoveries. However, the results are opposite. The explanation may be - at least in part - secondary oxidation of sulphides, as a result of long-term storage of analytical samples under conditions of air access.

The variability of the content of oxidized copper in all concentration products and in all concentrators was also analyzed, as well as variability of recovery and concentration ratio. Comparing these parameters with the analogous Cu_{tot} variations determined in the research carried out in previous years, the variability of results for oxidized copper turns out to be generally slightly higher than for total copper. It should be emphasized that comparing Cu_{ox} and Cu_{tot}, the least variability of such indexes as recovery and copper content in concentrate is usually obtained by Cu_{tot}. This is the result of the company's conscious policy of stabilizing the most important concentration indexes.

In order to determine the possible relationships between particular parameters, correlation coefficients were calculated for the relationship between the content of Cu_{tot}, Cu_{ox}, oxidation degree and recovery. In the vast majority of cases, the value of correlation coefficients turned out to be low or very low. The very low values of correlation coefficients between the Cu_{ox} content in the feed and the Cu recovery for all concentrators turned out to be particularly surprising. However, this is not a proof of the lack of significant dependencies between the analyzed variables, it only means that these relationships have not been revealed in the specific conditions of the analyzed population of results. The most important conditions include quite low level of oxidized Cu - in particular in flotation tailings and feed, behavior of Cu_{ox} in the flotation process and - first of all - accuracy of chemical analyzes and observed range of variability of analyzed parameters, especially Cu_{ox} content.

4.2. Behavior of oxidized copper in the flotation process – the results of sampling

In this subsection, the same abbreviations for balance products and concentration indexes as in subsection 4.1 were used.

4.2.1. Collection and analysis of concentration products. Due to the long-term storage of balance samples under conditions of free air access, it is possible that the results of the determination of the oxidized copper content are overstated. Therefore, a decision was made to carry out specialist sampling. Sampling balance products, i.e. feed, tailings and concentrate, in all concentrators of KGHM Polska Miedź S.A. was carried out from August 16 to 19, 2016, for a period of 9 shifts. The samples were then dried, averaged, reduced, ground and thoroughly mixed. The content of total Cu and CuO was determined in the prepared samples. The sampling and determination of the content of the listed components were carried out by the Center for Quality Research in Lubin

4.2.2. Results and their discussion. Using the obtained results of analyzes of industrial concentration products, Cu_{ox} content and its shares in particular concentration products in relation to Cu_{tot} were calculated as well as basic concentration indexes such as concentrate yield, CR (Concentration Ratio) and R (Recovery) for both total Cu, and oxidized. Obtained results of testing and results of calculations are given in table 3.

The Cu_{ox} contents found in all cases in samples taken during specialist sampling are clearly lower than for average annual samples. In turn, Cu_{ox} recoveries are smaller. This confirms the thesis about the oxidation process as a result of long-term stored analytical samples. What is surprising is the fact of Cu_{ox}'s high recoveries, as the technological flow-sheets of KGHM Polska Miedź SA's concentrators as well as the reagent regime in force are only suited to the recovery of sulphides. In the concentrator B, more than half of the oxidized copper contained in the feed is recovered, and in plant C almost 70%. Similar results, taking into account differences in the quality of the concentrates, were obtained during laboratory flotation experiments with the ore from plant C: after 30 minutes of reference flotation, using the same flotation reagents as in the industry, Cu_{ox} recovery was 80.7% with a Cu_{tot} yield of 91.7%.

Table 3. Average Cu, CuO and Cu_{ox} contents in the balance products of A, B and C concentrators and basic concentration indexes for Cu_{ox} and Cu_{tot} – sampling campaign.

Specification	Content, %			Yield,	CR	R,
	F	T	C	%		%
Concentrator A						
Cu, %	0.98	0.14	14.58	5.82	14.9	86.5
CuO, %	0.046	0.019	0.288	10.04	6.3	62.8
Cu_{ox}, %	0.0368	0.015	0.230	10.04	6.3	62.8
Share of Cu_{ox}, %	3.76	10.86	1.58	-	-	-
Concentrator B						
Cu, %	1.68	0.21	25.57	5.80	15.2	88.2
CuO, %	0.13	0.068	0.825	8.19	6.3	52.0
Cu_{ox}, %	0.104	0.054	0.66	8.19	6.3	52.0
Share of Cu_{ox}, %	6.19	25.90	2.58	-	-	-
Concentrator C						
Cu, %	1.66	0.20	26.49	5.55	16.0	88.6
CuO, %	0.098	0.034	0.895	7.43	9.1	67.9
Cu_{ox}, %	0.078	0.027	0.716	7.43	9.1	67.9
Share of Cu_{ox}, %	4.72	13.60	2.70	-	-	-

Table 4. The results of sieve analysis and contents of Cu_{tot}, CuO and Cu_{ox} in the individual particle classes of flotation tailings in concentrators A, B and C –sampling campaign.

Particle size +, mm	Yield, w/w%	Cum. yield passing, w/w%	Content			Share of Cu_{ox}, %	Distribution, %		
			Cu, %	CuO, %	Cu_{ox}, %		Cu,	CuO	Cu_{ox}
Concentrator A									
0.125	16.7	83.3	0.22	0.015	0.012	5.5	29.3	13.0	13.0
0.100	35.7	47.6	0.07	0.008	0.006	9.1	19.9	14.8	14.8
0.075	10.2	37.4	0.08	0.012	0.010	12.0	6.5	6.3	6.3
0.036	11.0	26.4	0.12	0.017	0.014	11.3	10.5	9.7	9.7
-0.036	26.4	-	0.16	0.041	0.033	20.5	33.7	56.1	56.1
F	100.0	100.0	0,13	0,019	0,015	12,3	100,0	100,0	100,0
Concentrator B									
0.125	13.7	86.3	0.15	0.020	0.016	7.3	8.7	4.6	4.6
0.100	17.2	69.1	0.27	0.032	0.026	9.5	19.7	9.2	9.2
0.075	9.7	59.4	0.32	0.043	0.034	10.8	13.2	7.0	7.0
0.036	15.3	44.1	0.32	0.040	0.032	10.0	20.8	10.3	10.3
-0.036	44.1	-	0.20	0.093	0.074	37.2	37.5	68.9	68.9
F	100.0	100.0	0.24	0.060	0.048	20.3	100.0	100.0	100.0
Concentrator C									
0.125	31.3	68.7	0.22	0.015	0.012	5.5	32.8	14.8	14.8
0.100	37.1	31.6	0.13	0.017	0.014	10.5	23.0	19.8	19.8
0.075	9.5	22.1	0.25	0.026	0.021	8.3	11.3	7.8	7.8
0.036	8.5	13.6	0.30	0.038	0.030	10.1	12.2	10.2	10.2
-0.036	13.6	-	0.32	0.111	0.089	27.8	20.7	47.5	47.5
F	100.0	100.0	0.21	0.025	0.032	15.2	100.0	100.0	100.0

Most strikingly, the Cu_{ox} concentration ratios given in table 3 for all concentrators are high, amounting to 6.3 (concentrator A and B), and even very high - 9.1 (plant C).

All these results show clearly better concentration of Cu oxide minerals found in Polish copper ores than in other ores described in the literature [23].

To assess the possibility of Cu recovery from flotation tailings, one of the most important issues is: the form of copper occurrence and its distribution in individual particle classes. Table 4 presents the results of sieve analysis and the content and distribution of selected components (Cu, CuO Cu_{ox}) in individual particle classes of flotation tailings from all concentrators.

Depending on the plant, the largest contents of Cu_{tot} in flotation tailings are observed in different classes: in the case of concentrator A it is the thickest class, in the case of plant B - these are intermediate classes, and in the case of plant C - it is the smallest class. The distribution of total copper in individual particle classes generally shows a fairly good correlation with the yield of a given class.

In the case of Cu_{ox}, its largest content, for all plants, is observed in the smallest class of flotation tailings - below 36 μm. The Cu_{ox} distribution is also similar: for all concentrators, oxidized copper accumulates mainly in the class below 36 μm. For concentrators A and B, in the smallest class, it collects more than twice as much Cu_{ox} than the outgoings of this class, and in the case of plant C - more than three times more. As a result, almost ¾ of oxidized copper accumulates in the smallest particle class - in the case of plant B, and more than half or nearly half of this form of copper - in plants C and A, although the share of the smallest class of flotation tailings in these plants is only 26.4 (A), 44.1 (B) and 13.6% (C).

The highest content of oxidized copper is observed in final concentrates (see table 3). They range from 0.23% Cu - for plant A, through 0.66% - for plant B, to 0.72% - for plant C. However, due to the high copper content of the concentrates, the Cu_{ox} shares in the concentrates are clearly smaller than in tailings or in the feed, which is consistent with the oxide flotation model.

5. Perspectives for the recovery of oxidized copper from flotation tailings

The observed, very low content of oxidized copper in flotation tailings determines the maximum possible level of income from recovery of oxidized copper and the type of concentration methods and equipment that could be used for this purpose.

Table 5 shows the value of recoverable oxidized and total copper from flotation tailings of individual concentrators.

Table 5. The value of recoverable oxidized and total copper included in flotation tailings of concentrators A, B, and C.

Concentrator	Ore throughput dry mass		Tailing yield, (%)	Cu content in tailings, (%)		Recoverable copper mass in tailings, (Mg)		Value of recoverable Cu, (mln USD)	
	(Mg/d)	(Thous. Mg/rok)		Cu_{ox}	Cu_{tot}	Cu_{ox}	Cu_{tot}	Cu_{ox}	Cu_{tot}
A	22 450	8 194	94,21	0,015	0,13	1 229	10 653	7,3	62,8
B	25 180	9 191	94,06	0,054	0,20	4 963	18 381	29,3	108,5
C	45 580	16 637	93,99	0,027	0,18	4 492	29 946	26,5	176,7
Total	93 210	34 022	-	-	-	10 684	58 980	63,0	348,0

The calculations involved ore throughputs and contents of Cu_{tot} and Cu_{ox} in flotation tailings actually produced in individual concentrators. The copper price was adopted at USD 5.900 per Mg. In the absence of appropriate studies, it was assumed optimistically that the recovery of oxidized Cu from flotation tailings will amount to 60%.

The estimation of the value of oxidized copper, which can theoretically be recovered from flotation tailings, is quite attractive at first sight, especially in the case of plants B and C, but the remaining aspects of the possible recovery process should be considered, especially the assumed level of capital and operating costs.

To further specify the above estimates, it is necessary first and foremost to perform laboratory tests to determine the actual indicators for the concentration of oxidized copper. Based on the literature information [9, 14-17, 19-20, 24], it can be predicted that the Cu contents in the flotation concentrate will be of the order of at most a few percent, if a satisfactory recovery is to be achieved. This is too low a value for this type of product to be considered as a qualified concentrate. After applying an appropriate sales formula to such a concentrate, the value of recovered copper will significantly decrease in relation to the values given in table 6, even if the recovery is maintained at 60%.

Another unfavorable factor is the low content of Cu_{ox} in flotation tailings in all concentrators as well as properties of the slurry of produced flotation tailings. It can be anticipated, the required number of flotation cells will be significant, which will translate into high capital and operating costs. One should also expect additional, significant investment outlays to increase the milling power. With interesting but not very high incomes (see table 5), it is essential to invest very economically, for example by shortening the flotation time of oxides through e.g. the intensification and/or implementation of oxide flotation in the final phase of the main sulphide flotation or by other ways. Confirmation of such possibilities also requires conducting appropriate tests.

Due to the low Cu_{ox} content in the flotation tailings, the technology that guarantees the simultaneous recovery of sulphide and oxide copper would be much more rational. This would significantly increase the level of income. Table 5 shows the mass of recoverable total (sulphide and oxide) copper found in the tailings and its value, assuming the same recovery for both type of copper as before. In this case, the revenues are about 6 times higher than in the recovery of only oxidized copper. It means that the margin for investment is clearly larger. Even higher revenues could be obtained by using hydrometallurgical methods, which would allow the recovery of both sulphide and oxidized copper, with higher recoveries than in the case of flotation only.

It seems, therefore, that further research should be carried out in this direction, despite the fact that previous experience with the recovery of copper from tailings by flotation, hydrometallurgical method or hybrid method indicates that even for such a technology obtaining the profitability of an investment will be difficult [10].

6. Summary and conclusions

1. Based on the previous diagnosis, oxidized copper contained in the feed processed at individual concentrators of KGHM Polska Miedź S.A. occurs primarily as cuprite, tenorite, azurite, malachite and chrysocolla. The presence of other Cu_{ox} minerals is so far doubtful.
2. The share of oxidized copper - determined on the basis of long-term average annual results - in the feed of all concentrators is slightly more than 10% of total copper, in tailings - from 27.1% to 38.6%, and in concentrates not more than 4.1%. Shares of oxidized copper in samples of feed, tailings and concentrate collected during sampling on August 16-19, 2016, are clearly smaller than the data cited above, which is understandable due to the chemistry of the oxidation processes.
3. On the basis of specialist sampling, it was found that the content of oxidized copper in the flotation tailings ranges from 0.015-0.054%, and in the concentrate from 0.23 to 0.72%. The share of oxidized copper is highest in the case of flotation tailings and varies from 10.7 to 25.7%, and in concentrates it is only a few percent. For the multiannual results, the contents of oxidized copper and its shares in enrichment products for all concentrators are clearly higher.
4. Achieving high Cu_{ox} recoveries (over 50%) in technological circuits of KGHM Polska Miedź S.A. concentrators, which are not adapted to the recovery of oxidized copper, requires further research in order to determine the reasons for this behavior.
5. The low content of Cu_{ox} in the concentrator feed, its significant (up to approx. 70%) recovery during sulfide flotation and rather fine particle of oxide Cu minerals make the prospects of

economic recovery of the remaining oxidized copper part for flotation tailings not very successful.

6. Taking into consideration the way of occurrence of Cu_{ox} and Cu_{tot} in flotation tailings as well as available copper recovery techniques, only hydrometallurgical methods, including bio-methods, may be effective, however the issue of economic profitability of such a project is questionable under the current conditions.

References

[1] Bolewski A 1982 *Mineralogia szczegółowa wydanie trzecie* (Warszawa: Wydawnictwo Geologiczne)
[2] Opracowanie zbiorowe 2007 *Monografia KGHM Polska Miedź S.A.* (Lubin: CBPM Cuprum)
[3] Grotowski A and Witecki K 2015 Etap II Badania w skali półtechnicznej nad mieleniem rudy miedzi w doświadczalnym młynie pierścieniowo-walcowym część II Wykonanie badań laboratoryjnych produktów mielenia i opracowanie wyników prób mielenia w doświadczalnym młynie Loesche *Raport KGHM Cuprum sp. z o.o. CBR* Wrocław
[4] Garbacki M 2016 *Raport mineralogiczny O/ZWR KGHM PM S.A. April 2016* Polkowice
[5] Lascelles D, Peters O and Caldwell R 2009 The Impact of Crushed Ore Ageing on Metallurgical Performance SGS Mineral Services *Technical paper 2009-07* Lakefield
[6] Drzymała J and Kozłowski A 2002 Wpływ czasu przechowywania rudy miedzi Lubin na jej ksantogenianową flotację *Prace Naukowe Instytutu Górnictwa Politechniki Wrocławskiej* 102 *Studia i Materiały* nr 29 Wrocław
[7] Gaudin A M 1957 *Flotation* (New York: McGraw-Hill Book Company Inc.)
[8] Castro S, Goldfrab J and Laskowski J 1974 Sulphidisation reactions in the flotation of oxidized copper minerals I Chemical factors in the sulphidisation of copper oxide *Int. J. Miner. Proc.* pp 141-9
[9] Dong-Hyun, Lee Dae, Sung Kim and Seung-Ho Lee 2015 Effect of Collectors for Malachite Flotation from Carbonate Ore Mixed with Malachite, Dolomite and Calcite *J. Korean Soc. Miner. Energy Resour. Eng.* **52** pp 1-8
[10] Kiraz E 2014 Recovery of Copper from Oxide Copper Ore by Flotation and leaching *Master of Science Thesis The Graduate School of Natural and Applied Sciences of Middle East Technical University*
[11] Grotowski A 2007 Możliwości i perspektywy zastosowania metod hydrometalurgicznych w KGHM PM S.A. *VIIIth Int.. Conf. On Non-ferrous Metal Ore Proc.* 21-23 May Wojcieszyce pp 29-46
[12] D`Hugues P, Norris P R, Hallberg K B, Sanchez F, Langwaldt J, Grotowski A, Chmielewski T and Groudev S 2008 Bioshale FP6 European project Exploiting black shale ores, using biotechnologies? *Mining Engineering* **21** pp111-20
[13] Chmielewski T 2015 Development of a hydrometallurgical technology for production of metals from KGHM Polska Miedź S.A. concentrates *Physicochem. Probl. Miner. Proc.* **51(1)** pp 335-50
[14] Szubert A and Grotowski A 2011 Elementarne czynniki kontroli optymalizacji procesów bioługowania siarczkowych rud miedzi *Copper Metallurgy Conference - 50th Anniversary of KGHM Polska Miedź S.A.* May 26-28 Kraków pp 319-30
[15] Singh R, Chandrashekhar S, Rath R K. Nayak B and Bhattacharyya K K 2008 Designing Process for Concentration of a Low Grade Copper Ore *Int. Seminar on Min. Proc. Technology (MPT-2008)* April 22-24 Trivandrum
[16] Gush J C 2005 Flotation of Oxide minerals by Sulphidisation – the Development of Sulphidisation Control System for Laboratory Testwork *The Journal of The South African Institute of Mining and Metallurgy* **105** pp 193-98
[17] Wenbin Z 1993 Thirty-Year Plant Practice of Sulphidisation - Flotation of Copper Oxidised Ores *XVIII Int. Min. Proc. Congress* Sydney pp 23-8

[18] Chabuka C and Witika L K 2001 Optimisation of the Baluba East ore treatment *African Journal of Science and Technology* **1(4)** pp 36–42
[19] Wills B A and Napier-Munn T 2006 *Wills' Mineral Processing Technology (7th ed.)* (Oxford: Elsevier Science & Technology Books)
[20] Ferron C J and Manu N N 1994 Recovery of copper oxide minerals by sulphidisation flotation in *Mulukutla P S (Ed.) Reagents for better metallurgy p 156–64 Littleton*: Society for Mining, Metallurgy and Exploration Inc.
[21] Ziyadanogullari R and Aydin F 2005 A New Application For Flotation Of Oxidized Copper Ore *Journal of Minerals & Materials Characterization & Engineering* **4/2** pp 67-73
[22] Lee K. et al. 2008 Flotation of mixed copper oxide and sulphide minerals with xanthate, *Miner. Eng.*
[23] Tebogo P P and Edison M 2010 A Multistage Sulphidisation Procedure for a Low Grade Malachite Copper Ore *Int. J. Chem. Molec. Nucl. Materials and Metallurgical Eng.* **4/9** pp 580-6
[24] Lee J and Nagaraj D R and Coe J E 1998 Practical aspects of oxide copper recovery with alkyl hydroxamates *Miner. Eng.* **11(10)** pp 929-39
[25] Kalichini M S 2015 A study of the characteristics of a complex copper ore *Master Thesis University of Cape Town*

IOP Conf. Series: Materials Science and Engineering **427** (2018) 012038 doi:10.1088/1757-899X/427/1/012038

Analysis of improving the quality of coal slurry based on the use of regression models

A Manowska[1] and M Mazurek[2]

[1]Faculty of Mining and Geology, Silesian University of Technology, Akademicka 2, 44-100 Gliwice, Poland

[2]JSW Innowacje S.A., Paderewskiego 41, 40-282 Katowice, Poland

anna.manowska@polsl.pl and michalmazurek@jswinnowacje.pl

Abstract. The article presents the results of research on the enrichment of fine-grained coal grades, also known as coal slurries. The research was carried out on a semi-technical scale, using the method of gravity flow enrichment on a spiral separator type Reichert LD-4. The research material used in the research were three types of coal slurries, diversified in qualitative terms that were taken from technological process of the preparation plants. The enrichment efficiency tests were carried out in relation to changes in the density of the feed directed to the separator. The best distribution results for all tested coal slurries were obtained for the feed density of $\beta = 300$ g/dm^3, where weighted averages of ash content in the obtained concentrates reached values of several percent. A statistical analysis was also performed, to which the multiple regression model was used. Using the constructed models, it is possible to determine which technological parameters affect the process of enriching coal sludge.

1. Introduction

Coal preparation plants are intermediaries between the mining and distribution stages. The mined raw material is a mixture of grains of various sizes, which consist of both organic and mineral matter and are described by a number of quality parameters. In the case of steam coal, such parameters are ash, sulphur, moisture content and heating value. In the beneficiation process, the raw coal is separated into density fractions and after reaching the appropriate level of enrichment, the stage of creating commercial assortments takes place. Taking into account the environment protection conditions, it is increasingly important to provide products with low emissivity [1,2]. This in turn is ensured by the correct course of the technological process of the coal preparation plant, which also improves the economic effects of its production. Thus, the article focuses on the review of mathematical models used in coal preparation plants, which are used in the simulation of the technological process of the technique of enriching minerals.

2. Source and classification of hard coal tailings

Output coal mined on the surface of the mine based on the used deposit exploitation technology, is a mixture of grains of various sizes ranging from several dozen millimetres to several microns. This mixture consists of coal grains, gangue, coal and dirt bands and carbonaceous shale. Thus, the generation of tailings is inseparably connected with the activity of hard coal mining. It is assumed that the extraction of one ton of coal is accompanied by the formation of about 400 kg of tailings [4], and according to other estimates, the amount of waste generated can range from 200 to 400 kg [5,6]. The ratio of the amount of tailings to coal production in individual mines varies, which results from the different geological structure and technical conditions of coal mining and processing.

At the present time, the utility value represented by the mined rock excavated from the bottom does not meet any form of interest on the part of potential buyers, due to the relatively high content of gangue in relation to the useful (coal) grains. Excessive amount of gangue that goes into the mined rock may result from ripping the roof and the floor when mining the deposit. The waste rock from the useful grains is separated using the enrichment processes.

Useful grains are separated from waste rock using the enrichment processes. These processes can be carried out in an aqueous and air medium, however, considering the very low accuracy of aerodynamic enrichment (high values of imperfection index), in domestic technological gravitational enrichment the main medium in which the separation process takes place is the aqueous medium [3]. The smallest grains in the <1mm class, commonly referred to as slurries, arise as a result of coal enrichment processes. Coal slurries by definition fall within lower quality parameters, i.e. low calorific value and high ash content. Considering the fact that the slurry products have a certain energy potential, it seems reasonable to undertake all actions aimed at its complete recovery [11,12]. The process of improving the quality of coal slurry can be carried out through the use of selected methods of gravity enrichment, i.e. enrichment on shaking tables, spiral separators, enrichments using centrifugal force or flotation enrichment [7,8,9,10]. Taking into account the economic conditions of the enrichment process for coal sludges, the cheapest of the abovementioned separation methods is enrichment on spiral coils.

3. Methodology and Data

The research on the possibility of improving the quality parameters of coal slurries was carried out on a spiral separator Reichert LD4. Three types of slurries diversified in terms of quality were analysed. The slurries analysed were named respectively Slurry 1, Slurry 2, Slurry 3. Prior to being directed to the concentrator unit, the slurries were subjected to a quantitative and qualitative analysis, the obtained results are presented in table 1.

Table 1. Quantitative and qualitative analysis of raw slurries.

	Slurry 1		**Slurry 2**		**Slurry 3**	
Size grade [mm]	Yield Y (%)	Ash content A^a(%)	Yield Y (%)	Ash Content A^a(%)	Yield Y (%)	Ash content A^a(%)
>0.75	41,32	21,17	55,89	44,06	52,21	23,01
0.7-0.5	32,11	32,50	7,84	44,98	7,77	16,14
0.5-031.5	31,88	21,65	7,42	41,51	5,82	19,35
0.315-0.2	33,32	13,27	19,61	40,83	14,02	20,14
0.2-0.1	30,17	17,41	3,73	39,81	6,47	23,59
< 0.1	27,73	14,01	5,51	48,00	13,71	32,94

source: own elaboration

Analysing the data contained in table 1, it can be noticed that Slurry 2 in relation to the other two materials is characterized by the worst qualitative parameters (the highest ash content) in all analysed

IOP Conf. Series: Materials Science and Engineering **427** (2018) 012038 doi:10.1088/1757-899X/427/1/012038

size grades. Ash content of individual size grades of Slurry 1 and Slurry 3 occurs in similar values, except for size grades 0.2-0.1mm and <0.1mm Slurry 1, where the ash content reaches a few percent.

Knowing the ash content in individual size grades in the tested raw slurries, they were directed to enriching on a spiral coil concentrator type Reichert LD4. The beneficiation of slurries was performed depending on the changing density of the feed successively β=350 g/dm^3 and β= 300 g/dm^3. Such selection of density of the feed was influenced by the conducted trial series aimed at the optimal selection of technological parameters of the separation. The feed for Reichert LD4 spiral was directed without pre-desliming (removing of clay) to check the effect of the increased amount of clays on the separation process. The feed for the concentrator was fed from a sender tank equipped with a mechanical stirrer and an aerator. As a result of enriching the feed, three separation products were obtained, i.e. concentrate, semi-finished product and tallings, from which samples were taken for technical analysis. Quantitative and qualitative analysis of enriched slurries Slurry 1, Slurry 2, Slurry 3 for density β=350 g/dm^3 and β= 300 g/dm^3 are presented in figures 1-6.

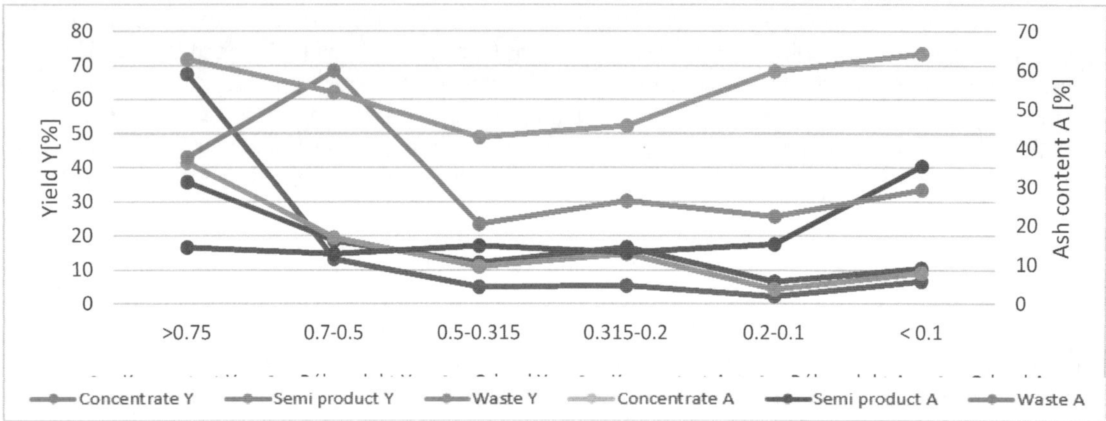

Figure 1. Quantitative and qualitative analyses of products for enrichment Slurry 1 density feed β=350 g/dm^3.

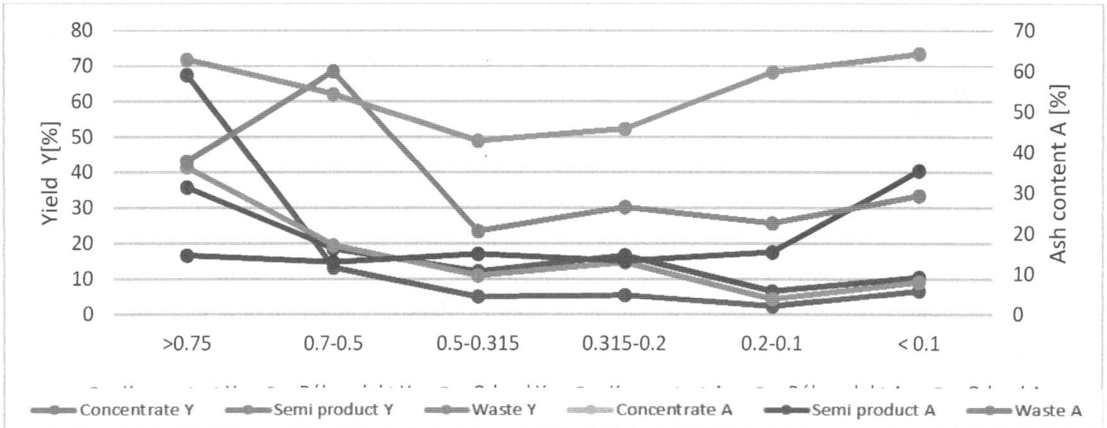

Figure 2. Quantitative and qualitative analyses of products for enrichment Slurry 1 density feed β=300 g/dm^3.

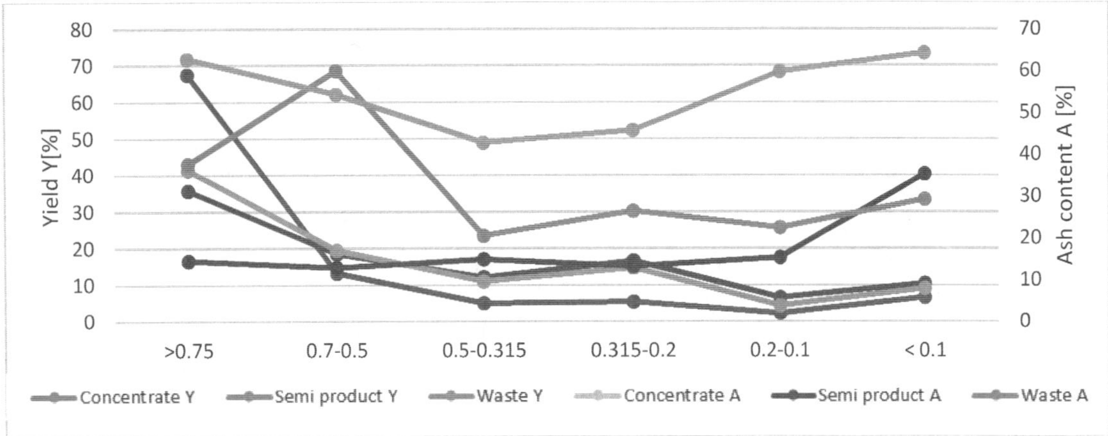

Figure 3. Quantitative and qualitative analyses of products for enrichment Slurry 2 density feed β=350 g/dm^3.

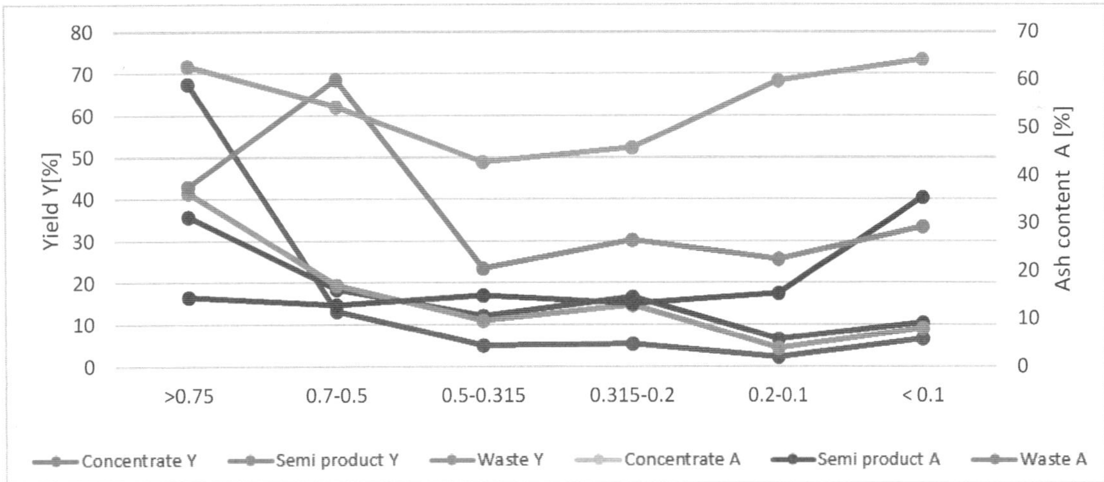

Figure 4. Quantitative and qualitative analyses of products for enrichment Slurry 2 density feed β=300 g/dm^3.

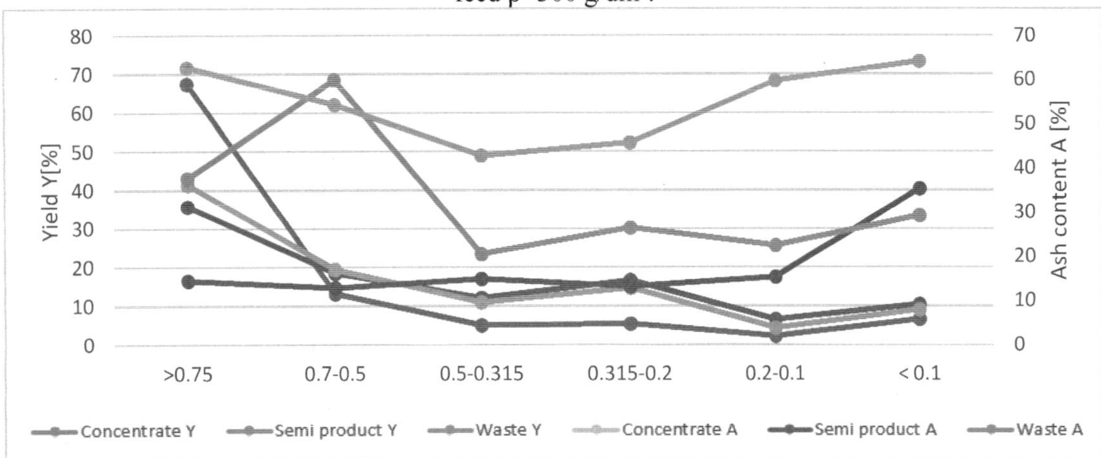

Figure 5. Quantitative and qualitative analyses of products for enrichment Slurry 3 density feed β=350 g/dm^3.

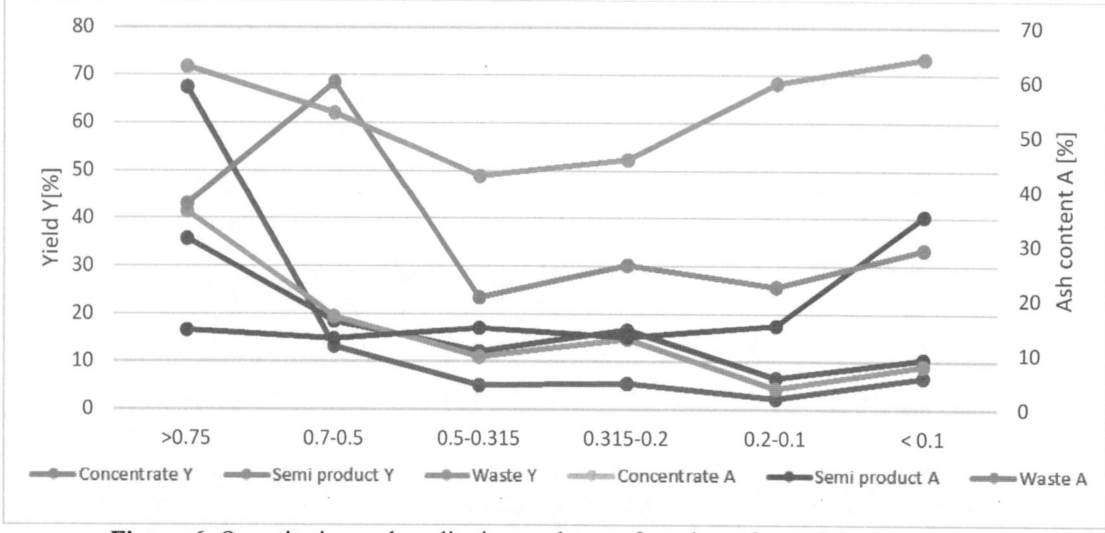

Figure 6. Quantitative and qualitative analyses of products for enrichment Slurry 3 density feed β=300 g/dm^3.

The analysis of the enrichment process of the three studied slurries at given densities of the feed shows that the best results were obtained for the Slurry 1 concentrate and the density of the feed β=300 g/dm^3. Similar results were obtained for the Slurry 3 concentrate and the density of the feed β=300 g/dm^3. The concentrates of Slurry 1 and Slurry 3 for density β=350 g/dm^3 are characterized by a higher content of ash in individual size grade. This phenomenon is most probably affected by too much solid phase in the feed and is connected with the unfavourable course of the grain separation process that occurs as a result of the impact of the restrained settling velocity. Ash content in semi-finished products of Slurry 1 and (Slurry 3) suggests that it is necessary to subject these products to secondary enrichment due to some content of useful grains in them. In the case of Slurry 2, a very small improvement in the value in use was found for both feed densities. Apparently, in the case of this material, the enrichment processes previously carried out on it were so accurate that only grains of gangue remained in the size grades of this slurry.

4. Research method - statistical analysis

In order to verify the assumed H_0 hypothesis that the enrichment process does not change the quality parameters of the analysed slurries, in the face of the alternative H_1 hypothesis that there is a change in the quality parameters of the analysed slurries, a multiple regression model was built. The general purpose of multiple regression is to quantify the relationship between many independent variables (explanatory) and dependent variable (criterial). The changeable variable is the yield and ash content of raw slurries, while the explanatory variables are the content of yield and ash after the enrichment process. In multiple regression, the rest of the model is assumed to be normally distributed, and again, despite that most tests (in particular, test F) are quite resistant to deviations from this assumption, it is always a good habit before drawing final conclusions to check how the distributions of major variables that are subject of the study look.

5. Results of research - statistical analysis of the correctness of the slurries enrichment process on spiral separator.

The first stage of statistical analysis is to determine basic descriptive statistics. This stage was made using the box plots chart (shown in figure 7) to show how the measures of the central position and the measure of variability of non-enriched products change (Zmn 1 and Zmn 2) against the products obtained after the enrichment process (Zmn 3 to Zmn 14) . The following variable designations have been adopted:

IOP Conf. Series: Materials Science and Engineering **427** (2018) 012038 doi:10.1088/1757-899X/427/1/012038

- Zmn 1 – amount of yield in raw slurries
- Zmn 2 – ash content in the raw slurries
- Zmn 3 – amount of yield in product "concentrate" for the density of the feed β=350 g/dm^3
- Zmn 4 – ash content in the product "concentrate" for the density of the feed β=350 g/dm^3
- Zmn 5 – amount of yield in product "concentrate" for the density of the feed β=300 g/dm^3
- Zmn 6 – ash content in the product "concentrate" for the density of the feed β=300 g/dm^3
- Zmn 7 - amount of yield in product "semi-finished product" for the density of the feed β=350 g/dm^3
- Zmn 8 – ash content in the product "semi-finished product" for the density of the feed β=350 g/dm^3
- Zmn 9 - amount of yield in product "semi-finished product" for the density of the feed β=300 g/dm^3
- Zmn 10 - ash content in the product "semi-finished product" for the density of the feed β=300 g/dm^3
- Zmn 11– amount of yield in product "tailings" for the density of the feed β=350 g/dm^3
- Zmn 12– ash content in the product "tailings" for the density of the feed β=350 g/dm^3
- Zmn 13– amount of yield in product "tailings" for the density of the feed β=300 g/dm^3
- Zmn 14 – ash content in the product "tailings" for the density of the feed β=300 g/dm^3

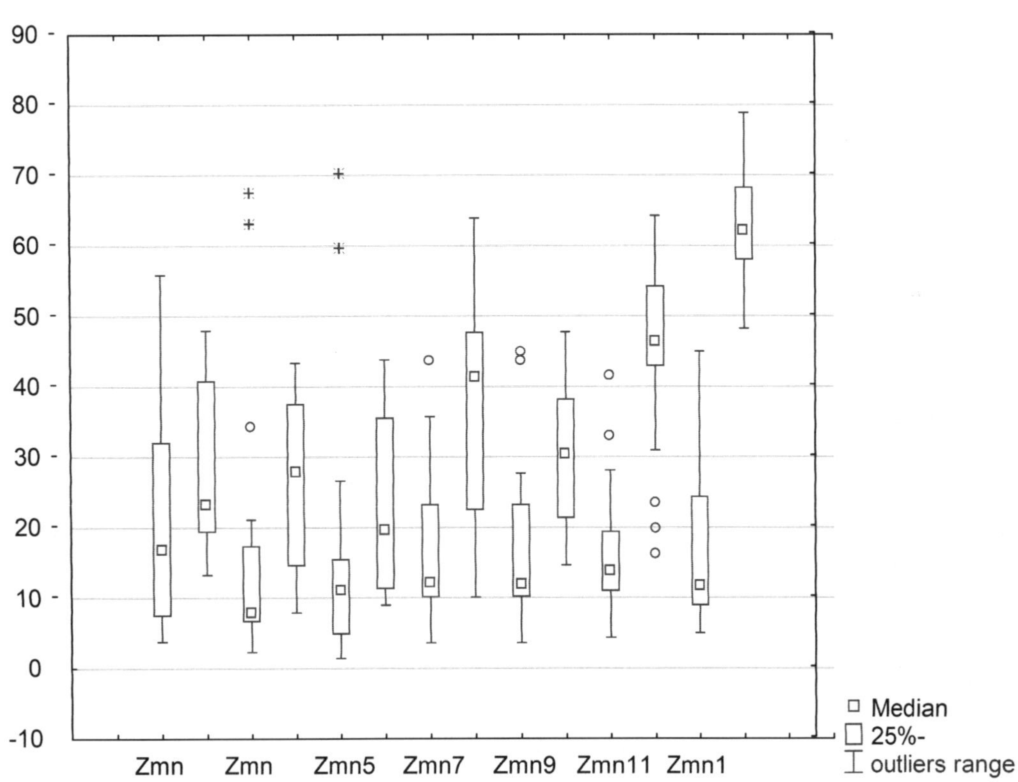

Figure 7. Box plots chart for the analysed variables.

Visual analysis shows that the variables differ significantly from each other. It means that in the enrichment process a product with different parameters can be obtained.

IOP Conf. Series: Materials Science and Engineering **427** (2018) 012038 doi:10.1088/1757-899X/427/1/012038

In order to verify the reliability of the adopted research methodology, the normality of the analysed variables was examined, as shown in figure 8.

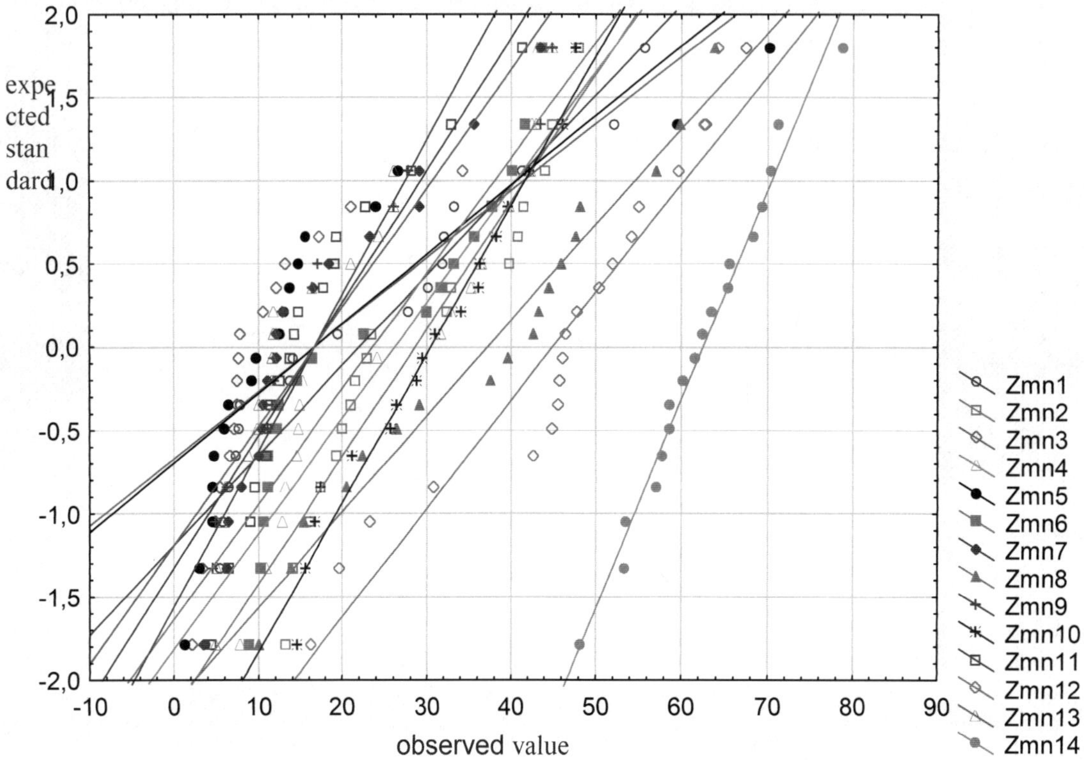

Figure 8. Normality diagram of the analysed variables.

From the analysis, it can be seen that a multiple regression model can be built for each product. The following assumptions have been adopted in the research methodology:

- multiple regression models were created separately for the yield and ash content of raw slurries,
- the explanatory variables were respectively: concentrate, semi-finished product and tallings for the density of the feed β=350 g/dm^3 and β=300 g/dm^3,
- in the multiple regression model, both yield and ash content were introduced,

With red, parameters which are statistically significant according to the assumed Ho hypothesis were marked, which means that raw slurry has similar quality parameters as the product after the enrichment process. The results of calculations are included in table 2-7.

Table 2. A multiple regression model for the yield of raw slurries and concentrate.

Effect	One-dimensional significance tests for Zmn1 Parameterization with sigma-restrictions Decomposition of effective hypotheses; Standard evaluation error: 12.8340				
	SS	Degrees of freedom	MS	F	p
Constant term	905,107	1	905,1071	5,495101	0,035611
Zmn1	205,527	1	205,5269	1,247798	0,284194
Zmn2	544,294	1	544,2939	3,304526	0,092199
Zmn3	81,400	1	81,4000	0,494197	0,494459
Zmn4	795,421	1	795,4207	4,829171	0,046708

Table 3. A multiple regression model for the ash content of raw slurries and concentrate.

Effect	One-dimensional significance tests for Zmn2 Parameterization with sigma-restrictions Decomposition of effective hypotheses; Standard evaluation error:7,4612				
	SS	Degrees of freedom	MS	F	p
constant term	307,8048	1	307,8048	5,529201	0,035133
Zmn1	0,0059	1	0,0059	0,000107	0,991922
Zmn2	15,8331	1	15,8331	0,284416	0,602820
Zmn3	0,3245	1	0,3245	0,005829	0,940303
Zmn4	171,8997	1	171,8997	3,087892	0,102386
Error	723,6962	13	55,6689		

Table 4. A multiple regression model for yield of raw slurries and semi-finished products.

Effect	One-dimensional significance tests for Zmn1 Parameterization with sigma-restrictions Decomposition of effective hypotheses; Standard evaluation error:13,0103				
	SS	Degrees of freedom	MS	F	p
Constant term	481,420	1	481,4204	2,844133	0,115543
Zmn1	0,426	1	0,4257	0,002515	0,960767
Zmn2	187,762	1	187,7621	1,109260	0,311426
Zmn3	860,191	1	860,1908	5,081832	0,042073
Zmn4	683,329	1	683,3286	4,036966	0,065745
Error	2200,482	13	169,2679		

Table 5. A multiple regression model for the ash content of raw slurries and semi-finished products.

Effect	One-dimensional significance tests for Zmn2 Parameterization with sigma-restrictions Decomposition of effective hypotheses; Standard evaluation error:7,6280				
	SS	Degrees of freedom	MS	F	p
Constant term	8,2323	1	8,2323	0,14148	0,712878
Zmn3	22,8318	1	22,8318	0,39239	0,541888
Zmn4	234,0743	1	234,0743	4,02286	0,066160
Zmn5	50,0208	1	50,0208	0,85967	0,370726
Zmn6	780,5730	1	780,5730	13,41514	0,002867
Error	756,4179	13	58,1860		

Table 6. A multiple regression model for yield of raw slurries and tallings.

Effect	One-dimensional significance tests for Zmn1 Parameterization with sigma-restrictions Decomposition of effective hypotheses; Standard evaluation error:14,8908				
	SS	Degrees of freedom	MS	F	p
Constant term	14,828	1	14,8283	0,066874	0,799993
Zmn7	32,698	1	32,6977	0,147463	0,707179
Zmn8	159,725	1	159,7255	0,720344	0,411387
Zmn9	617,620	1	617,6195	2,785395	0,119018
Zmn10	14,239	1	14,2387	0,064215	0,803917
Error	2882,555	13	221,7350		

Table 7. A multiple regression model for the ash content of raw slurries and tallings.

Effect	One-dimensional significance tests for Zmn2 Parameterization with sigma-restrictions Decomposition of effective hypotheses; Standard evaluation error:10,9619				
	SS	Degrees of freedom	MS	F	p
Constant term	1273,725	1	1273,725	10,59995	0,006259
Zmn11	117,461	1	117,461	0,97751	0,340859
Zmn12	84,926	1	84,926	0,70675	0,415719
Zmn13	186,112	1	186,112	1,54882	0,235279
Zmn14	795,289	1	795,289	6,61840	0,023178
Error	1562,123	13	120,163		

The statistical analysis carried out shows that the enrichment process has improved the quality parameters of coal products. Therefore, the accepted methodology in the article is justified. It is also possible to detect a human error during the beneficiation process, which significantly affected the deterioration of this process. This is evident by the statistical significance of the multiple regression model coefficient. A well-conducted enrichment process is characterized by a lack of correlation between the quality parameters of raw slurries and the products obtained, as confirmed in scientific articles [9,10,11].

6. Summary

The process of slurries beneficiation in a spiral concentrator has shown that it is possible to improve the utility value of products (slurries) often considered as unprofitable waste. This possibility depends on the initial state of the feed, i.e. the ash content and the amount of clays in the material as well as the correctly selected technological parameters. The subject of statistical analyses were data defining the quality of feed, concentrates, semi-finished products and tallings. They were analysed in terms of ash content and yield. The statistical analysis carried out confirmed the relationship between the technological parameters of the enrichment process and the obtained products. The selected, optimal parameters of the technological process will allow to obtain concentrates from slurries in an economically feasible way. The proposed method can also capture errors that arise in this process, which in turn will have a positive effect on reducing the costs of the beneficiation process.

References

[1] Gawlik L, and Mokrzycki E 2014 *Przegląd Górniczy* **70** pp. 1-8
[2] Lorenz U 1999 Metoda oceny wartości węgla kamiennego energetycznego uwzględniająca skutki jego spalania dla środowiska przyrodniczego. *Studia, Rozprawy, Monografie* **64**

[3] Blaschke W 2009 Przeróbka węgla kamiennego – wzbogacanie grawitacyjne *Wydawnictwo Instytutu Gospodarki Surowcami Mineralnymi i Energią PAN*
[4] Strzyszcz Z and Harabin Z 2004 Rekultywacja i biologiczne zagospodarowanie odpadów górnictwa węgla kamiennego ze szczególnym uwzględnieniem centralnych zwałowisk. *Prace i Studia* **61**
[5] Ostrowski J 2001 Ochrona środowiska na terenach górniczych *Wydawnictwo IGSMIE PAN Kraków* p 10
[6] Ostrowicz H 1969 Gospodarka odpadami górniczymi i hutniczymi *Wyd. Śląsk, Katowice* p 53
[7] Manowska A and Mazurek M, 2016 Laboratory and statistical efficiency of fine coal enrichment in the spiral separator Reichert LD-4 *Syst. Wspomag. Inż. Prod.* **6** **(3)**, pp 136-149
[8] Jendruś R, Manowska A and Mazurek M 2016 Influence of concentrating table deck inclination angles on energy reclamation potential from fine coal particles based on laboratory and statistical studies *Inżynieria Mineralna J Polish Min Proc Soc. Miner* **2(38)** pp 191-200
[9] Jendruś R, Manowska A, and Mazurek M 2016 *Gosp. Sur. Miner.* **32**(**3**) pp 173-194
[10] Baic I, and Blaschke W, Szafarczyk J 2010 *Przegląd Górniczy* **66**(**1–2)** pp 73-77
[11] Sobko W, and Baic I 2011 *Przegląd Górniczy* **67** pp 32-35
[12] Lutyński A, and Szpyrka J 2012 *Czasopismo Techniczne* **150**

IOP Conf. Series: Materials Science and Engineering **427** (2018) 012039 doi:10.1088/1757-899X/427/1/012039

Sorption rate of CH_4 and CO_2 in coal at different pressure ranges

M Lutynski[1] **and L Zavanella**[2]

[1] Silesian University of Technology, Faculty of Mining and Geology, ul. Akademicka 2, 44-100 Gliwice, Poland
[2] Brescia University, Department of Mechanical and Industrial Engineering, via Branze 38, 25123 Brescia, Italy

E-mail: marcin.lutynski@polsl.pl

Abstract. The aim of this study was to verify the dynamic factor, that is the diffusion rate, which can directly affect the efficiency of CO_2 injection and as a consequence – storage. A manometric setup was used for experiments on two hard coals from Upper Silesian Coal Basin in Poland. A model combining two firs-order rate functions with different rate constants was used to plot normalized equilibration curves. Diffusion curves were plotted at three pressure ranges 5-6 MPa, 3.5-4 MPa and 1.5-2 MPa. Result show that fast adsorption rate is higher at 5.5-6 MPa than at lower pressure range with highest fast adsorption rate fraction both for CH_4 and CO_2. Lower (1.5-2 MPa) pressure range allows achieving sorption equilibrium in less time for both gases. Diffusion rates are lower for CO_2 than for methane the CH_4 desorption rate has a slight impact on the CO_2 adsorption and as a consequence CO_2 storage capacity.

1. Introduction

Recent regulations related to carbon dioxide emission force power companies, cement and metal manufacturers to seek for different methods of CO_2 emission reductions. Although there are numerous technologies which can be used to reduce CO_2 emissions there is still lack of sufficient economic factor that can foster their implementation and meet targets of 2050 Energy roadmap in European Union [1]. One of the solutions that could work for fossil fuel based economies is the Carbon Capture and Storage (CCS). In this technology CO_2 is captured in stationary emission source, transported and permanently stored underground in geological traps. This technology is well known and although not widely implemented to date, it has a huge potential and when taking into consideration the dominant role that fossil fuels continue to play in primary energy consumption, the deployment of CCS technologies is becoming increasingly urgent [2]. Concerning Europe, a considerable amount of studies has been conducted about the implementation of CCS in unmineable coal seams, with particular attention to Poland (particularly in the Upper Silesian Basin) where there are still considerable amounts of Coalbed Methane reservoirs. In order to assess CO_2 storage capacity the following parameters have to be taken into consideration:

- Sorption capacity of coal (Langmuir parameters)
- Water saturation of coal
- Coal porosity

IOP Conf. Series: Materials Science and Engineering **427** (2018) 012039 doi:10.1088/1757-899X/427/1/012039

In order to assess the dynamic properties of CBM reservoir, other parameters, such as permeability diffusion coefficient, are needed. In the previous study, a petrophysical static model was applied in order to assess the total storage capacity of coalbed [3]. This research is focused on the dynamic factor, that is the sorption rate which can directly affect the efficiency of CO_2 injection and as a consequence – storage.

2. Materials and methods

In the study two coal samples from the Upper Silesia Coal Basin (USCB) in Poland were used. They represented two types of steam coal i.e. good quality coal with low ash content (Coal A) and low quality coal with high ash content (Coal B). Experiments were carried in a manometric sorption setup. Detailed description of the setup and experimental procedure is given in [3,4]. Proximate analysis of coals are presented in Table 1.

Table 1. Properties of coal used in the study.

Parameter	Coal A	Coal B
Ash content, %	4.66	11.33
Moisture content (as received), %	0.2	2.1
Moisture content (air dried), %	1.38	7.46
Higher heating value, KJ/kg	33376	22060

The diffusion curves of gas in coal can be assumed as equal both in adsorption and desorption process [5,6]: therefore, diffusion curves of CO_2 can give a reference for the variation of the stored quantities in time, while the ones of methane can give a reference for the variation of the produced quantities. The diffusion processes of methane and carbon dioxide have been analyzed at different pressure levels to assess possible difference and changings in the process due to the injection pressure, which is the initial pressure of the diffusion process. Three pressure ranges for carbon dioxide and methane were selected i.e.: 5-6 MPa, 3.5-4 MPa and 1.5-2 MPa. This indicates the pressure step at which adsorption starts. The time at which the equilibrium was attained varied from 0.6 days up to almost two weeks.

To estimate diffusion rate, a numerical diffusion model proposed by Busch and Gensterblum [5] combining two firs-order rate functions with different rate constants was used. The model expresses the quantity of gas diffused at a certain moment of time in terms of the residual (unoccupied) sorption capacity $Q_{residual}(t)$:

$$Q_{residual}(t) = \frac{P(t) - P_\infty}{P_0 - P_\infty}$$

Where:
$P(t)$ - system pressure at time t;
P_∞- system equilibrium pressure; and
P_0- initial system pressure.

The formula above allows the calculation of $Q_{residual}(t)$ for discrete values of pressure and time derived from experimental data. To extend the calculation of $Q_{residual}(t)$ and create a continuous function, the model proposes to calculate the residual sorption capacity as:

$$Q_{residual}(t) = Q_0' \times \exp(-k' \times t) + Q_0'' \times \exp(-k''t)$$

Where:
Q_0'- fraction of fast sorption;
k'- fast sorption first order rate constant;
Q_0'' - fraction of slow sorption;
k''- slow sorption first order rate constant.

During diffusion, two characteristic speed rates are present: the first one, which is the fastest, occurs in the first periods of time of the process, while the second one, which is slower, rules the diffusion in the following time. Fast and slow sorption fractions therefore refer to the fractions of gas adsorbed during these two sorption processes. The two first rate constant are used instead to calculate the time of fast and slow sorption:

$$t_{Q_0'/''} = \frac{\ln(2)}{k'/''}$$

The calculation of the fractions and rate constants value is obtained from the minimisation of the standard deviations between $Q_{residual}(t)$ values calculated with the first formula from the sorption experimental data.

3. Results and discussion

In Table 2 results of diffusion rate calculated with numerical model for the two types of coal at different pressure ranges have been shown.

Table 2. Results of CH_4 and CO_2 diffusion rate calculation at different pressure range.

	Pressure Range	Coal sample	Fast sorption fraction	Slow sorption fraction	Fast sorption time (s)	Slow sorption time (s)	Diffusion time (s)	Diffusion time (d)
CH_4	5.5-6 MPa	A	39%	61%	417	20090	607000	7.0
		B	48%	52%	65	13269	398000	4.6
	3.5-4 MPa	A	22%	75%	439	39138	1195000	13.8
		B	41%	59%	340	35322	1065000	12.3
	1.5-2 MPa	A	29%	71%	530	18214	447000	5.2
		B	46%	54%	378	14887	447000	5.2
CO_2	5.5-6 MPa	A	52%	48%	216	1712	52000	0.6
		B	61%	39%	382	8531	252000	2.9
	3.5-4 MPa	A	53%	47%	477	4315	129000	1.5
		B	69%	31%	449	12750	373000	4.3
	1.5-2 MPa	A	41%	59%	404	4939	149000	1.7
		B	53%	47%	254	4035	121000	1.4

Results show that for coal A the fast adsorption rate is higher at 5.5-6 MPa than at lower pressure range with highest fast adsorption rate fraction both for CH_4 and CO_2, 39% and 52% respectively. The 1.5-2 MPa pressure range allows achieving sorption equilibrium in less time for both gases. At higher pressure (5.5-6 MPa) it is possible to achieve a larger amount of gas stored in a comparable amount of time (7 days against 5.2 days) for sample A and CH_4. Comparison of normalized CH_4 and CO_2 sorption equilibration rates for coal A and B are shown in Figure 1, 2, 3 and 4.

IOP Conf. Series: Materials Science and Engineering **427** (2018) 012039 doi:10.1088/1757-899X/427/1/012039

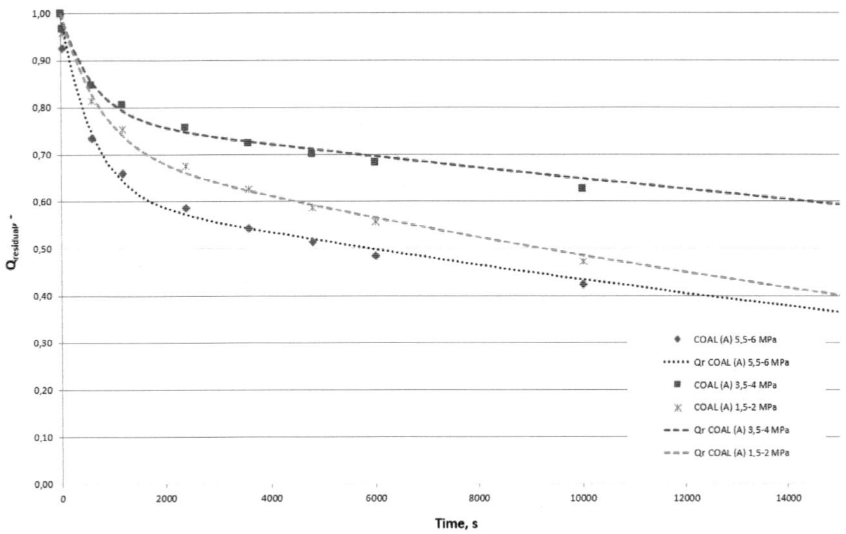

Figure 1. Normalized CH_4 sorption equilibration rate at 5,5-6 MPa, 3.5-4 MPa and 1,5-2 MPa on coal A.

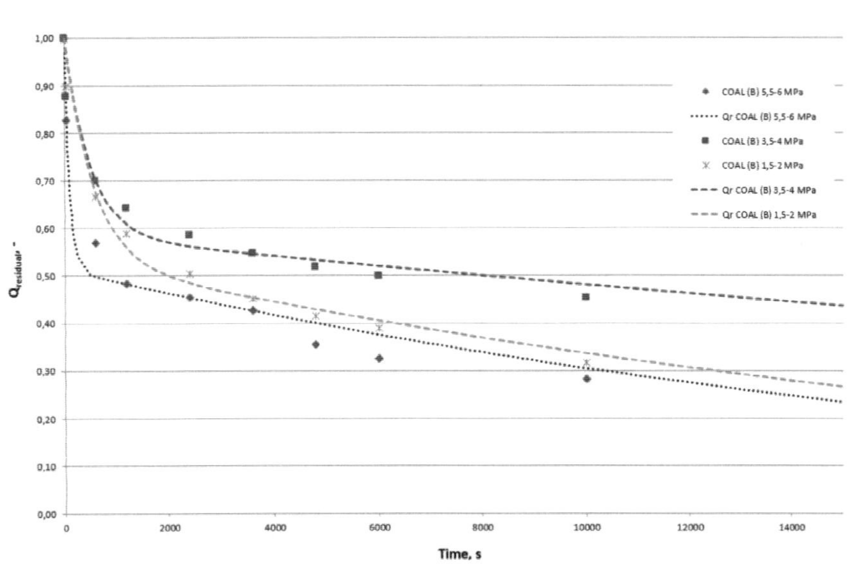

Figure 2. Normalized CH_4 sorption equilibration rate at 5,5-6 MPa, 3.5-4 MPa and 1,5-2 MPa on coal B.

IOP Conf. Series: Materials Science and Engineering **427** (2018) 012039 doi:10.1088/1757-899X/427/1/012039

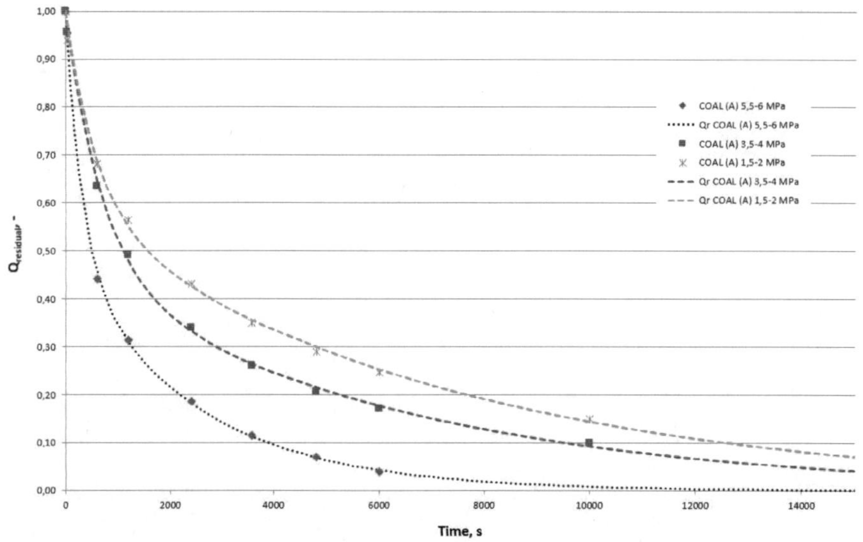

Figure 3. Normalized CO_2 sorption equilibration rate at 5,5-6 MPa, 3.5-4 MPa and 1,5-2 MPa on coal A.

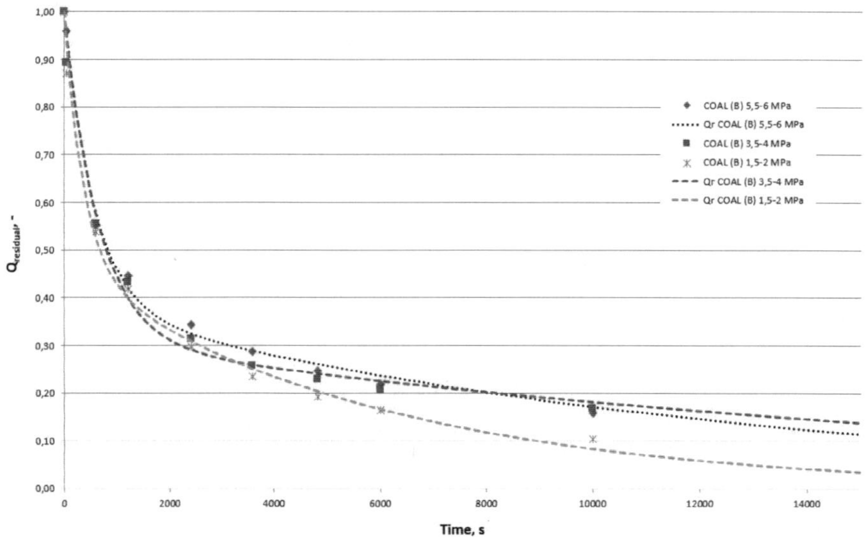

Figure 4. Normalized CO_2 sorption equilibration rate at 5,5-6 MPa, 3.5-4 MPa and 1,5-2 MPa on coal B.

When concerning coal B, results show that both fast and slow adsorption rates are faster at 5.5-6 MPa than at two other pressure ranges (except for CO_2). Hence, shortest diffusion time (4.6 days) for methane is at the highest pressure. This is not the case for CO_2 where diffusion times are considerably shorter.

Comparing the two types of coal, the 3.5-4 MPa pressure range accounts for the slowest diffusion time and quantity. Diffusion times are generally faster in coal B with also higher quantities of adsorbed gas in the fast adsorption rate. Coal A requires a longer diffusion process since this type of coal adsorbs more methane than coal B as shown in the previous study [3]. Experimental results show that diffusion rates are lower for CO_2 than for methane which is in agreement with other studies [5–7]. A slightly higher water content of coal B might have an impact on the diffusion time which was longer in case of CO_2.

The fraction of fast adsorption is considerably higher in CO_2 than in CH_4 by 15%-20% for the same pressure range and type of coal.

4. Diffusion rate and storage capacity

Laboratory experiments are not a direct representation of real CBM reservoir. The CBM reservoir is always filled with CH_4 and 100% recovery rates are rarely achievable. In case of ECBM process the CO_2 is injected into the reservoir saturated with CH_4. Aa result, the storage capacity varies in time, because the carbon storage capacity increases with time (injection) and a with desorption of methane. In fact longer time for CO_2 storage is required because the equilibrium will be reached when all the methane inside the coal will be desorbed.

This could be considered as a starting point for the application of simple management optimization models for the storage and production processes of CO_2 and CH_4. When analysing and comparing the diffusion functions of carbon dioxide and methane in coal at a certain pressure level it is possible to assess the storage and production processes and how they vary with the variation in pressure which is the only parameter than can readily be regulated.

In Table 3 a summary of the results obtained for the calculation of storage capacity in two cases: CO_2 is stored in empty reservoir (Case 1) and (Case 2) CO_2 is stored when methane is present inside the coal. These values show how different pressure levels correspond to different times of CO_2 storage and CH_4 production. The storage capacity variation in time has been calculated as the difference between the amount of CO_2 stored and the amount of CH_4 produced at a certain instant of time. Example of effective variation of the storage capacity at a certain pressure level in time in the first moments of the storage process are shown in Figure 5.

Table 3. Results of the storage capacity assessment in the function of time.

Coal A	5,5-6 MPa		3,5-4 MPa		1,5-2 MPa	
	Case 1	Case 2	Case 1	Case 2	Case 1	Case 2
CO_2 storage capacity (sm³/t)	30,97		27,96		22,68	
%variation from the max capacity	0%		-10%		-27%	
CH_4 production capacity (sm³/t)	5		5		5	
Time of Storage (s)	52000	607000	129000	1195000	149000	554000
Time of Storage (d)	0,6	7,0	1,5	13,8	1,7	6,4
%variation from the min storage time	10%		116%		0%	
Coal B	5,5-6 MPa		3,5-4 MPa		1,5-2 MPa	
	Case 1	Case 2	Case 1	Case 2	Case 1	Case 2
CO_2 storage capacity (sm³/t)	42,34		39,32		32,26	
%variation from the max capacity	0%		-7%		-24%	
CH_4 production capacity (sm³/t)	4		4		4	
Time of Storage (s)	252000	398000	373000	1065000	121000	447000
Time of Storage (d)	2,9	4,6	4,3	12,3	1,4	5,2
% variation from the min storage time	0%		120%		9%	

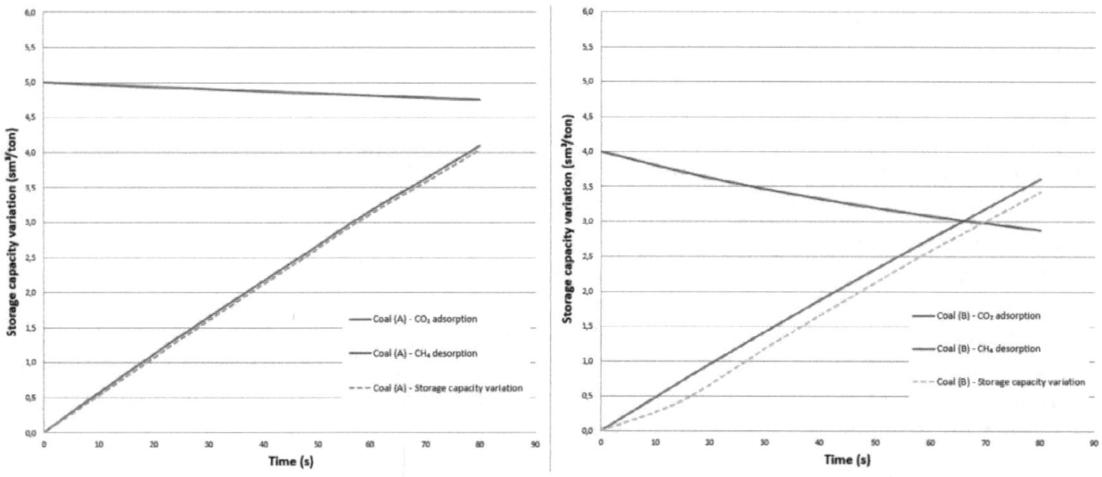

Figure 5. Graphical representation of CO storage, CH4 production and storage capacity variation trend in the first moments of the process in coal A and coal B at 5,5-6 MPa.

The selected diffusion model represents the adsorbed and desorbed quantities of gas in time as a fraction of the total capacity: cross-referencing those results with the values of the total storage capacity of CO_2 obtained through the storage capacity model, a quantification of the values becomes possible.

Concerning the production capacity of methane of the two coal samples, these values are site specific and their evaluation was not possible with the available equipment: therefore, an assumption on these values has been made based on the common values of similar coal types. It's important to notice that these values don't depend on the pressure of injection but on the characteristics of the reservoir; in these cases, the pressure of injection only rules the diffusion process and the time required to achieve the complete desorption of methane. From the Figure 5 it is evident that diffusion processes have an impact on the CO_2 injection time and CH_4 production.

5. Conclusions

The aim of this study was to verify the dynamic factor, that is the sorption rate which can directly affect the efficiency of CO_2 injection and as a consequence – storage. When aassessing the diffusion processes of CO_2 and CH_4 at different pressure levels, it was possible to understand how the CO_2 storage capacity of the coals varies in time due to the consequent desorption of methane which frees space for the storing of carbon dioxide. This phenomenon doesn't involve the creation of a higher storage capacity, but it shows how diffusion rate can have an impact on the injection time. The following conclusions can be drawn from this study:

- fast adsorption rate is higher at 5.5-6 MPa than at lower pressure range with highest fast adsorption rate fraction both for CH_4 and CO_2.
- lower (1.5-2 MPa) pressure range allows achieving sorption equilibrium in less time for both gases.
- diffusion rates are lower for CO_2 than for methane
- the CH_4 desorption rate has a slight impact on the CO_2 adsorption and as a consequence CO_2 storage capacity. The time to attain sorption equilibrium is in the range of a few hours.

Acknowledgments
Authors of this study would like to thank Sebastiano Corradini for his great contribution to this work.

References

[1] Anon 2018 *Global Energy Transformation: A Roadmap to 2050* (International Renewable Energy Agency)

[2] Alcalde J, Flude S, Wilkinson M, Johnson G, Edlmann K, Bond C E, Scott V, Gilfillan S M V, Ogaya X and Stuart H 2018 Estimating geological CO2 storage security to deliver on climate mitigation *Nat. Commun.* **9**

[3] Lutynski M and Zavanella L 2018 Application of petrophysical shale gas model for CO2 storage capacity assessment of coals IOP Conference Series: Earth and Environmental Science vol 174

[4] Lutynski M and González G 2016 Characteristics of carbon dioxide sorption in coal and gas shale - The effect of particle size *J. Nat. Gas Sci. Eng.* **28** 558–65

[5] Busch A, Gensterblum Y, Krooss B M and Littke R 2004 Methane and carbon dioxide adsorption–diffusion experiments on coal: upscaling and modeling *Int. J. Coal Geol.* **60** 151–68

[6] Charrière D, Pokryszka Z and Behra P 2010 Effect of pressure and temperature on diffusion of CO2 and CH4 into coal from the Lorraine basin (France) *Int. J. Coal Geol.* **81** 373–80

[7] Gruszkiewicz M S, Naney M T, Blencoe J G, Cole D R, Pashin J C and Carroll R E 2009 Adsorption kinetics of CO2, CH4, and their equimolar mixture on coal from the Black Warrior Basin, West-Central Alabama *Int. J. Coal Geol.* **77** 23–33

IOP Conf. Series: Materials Science and Engineering **427** (2018) 012040 doi:10.1088/1757-899X/427/1/012040

The testing of waste from the installation of salt debris leaching

T Suponik, P Pierzyna, D Franke and Z Adamczyk

Silesian University of Technology, Faculty of Mining and Geology, Akademicka 2, 44-100 Gliwice, Poland

tomasz.suponik@polsl.pl

Abstract. Although the Salt Mine in Bochnia is designed for tourism, it still produces waste. In order to reduce the mass of waste and its harmful impact on the environment, the leaching method was proposed. The authors of the article have presented in their earlier works the method of leaching a salt debris as well as the large-scale laboratory installation used for leaching. The waste generated in the leaching installation are to be used e.g. for the production of building materials or, in the worst case, stored in landfill. The aim of the article was to assess the possibility of their storing at the landfill of inert waste or using them as a secondary raw material in the environmental aspects. The analysis of the mineral and chemical composition of salt debris and waste after leaching in the installation as well as the chemical composition of the leachate after the leaching tests of these waste were performed. Halite, anhydrite and gypsum are the dominant mineral in salt debris, whereas only anhydrite and gypsum in waste after leaching in a large-scale laboratory installation. The remaining, in smaller amounts, minerals are illite and quartz. Waste from the installation of salt debris leaching still releases small amounts of chlorides into the water. Therefore waste still poses a threat to the environment. In order to reduce the amount of chlorides released into the environment, the pump in the leaching plant should work longer than the assumed 5 minutes or the leaching temperature should be higher than 15°C.

1. Introduction

The Bochnia Salt Mine produces large amounts of waste in the form of salt debris annually. This is to maintain tourist routes and create new underground heading elements. In the case of active mines, the debris are used in liquidation works [1].

The Salt Mine in Bochnia is not active, so there are no new locations to store salt debris. In order to reduce the mass of waste and its harmful impact on the environment, the leaching method in 2017 has been proposed. Pierzyna et al. presented in their paper [2] the method of leaching a salt debris as well as the large-scale laboratory installation used for leaching. The saturated brine and waste will be created as a result of this operation. In the article, this waste is called waste after leaching. Saturated brine is to be pumped into inactive underground excavations, salt caverns tightly isolated from groundwater, while the waste from the leaching installation either is to be used as a secondary raw material or disposed at a landfill for inert waste. This article answers the question about the possibility of such waste management, in the environmental aspects. Thus, the aim of the article is to assess the possibility of their storing at the landfill of inert waste or using them as a secondary raw material in the environmental aspect. The following studies were carried out:

- the analysis of mineral and chemical composition of salt debris and waste after leaching,
- the analysis of chemical composition of the leachate after the leaching test for both salt debris and waste after leaching,
- the analysis of weight loss during the roasting process of the salt debris and waste after leaching.

2. Material and methods

The waste for the environmental tests was obtained from the laboratory installation for waste leaching located at the technological hall of the Silesian University of Technology. The system operates as a closed cycle and is equipped with a PH65 Powen's rotary pump, which serves the mixing, pumping and exchange of the mixture in the tank. The piping is equipped with a series of flap valves allowing for the control of the flow through the selected diameters of the pipes [2]. The tests were conducted for one grain classes: 0-6 mm. The leaching temperature was set at 15°C, whereas thy leaching time was 5 minutes, according to previous research and recommendations presented in the article [2]. The waste was then dried.

In order to identify the mineral and chemical composition of the samples (salt debris and waste after leaching), the X-ray diffraction (XRD) method and X-ray fluorescence (XRF) were used, respectively. For the analysis of crystalline substances, AERIS 1 with a $Cu_{K\alpha}$ lamp by PANalytical was used. Working conditions of the diffractometer: voltage 40kV, current intensity 8 mA, the step of 0.003° of 2θ angle, time 4.84 s, the range of the 2θ angle 7-77°. The identification of elements and determination of atomic concentrations of samples were performed using Epsilon 1 Spectrometer by PANalytical.

A roasting was used to check the total weight loss of the samples at high temperature. The samples were comminuted and sieved to a size < 0.1 mm. Approximately 2 grams of both salt debris and waste after leaching were roasted for 3 hours in a laboratory furnace at 850°C. The temperature in the furnace increased to 500°C for 40 minutes and to 850°C for a further 20 minutes. Then the temperature was maintained at 850 °C for 3 hours. The samples were weighed before and after roasting, and then the total weight loss was calculated as a percentage.

In order to assess the possibility of storing waste at the landfill of inert waste, according to the Regulation of the Minister of Economy [3], and their impact on the environment, the waste was leached in a ROTAX 6.8 shaker in accordance with the PN-EN standard [4]. The liquid/waste ratio (L/S) in the tests was 10:1, while the shaking time was 24 hours. The leaching test was carried out at 22°C. Determination of the content of chemical substances and parameters in the eluates formed during the leaching tests was performed using: JY2000 spectrometer (ICP-AES) for elements, DR5000 UV-Vis spectrophotometer for chlorides, fluorides, sulfates, phenol indicator, Shimadzu $TOC\text{-}V_{CSH}$ Analyser and Shimadzu ASI-C sampler for dissolved organic carbon and Knick PORTAMESS meters for Total Dissolved Solids.

Four samples of salt debris and four samples of waste after leaching were taken for roasting and leaching tests, and the results were presented as an arithmetic average.

3. Results and discussion

3.1. X-ray diffraction (XRD)

Based on the diffraction patterns presented in figure 1 and 2 it can be noticed that the mineral composition of the samples, i.e. salt debris and waste after leaching, was practically the same. Both of the samples consist the following minerals:

- halides, represented by halite (NaCl) and sylvine (KCl),
- calcium sulphates, represented by anhydrite ($CaSO_4$) and gypsum ($CaSO_4$ x$2H_2O$),
- mineral from the silica group, represented by quartz (SiO_2),
- clay minerals, represented by illite ($K_{0.6\text{-}0.85}Al_2(Si,Al)_4O_{10}(OH)_2$.

IOP Conf. Series: Materials Science and Engineering **427** (2018) 012040 doi:10.1088/1757-899X/427/1/012040

This was indicated by the characteristic reflections from these minerals, appearing on the diffraction patterns (figures 1 and 2).

- halite: 3.25, 2.82, 1.99, 1.70, 1.63, 1.41, 1.26Å,
- anhydrite: 3.88, 3.50, 3.12-3.11, 2.85, 2.47, 2.33, 2.21, 2.18, 2.08, 1.94, 1.87, 1.75, 1.65, 1.56, 1.52, 1.49Å,
- gypsum: 7.62-7.60, 4.28, 3.06, 2.68Å,
- quartz: 4.25, 3.34, 2.46, 2.28, 2.13, 1.81, 1.54Å,
- illite: 10.21-10.19, 4.47-4.46, 2.36Å,
- sylvine: 3.11, 2.89, 2.20, 2.04, 1.29Å.

The intensity of basic reflections indicated a clear quantitative differentiation of minerals in the samples:

- in salt debris the halite and anhydrite dominated, while illite, quartz and sylvine were identified in smaller quantities,
- in the waste after leaching anhydrite prevailed, but there were also, in much smaller quantities, illite, quartz, halite and sylvine.

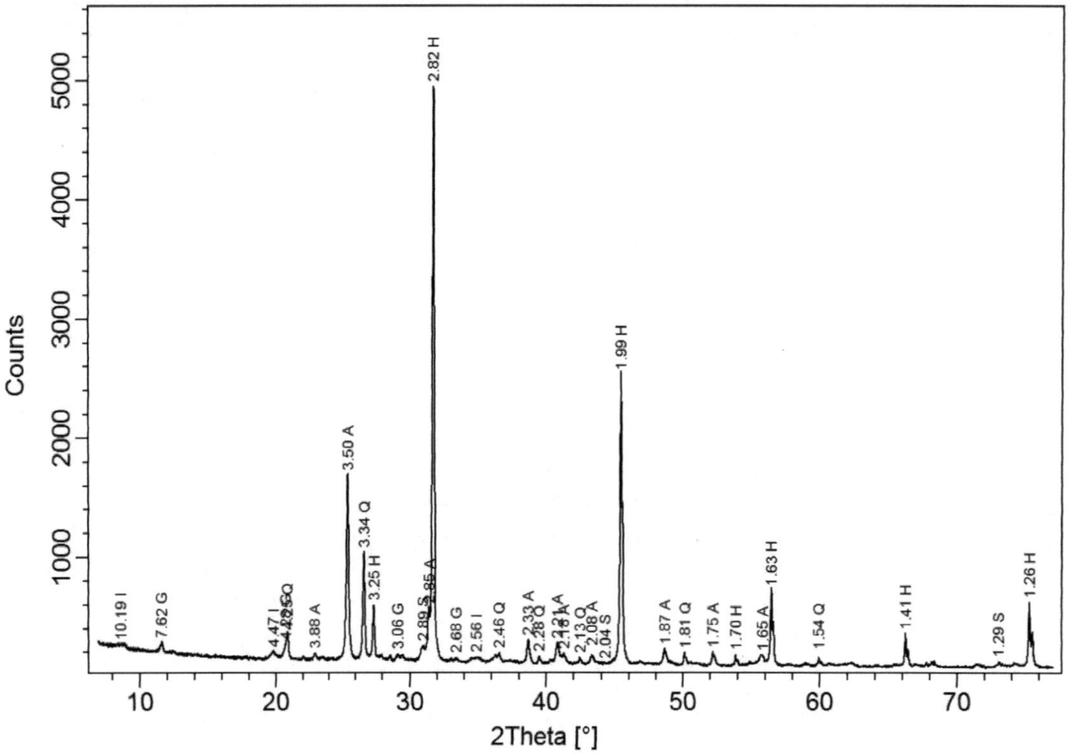

Figure 1. Diffractogram for the salt debris. A – anhydrite, G – gypsum, H – halite, I – illite, Q – quartz, S – sylvine.

IOP Conf. Series: Materials Science and Engineering **427** (2018) 012040 doi:10.1088/1757-899X/427/1/012040

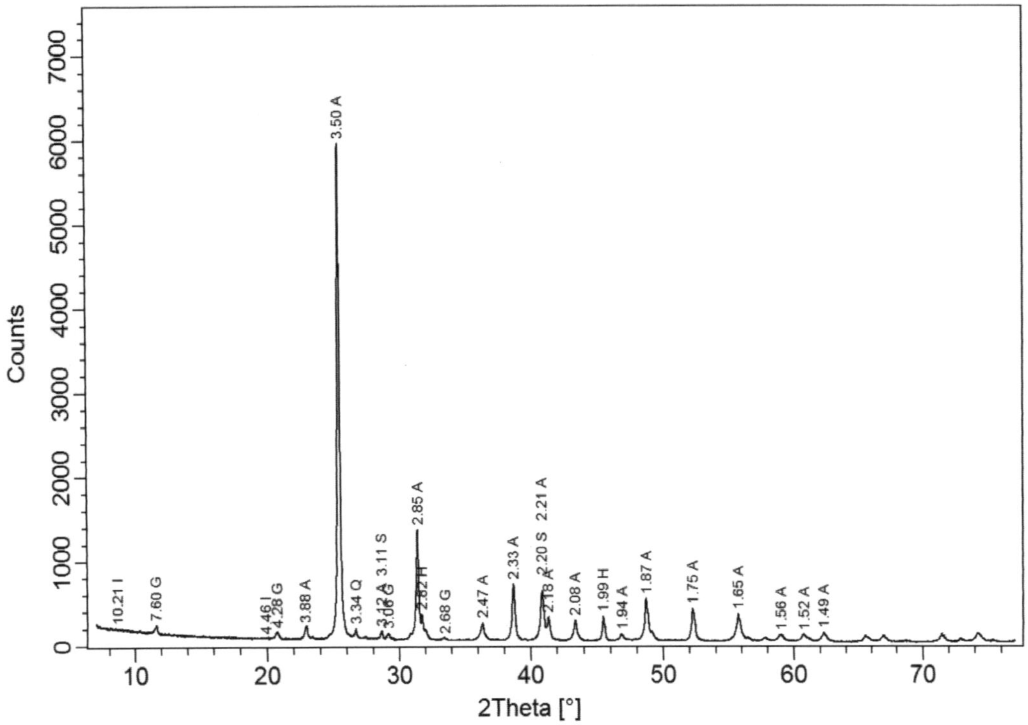

Figure 2. Diffractogram for waste after leaching. A – anhydrite, G – gypsum, H – halite, I – illite, Q – quartz, S – sylvine.

3.2. X-ray fluorescence (XRF)

The results of the tests using XRF spectrometer are presented in table 1. These results show that the chemical composition of both samples differs significantly:

- in salt debris Na_2O and Cl dominated, with contents of 22.65 and 26.48 % by mass, respectively. There were also SiO_2, CaO and SO_3, in an amount of over ten percent each,
- in waste after leaching there were SO_3 and CaO, together in an amount of over 89 % by mass. The other chemical components were present in much smaller quantities.

Table 1. Chemical composition of the tested samples (in % by mass).

Chemical ingredient	The name of the sample	
	Salt debris	Waste after leaching
SiO_2	15.08	5.84
TiO_2	0.25	0.03
Al_2O_3	5.32	2.09
Fe_2O_3	1.83	0.26
CaO	13.09	37.15
Na_2O	22.65	1.11
K_2O	1.22	0.13
SO_3	14.05	52.03
Cl	26.48	1.35
V_2O_5	0.01	0.00
Cr_2O_3	0.01	0.00
MnO	0.03	0.00
CuO	0.01	0.00
SUMA	100.00	100.00

Based of mineral and chemical composition the percentage of mineral components in both samples was calculated (anhydrite and gypsum were presented together as calcium sulphates). The results were presented graphically in the figure 3.

In salt debris, the dominant mineral was halite in the amount of 47% by mass, while the second component was calcium sulphate in the amount of 26% by mass (see figure 3). The content of illite and quartz was very similar and amounted to 12-13% by mass, while the content of sylvine was 2% by mass.

In the waste after leaching calcium sulphates (anhydrite and gypsum) were dominated (nearly 90% by mass). The content of the remaining minerals were: quartz and illite 4% by mass each, halite 2% by mass and sylvine 1% by mass.

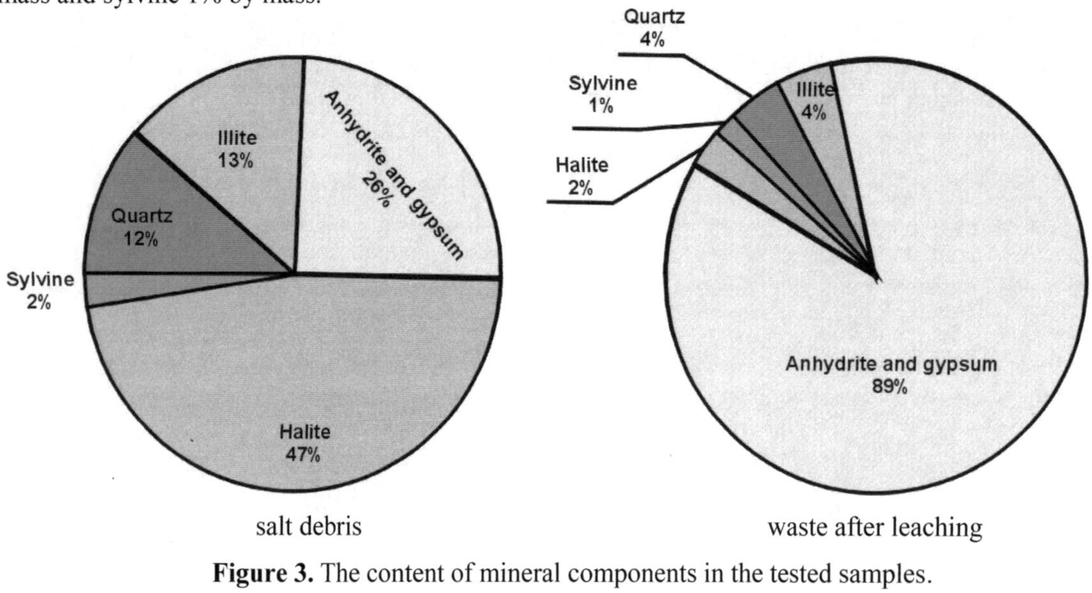

Figure 3. The content of mineral components in the tested samples.

3.3. Weight loss during roasting

The weight loss of samples as a result of roasting at 850oC, for salt debris and waste after leaching, was 1.6365 % and 0.3718 %, respectively. The loss of mass is much smaller (more than four times) for waste after leaching. This indicates that this material contains less minerals that undergo chemical changes that cause the loss of mass due to high temperature. Gypsum and especially anhydrite, which together make up 89% of the total, belongs to such minerals. As a result of raising the temperature to 150°C, the gypsum turns into burned gypsum, and up to 200°C in anhydrite. These changes cause loss of water from the sample, thereby reducing its weight to a small extent. At temperatures up to 850°C (and even up to 1180°C), the anhydrite changes slightly, which does not lead to weight loss [5].

3.4. Leaching tests

Based on the results of leaching tests (table 2), it can be claimed that waste from the installation of salt debris leaching can be stored in the landfill of waste other than hazardous and inert, according to the Regulation of the Minister of Economy [3]. Chlorides are the only substance that exceeds the permissible values of their storage at landfills for inert waste. The test results indicate elevated concentrations of copper and zinc in the eluates. These values, however, do not exceed the permissible level, according to the Regulation of the Minister of Economy, for both types of landfills.

Based on these observations, it is recommended that the pump in the leaching plant should work longer than the assumed 5 minutes or the leaching temperature should be higher than 15°C. If one or both of these conditions are met, the waste may be stored at the landfills of inert waste and will not be a problem for the environment. At the same time, it can be expected that the waste will be able to be used in engineering works (e.g. in the production of building materials). For this purpose, however, additional studies needs to be done.

In order to obtain a product with a high content of calcium sulphate and a suitable crystalline structure the waste should be free of impurities [5]. Then the waste can be thermally treated. Directions for the use of waste containing a very large amount of calcium sulphate have been presented in the following papers [6,7].

IOP Conf. Series: Materials Science and Engineering **427** (2018) 012040 doi:10.1088/1757-899X/427/1/012040

Table 2. The results of the leaching tests of waste vs the permissible values of substance/parameters that allow to storage them at the landfills of inert waste or at the landfills of waste other than hazardous and inert, in mg/kg of dry matter; L/S = 10 dm^3:1 kg.

Substance/parameter	The results of leaching tests for:		The permissible values* of substance/parameters that allow to storage waste at the landfills of:	
	salt debris	waste obtained from the installation of salt debris leaching	inert waste	waste other than hazardous and inert
Arsenic (As)	<0.04	<0.04	0.5	2
Barium (Ba)	7.11	1.94	20	100
Cadmium (Cd)	0.24	<0.002	0.04	1
Chromium tot. (Cr)	2.90	0.05	0.5	10
Copper (Cu)	12.44	1.35	2	50
Mercury(Hg)	0.04	<0.002	0.01	0.2
Molybdenum (Mo)	0.14	<0.006	0.5	10
Nickel (Ni)	4.45	0.09	0.4	10
Lead (Pb)	3.81	<0.03	0.5	10
Antimony (Sb)	<0.01	<0.01	0.06	0.7
Selenium (Se)	<0.02	<0.02	0.1	0.5
Zinc (Zn)	9,24	0.6	4	50
Chlorides (Cl^-)	307 646	7160	800	15 000
Fluorides (F^-)	1.21	<0.02	10	150
Sulphates (SO_4^{2-})	1 084	459	1000	20 000
Phenol indicator	<0.1	<0.1	1	-
Dissolved organic carbon	15	2	500	800
Total dissolved solids	>19990	14 225	4000	60 000

* - according to the Regulation of the Minister of Economy [3]

4. Conclusions

- The salt debris from the Salt Mine in Bochnia, which has not undergone any processes, releases to the aquatic environment chlorides, sulphates as well as chromium, copper, nickel, lead and zinc in ionic state, while waste from the installation of salt debris leaching only chlorides. Extending the leaching time in the leaching installation and/or increasing the process temperature (both factors to a small extent) will improve the efficiency of the process and will make the waste environment-friendly.
- The salt debris mainly consists of halite, anhydrite and gypsum, illite and quartz, while waste from leaching installation from anhydrite and gypsum. The latter material could be used, after prior preparation, for the production of building materials.

References

[1] Poborska-Młynarska K 2015 *Przeg. Górn.* **12** pp 140-146

[2] Pierzyna P, Popczyk M and Suponik T 2017 *Mineral Engineering Conf. MEC2017* vol 18 (E3S Web of Conferences) p 01033

[3] Journal of Lows (2015, item 1277)

[4] PN-EN 12457-4:2006

[5] Jarosiński A, Żelazny S, Nowak AK 2007 *Czasopismo Techniczne* **1** pp 24-28

IOP Conf. Series: Materials Science and Engineering **427** (2018) 012040 doi:10.1088/1757-899X/427/1/012040

[6] Zhuang LJ, Querol X, Font O and Moreno N 2018 *International Geology Review,* **60(5-6)** pp 671-716

[7] Ngernchuklin P, Yongpraderm N, Boornuang A, Kanchanasutha S, Laoauyporn P and Busabok C 2018 Upgrading of Waste Gypsum for Building Materials *Key Engineering Materials* vol 766 (Trans Tech Publications) pp 211-16